Lecture Notes Electrical Engineering

Volume 53

Alexander Barkalov and Larysa Titarenko

Logic Synthesis for FSM-Based Control Units

Springer

Prof. Alexander Barkalov
Institute of Informatics and Electronics
University of Zielona Gora
Podgorna Street 50
65-246 Zielona Gora
Poland
E-mail: a.barkalov@iie.uz.zgora.pl

Dr. Larysa Titarenko
Institute of Informatics and Electronics
University of Zielona Gora
Podgorna Street 50
65-246 Zielona Gora
Poland
E-mail: abar54@mail.ru

ISBN 978-3-642-26064-3 e-ISBN 978-3-642-04309-3

DOI 10.1007/978-3-642-04309-3

Softcover reprint of the hardcover 1st edition 2010

Typeset & Coverdesign: Scientific Publishing Services Pvt. Ltd., Chennai, India.

Printed in acid-free paper

9 8 7 6 5 4 3 2 1

springer.com

Acknowledgements

Several people helped us with preparation of this manuscript. Our PhD students Mr Jacek Bieganowski and Mr Sławomir Chmielewski worked with us on initial planning of this work, distribution of tasks during the project, and final assembly of this book.

We also thank Professor Marian Adamski for his support and special attention to this work. His guidelines in making this book useful for students and practitioners were very helpful in the organization of this book.

Contents

Symbols

$X = \{x_1, \ldots, x_L\}$	set of logical conditions
$Y = \{y_1, \ldots y_N\}$	set of microoperations
$Y_q \subseteq Y$	collection of microoperations (microinstruction)
Γ	graph–scheme of algorithm
b_0	start vertex of GSA
b_E	end vertex of GSA
B_1	set of GSA operator vertices
B_2	set of GSA conditional vertices
$E = \{\langle b_t, b_q \rangle\}$	set of GSA arcs
$a_m \in A$	internal state of FSM
$K(a_m)$	code of internal state $a_m \in A$
$A = \{a_1, \ldots, a_M\}$	set of FSM internal states
$T = \{T_1, \ldots, T_R\}$	set of FSM state variables
$A_m \in A$	conjunction of state variables corresponding to the state code $K(a_m)$
$\Phi = \{\varphi_1, \ldots, \varphi_R\}$	set of FSM input memory variables (excitation functions)
H	the number of structure table rows (lines)
$\Pi_A = \{B_1, \ldots, B_I\}$	set of the classes of pseudoequivalent states
$K(B_i)$	code of class of pseudoequivalent states $B_i \in \Pi_A$
$\alpha_g = \langle b_{g_1}, \ldots, b_{g_{F_g}} \rangle$	operational linear chain
M_i	matrix (AND- or OR-plane)
$S(M_i)$	area of matrix M_i
$F = \{F_1, \ldots, F_H\}$	set of FSM terms
$X(a_m)$	set of logical conditions determining transitions from the state $a_m \in A$
$p_g \in P$	additional variable used to replace logical conditions, where $\|P\| = G$, $G = \max(\|X(a_1)\|, \ldots, \|X(a_M)\|)$
$X(p_g)$	set of logical conditions written in the column p_g
$z_r \in Z$	additional variable used to encode the microinstructions
$K(Y_t)$	binary code of collection Y_t
τ	set of variables used to code classes $B_i \in \Pi_a$, where $\|\tau\| = R_0$ and $R_0 = \lceil \log_2 I \rceil$

BM	FSM block generating variables for logical condition replacement
BP	FSM block generating variables written in the rows of (transformed) structure table
BY	FSM block generating microoperations and implemented with embedded memory blocks
BD	FSM block generating microoperations and implemented with decoders
BF	FSM block generating variables corresponding to rows of (transformed) structure table
MX_g	multiplexer from block BM generating function $p_g \in P$
$K(y_n)$	code of microoperation $y_n \in Y^k$ from the class k of compatible microoperations
R_k	the number of bits in the code $K(y_n)$
$z_r \in Z^k$	additional variables used for encoding of microoperations $y_n \in Y^k$
DC_k	decoder from block BD generating microoperations from the class k of compatible microoperations
$K(F_h)$	binary code of row h of FSM structure table
R_F	the number of bits in code $K(F_g)$
$H(f)$	the number of terms for SOP of some function f
q	the number of terms for PAL-based macrocell
$n(f,q)$	the number of macrocells having q terms, necessary to implement the logic circuit for function f
NL_i	the number of FSM models having i levels
$V(\Gamma)$	graph-scheme of algorithm Γ after verticalization
I	set of identifiers for FSM with object codes transformation
$K(I_k)$	binary code of identifier $I_k \in I$ having $R_V = \lceil \log_2 K \rceil$ bits
$V = \{v_1, \dots, v_{R_V}\}$	set of variables used for encoding of identifiers $I_k \in I$
$CE = \{\alpha_1, \dots, \alpha_{G_E}\}$	set of elementary operational linear chains
R_E	the number of microinstruction address bits, where $R_E = \lceil \log_2 M_E \rceil$
M_E	the number of operator vertices in transformed GSA
O_g	output of EOLC $\alpha_g \in C_E$
$A(O_g)$	address of EOLC output O_g
I_j	input of EOLC $\alpha_j \in C_E$
$A(I_j)$	address of EOLC input I_j
G_E	the number of EOLC in GSA Γ
M_g	the number of components in EOLC $\alpha_j \in C_E$
Q_E	the maximal number of components in EOLC of GSA Γ

R_{EO}	the number of variables for encoding of EOLC, where $R_{EO} = \lceil \log_2 G_E \rceil$
R_{CO}	the number of variables for encoding of EOLC, where $R_{EO} = \lceil \log_2 Q_E \rceil$
$K(\alpha_g)$	code of EOLC $\alpha_g \in C_E$
$K(b_t)$	code of component $b_t \in B_1$ of EOLC $\alpha_g \in C_E$

Abbreviations

ASIC	application-specific integrated circuit
BAT	block of address transformer
BTC	block of code transformer
BM	block for logical condition replacement
BP	block forming input memory functions of FSM
BTC	block for code transformation
BY	block forming microoperations of FSM
CA	control automaton
CAD	computer-aided design
CAMI	counter of microinstruction address
CC	sequential circuit
CCS	state code transformer
CFA	circuit of address formation (sequencer)
CLB	configurable logic block
CM	control memory
CMCU	compositional microprogram control unit
CMO	circuit (block) of microoperation generation
CPLD	complex programmable logic devices
EAB	embedded array block
EPROM	erasable programmable read-only memory
EEPROM	electrically erasable programmable read-only memory
EOLC	elementary operational linear chain
FPLD	field-programmable logic devices
FSM	finite state machine
FPGA	field-programmable gate arrays
GFT	generalized formula of transition
GSA	graph- scheme of algorithm
HDL	hardware description language
LAB	logic array block
LE	logic element
LUT	look-up table

MCU	microprogram control unit
MX	multiplexer
OA	operational automaton
OLC	operational linear chain
PAL	programmable array logic
PLD	programmable logic device
PLA	programmable logic array
PLS	programmable logic sequencer
PROM	programmable read-only memory
RAM	random-access memory
RAMI	register of microinstruction address
RG	register
ROM	read-only memory
SBF	system of Boolean functions
SOP	sums of products
SPLD	simple programmable logic devices
ST	structure table
TMS	microoperation code transformer
TSM	state code transformer
VGSA	vertical graph- scheme of algorithm
VLSI	very large scale integration circuit

Introduction

Tremendous achievements in the area of semiconductor electronics turn microelectronics into nanoelectronics. Actually, we observe a real technical boom connected with achievements in nanoelectronics. It results in development of very complex integrated circuits, particularly the field programmable logic devices (FPLD). Up-to-day FPLD chips are so huge, that it is enough only one chip to implement a really complex digital system including a data-path and a control unit. Because of the extreme complexity of modern microchips, it is very important to develop effective design methods oriented on particular properties of logic elements. The development of digital systems with use of FPLD microchips is not possible without use of different hardware description languages (HDL), such as VHDL and Verilog. Different computer-aided design tools (CAD) are wide used to develop digital system hardware. As majority of researches point out, the design process is now very similar to the process of program development. It allows a researcher to pay more attention to some specific problems, where there are no standard formal methods of their solution. But application of all these achievements does not guarantee per se development of some competitive electronic product, especially in the acceptable time-to-market. This problem solution is possible only if a researcher possesses fundamental knowledge of a design process and knows exactly the mode of operation of industrial CAD tools in use. As it is known, any digital system can be represented as a composition of a date-path and a control unit. Logic schemes of data-path have regular structures; it allows use of standard library elements of CAD tools (such as counters, multibit adders, multipliers, multiplexers, decoders and so on) for their design. A control unit coordinates interplay of other system blocks producing a sequence of control signals that causes some operations in a data-path. As a rule, control units have irregular structures, which makes process of their design very sophisticated. In case of complex logic controllers, the problem of system design is reduced practically to the design of control units. Many important features of a digital system, such as performance, power consumption and so on, depend to a large extent on characteristics of its control unit. Therefore, to design competitive digital systems with FPLD chips, a designer should have fundamental knowledge in the area of logic synthesis and optimization of logic circuits of control units. As our experience shows, design methods used by standard industrial packages are, in case of complex control units design, far from optimal. It means that a designer may be forced to develop his own design methods, next to program them and at last to combine them with standard packages to get a result with desired characteristics. To help such a designer, this book is devoted to solution of the problems of logic synthesis and reduction of hardware amount in control units, when a control unit is represented using the model of finite state machine (FSM). The book contains some original design and optimization methods based on the structural decomposition of FSM model. Such an approach results in multilevel models of FSM, where regularity of the device increases in comparison with known single- and double-level models. Regular parts of these models can be implemented using such library elements as memory blocks, decoders and multiplexers. In the same time, an irregular part of the control units described by means of Boolean functions is reduced. It permits to decrease the total number of logic elements (PAL, GAL, PLA, or LUT macrocells) in comparison with logic circuits based on known models of FSM. This approach is especially fruitful when a control unit is implemented using up-to-day FPLD chips which include not only combinational macrocells, but also the embedded memory blocks. In our book, control algorithms are represented by graph-schemes of algorithms (GSA). This choice is based on obvious fact that this specification provides simple explanation of methods proposed by authors. The methods of synthesis and design presented in the book are not oriented to any particular FPLD chips, but to construction of tables describing the behaviour of FSM blocks. These tables are used to find the systems of Boolean functions, which can be used to implement logic circuits of particular FSM blocks. In order to implement corresponding circuits, this information should be transformed using data formats of particular industrial CAD systems. This step is beyond the scope of our book, in which the following information is presented:

Chapter 1 introduces such basic topics as principle of microprogram control and specification of control units by graph-scheme of algorithms. Such conceptions as microoperations (FSM output signals), logical conditions (FSM input signals), FSM states, interstate transitions, and FSM structure table are introduced. Next, some methods of control algorithms interpretation are discussed, such as finite state machines and microprogram control units. The FSM models of Mealy and Moore are introduced; the methods of transition from GSA to Mealy and Moore FSM graphs are shown. All FSM discussed in the book are specified either by GSA or by structure table of FSM. Last part of the chapter is devoted to organization principles of compositional microprogram control units, which can be viewed as a composition of Mealy finite-state machine addressing microinstructions and microprogram control unit with natural microinstruction addressing. These control units are Moore

FSMs using counter to represent their state codes; they can be used for interpretation of linear GSA.

Chapter 2 discusses some problems, connected with logic synthesis and optimization of FSM implemented with custom matrix integrated circuits. The primitive matrix implementation of FSM circuit is analyzed first. It is reduced to direct interpretation of FSM structure table and is characterized by considerable redundancy. Next, the methods of logical condition replacement and encoding of collections of microoperations are considered. These methods allow decrease for circuit redundancy due increase of the number of FSM model levels. Next, it is shown that the model of Moore FSM offers an additional possibility for its circuit optimization due to existence of the classes of pseudoequivalent states. Each such class corresponds to one state of the equivalent Mealy FSM. Optimization methods are introduced based on different approaches for state encoding, as well as on transformation of state codes into class codes. The last part of the chapter is devoted to optimization of the block generating microoperations.

Chapter 3 discussed contemporary field-programmable logic devices and their evolution, starting from the simplest programmable logic devices such as PROM, PLA, PAL and GAL, and finishing with very sophisticated chips such as CPLD and FPGA. This analysis shows particular features of different logic elements and permits to optimize the FSM logic circuits, in which some particular elements are used. The analysis is accompanied by some examples for systems of Boolean functions implementation using PROM, PLA and PAL chips. The principle of functional decomposition oriented on FPGA chips is analysed in the last part of the chapter.

Chapter 4 is devoted to the hardware amount reduction in the logic circuit of Mealy FSM. The methods of logical condition replacement are analyzed, as well as different methods of encoding of collections of microoperations (maximal encoding and encoding of the classes of compatible microoperations). Next, the methods of structure table rows encoding are discussed. Each of these methods produces double-level circuit of Mealy FSM. The main part of the chapter is devoted to joint application of these methods, the main advantage of whose is possibility of standard library cells use for implementation of logic circuits for some blocks of an FSM model. For example, the logical condition replacement allows application of multiplexers, whereas the encoding of collections of microoperations permits to use embedded memory blocks. Standard decoders can be used in case of encoding of the classes of compatible microoperations. It increases FSM logic circuit regularity and leads to simplification of its design process.

Chapter 5 is devoted to original synthesis and optimization methods oriented on Moore FSM logic circuit implemented with CPLD. These methods are based on results of joint investigations conducted by the authors and their PhD students Cololo S. (Ukraine) and Chmielewski S. (Poland). These methods deal with both homogenous and heterogeneous CPLD chips. In the first case, only PAL- or PLA- based macrocells are used for logic circuit implementation. In the second case, the logic circuit is implemented using both PAL-based macrocells and embedded memory blocks. The hardware amount reduction is based on use of several sources (up to three) to represent the codes of classes of pseudoequivalent states. The methods assume joint minimization of Boolean expressions for input memory functions and microoperations of Moore FSM. The last part of the chapter is devoted to joint application of proposed methods and logical condition replacement.

Chapter 6 is devoted to design methods based on transformation of an interpreted graph-scheme of algorithm. The methods of decrease for the number of logical conditions per FSM state are discussed. In extreme case, all FSM transitions depend on single logical condition; it allows use of embedded memory blocks for implementation of FSM input memory functions. In this case all FSM blocks are implemented using standard library cells (not just macrocells of a particular FPLD chip). The second part of the chapter is devoted to hardware optimization for block of microoperations, based on verticalization of an interpreted GSA. It permits to decrease the number of decoders (up to 1) and bit capacity of microinstruction word, but this optimization is connected with increase for the number of cycles required for a control algorithm interpretation. At last, the models based on joint application of these methods are discussed.

Chapter 7 is devoted to original optimization methods oriented on decrease of the number of outputs for FSM block generating input memory functions. These methods are based on the object code transformation. The FSM objects are either states or collections of microoperations. Sometimes, some additional identifiers are needed for one-to-one representation of different objects. Such optimization methods are discussed for both Mealy and Moore finite state machines. At last, the multilevel models of FSM with object code transformation, logical condition replacement and encoding of collections of microoperations are discussed. This chapter is written together with employee of "Nokia-Siemens Network" Alexander Barkalov (Ukraine).

Chapter 8 is devoted to original methods oriented on optimization of Moore FSM interpreting graph-schemes of algorithms with long sequences of operator vertices having only one input. These sequences are named elementary operational linear chains (EOLC). These FSM models include the counter keeping, either microinstruction addresses or code of EOLC component. In the beginning the Moore FSM models with code sharing are analysed, where the register keeps EOLC codes. The methods of EOLC encoding and transformation are discussed; these methods permit to decrease the number of macrocells in the block generating input memory functions. The second part of the chapter is devoted to reduction of the number of embedded memory blocks in the FSM block generating microoperations. These methods are based on transformation of microinstruction address represented as concatenation of EOLC code and code of its component into either linear microinstruction address or code of collection of microoperations. The last part of the chapter discusses synthesis methods for multilevel FSM models with EOLC.

We hope that our book will be interesting and useful for students and postgraduates in the area of Computer Science, as well as for designers of modern digital devices. We think that proposed FSM models enlarge the class of models applied for implementation of control units with modern CPLD and FPGA chips.

Chapter 1
Hardwired Interpretation of Control Algorithms

Abstract. The chapter introduces such basic topics, as principles of microprogram control and specification of the control unit behavior using the graph-scheme of algorithm. Next, some methods of control algorithm interpretation, such as finite-state machines (FSM) and microprogram control units (MCU), are discussed. Last part of the chapter is devoted to the organization principles of compositional microprogram control units, which can be viewed as compositions of finite-state machine and microprogram control unit. These control units provide efficient interpretation of the so-called linear GSA, in which long sequences of operator vertices can be found. These sequences are called operational linear chains (OLC). Microinstructions corresponding to the components of OLC are addressed using the principle of natural microinstruction addressing. It permits to use the counter to keep microinstruction addresses and to simplify the combinational part of control unit, as compared with the classical Moore FSM. The Mealy FSM is used in CMCU to address microinstructions. It permits to calculate the transition address during one cycle of control unit's operation. Due to this feature, performance of the CMCU (proportional to the number of cycles needed to execute the control algorithm) is better than performance of the equivalent MCU with natural microinstruction addressing.

1.1 Principle of Microprogram Control

The principle of microprogram control was proposed by M. Wilkes in 1951 [68,69] and was developed by V. Glushkov [1]. According to this principle, any complex operation executed by a digital system is represented as a sequence of elementary operations of information processing. These elementary operations are named *microoperations*. An ensemble of microoperations executed during one cycle of a digital system operation is named *microinstruction*. Special *logical conditions* (status signals or flags) are used to control the order of execution of microoperations. Their values are calculated as some Boolean functions depending on the values of operands. An algorithm of execution of some operation is represented in terms of microinstructions and logical conditions is named *microprogram* [22]. A digital

A. Barkalov and L. Titarenko: Logic Synthesis for FSM-Based Control Units, LNEE 53, pp. 1–28.
springerlink.com

system with microprogram control is represented by an operational unit. The operational unit is the composition of operational automaton (OA), which is a data-path of the system, and control automaton (CA), which coordinates the interplay of all system blocks (Fig. 1.1) [1,8,9].

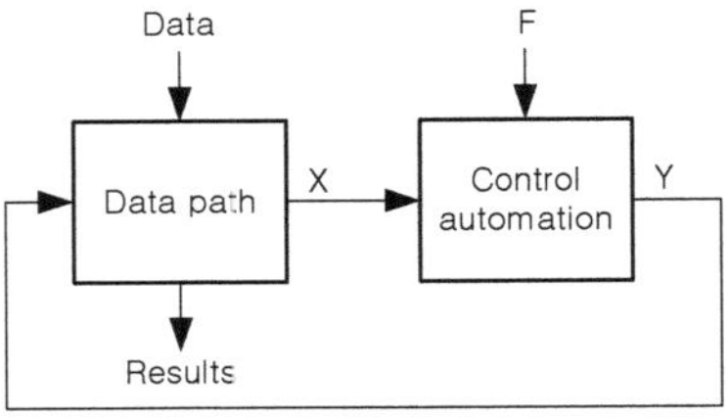

Fig. 1.1 Structural diagram of operational unit

In the operational unit, CA analyses the code of operation together with values of logical conditions from the set $X = \{x_1, \ldots, x_L\}$. Microinstructions $Y_q \subseteq Y$ are executed on the base of this analysis, where $Y = \{y_1, \ldots, y_N\}$ is a set of microoperations, which initialize operand processing and obtaining of intermediate and final results of operations, executed by the data-path. An algorithm of operational unit's operation is represented using one of the formal methods [8]. In this book we use the language of graph- schemes of algorithm (GSA) which is very popular in design practice [8,9].

Graph-scheme of algorithm Γ is the directed connected graph, characterized by a finite set of vertices, namely (Fig. 1.2): start (initial) vertex, end (final) vertex, operator and conditional vertices.

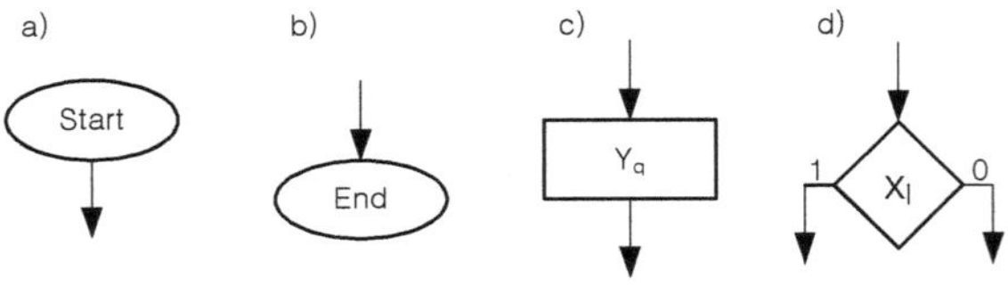

Fig. 1.2 Types of vertices of GSA

The start vertex, denoted here by the symbol b_0, corresponds to the beginning of control algorithm to be interpreted and has no input. The end vertex, denoted here by the symbol b_E, corresponds to the end of control algorithm and has no output. The operator vertex $b_t \in B_1$, where B_1 is a finite set of operator vertices of GSA Γ, contains a collection of microoperations $Y_q \subseteq Y$ which are executed in parallel. The conditional vertex $b_t \in B_2$, where B_2 is a set of conditional vertices of GSA Γ, contains single element $x_l \in X$. It has two outputs, first corresponding to value "1" and second to value "0" of the logical condition to be checked. Thus, GSA Γ is characterized by a finite set of vertices $B = B_1 \cup B_2 \cup \{b_0, b_E\}$. The vertices $b_t \in B$ are connected by arcs from a finite set $E = \{\langle b_t, b_q \rangle\}$, where $b_t, b_q \in B$.

For example, the GSA Γ_1 (Fig. 1.3) is characterized by the following sets:

- the set of vertices $B = \{b_0, b_1, \ldots b_6, b_E\}$;
- the set of arcs $E = \{\langle b_0, b_1\rangle, \langle b_1, b_2\rangle, \ldots, \langle b_6, b_E\rangle\}$;
- the set of microoperations $Y = \{y_1, \ldots, y_4\}$;
- the set of logical conditions $X = \{x_1, x_2\}$.

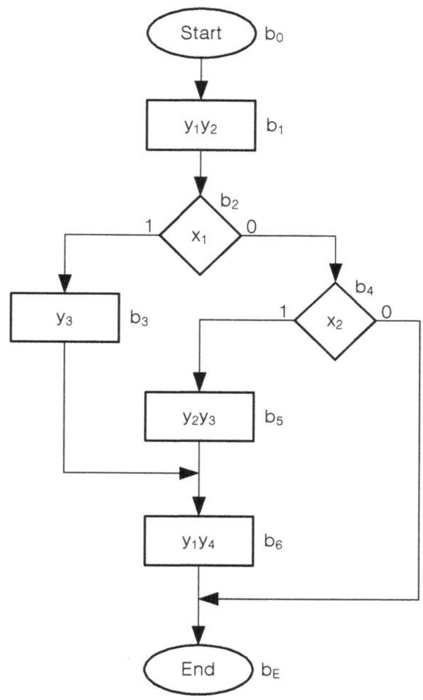

Fig. 1.3 Graph-scheme of algorithm Γ_1

A control algorithm can be implemented either as a program (program interpretation) or as a network of logic elements connected in a particular way (hardwired interpretation). In this book we discuss the methods of hardwired interpretation for control algorithms represented by GSAs. These methods could be based either on a model of a finite state machine (automaton with hardwired logic) or on the principle of keeping the microprogram in a special control memory (automaton with programmed logic) [1].

Methods of data-path design are not discussed in this book. They could be found, for example, in [1, 6, 8, 35, 47]. Let us discuss the classical methods of control units design.

1.2 Control Algorithm Interpretation with Finite State Machines

The finite state machine, called in [9] as microprogram automaton, represents a control algorithm by a classical model, which is the composition of sequential circuit CC and register RG (Fig. 1.4).

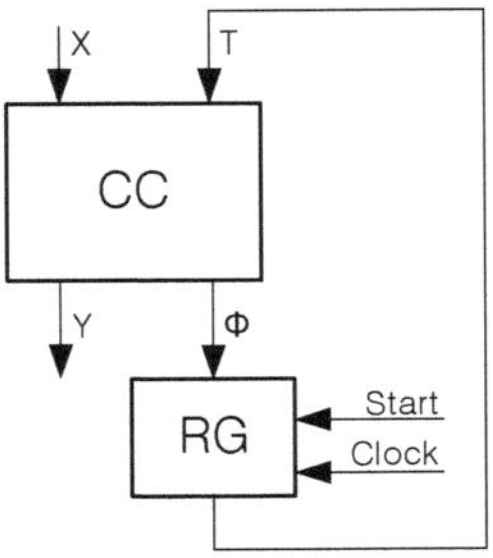

Fig. 1.4 Structural diagram of finite state machine

Presence of the register RG can be explained in the following way. The FSM produces as its output information a time-distributed microinstruction sequence $Y(0), Y(1), \ldots, Y(t)$, where t is the automaton time determined by synchronization pulse Clock. The initial instant $t = 0$ is determined by a single-shot pulse "Start". To produce such a sequence, some information about prehistory of the system operation is needed. This sequence is determined by input signals $X(0), \ldots, X(t-1)$ for previous time intervals. Thus, output signal $Y(t)$ at time t is determined by the following expression:

$$Y(t) = f(X(0), \ldots, X(t-1), X(t)). \tag{1.1}$$

Expressions of this kind are very complex and could not be easy realized in hardware, especially if they contain cycles with unpredictable number of iterations. The internal states of FSM are used to represent the prehistory of its operation. The states $a_m \in A$, where $A = \{a_1, \ldots, a_M\}$ is a set of internal states, are encoded by binary codes $K(a_m)$ having

$$R = \lceil \log_2 M \rceil \tag{1.2}$$

bits, where $\lceil A \rceil$ is the least integer, greater than or equal to A. This function is known as a ceil function or ceiling [48]. Elements of the set of state variables $T = \{T_1, \ldots, T_R\}$ are used to encode the states of FSM. The code of current state is kept in register RG, which includes R flip-flops, and common timing signal Clock is used for their synchronization. The code of initial state $a_1 \in A$ is loaded into register using pulse "Start", the content of RG can be changed by pulse Clock on the base of input memory (excitation) functions, which form the set $\Phi = \{\phi_1, \ldots, \phi_R\}$. As a rule, the register RG is implemented using D flip-flops [26, 45].

The combinational circuit CC produces both input memory functions

$$\Phi = \Phi(T,X) \tag{1.3}$$

and the output functions Y, which depend strongly on the FSM model in use [1]. In case of the Mealy FSM, system Y is represented as

$$Y = Y(T,X), \tag{1.4}$$

whereas in the case of Moore FSM these functions depend only on the states:

$$Y = Y(T). \tag{1.5}$$

The method of FSM synthesis on the base of GSA Γ includes the following steps [9]:

- construction of marked GSA Γ;
- encoding of the internal states (state assignment);
- construction of the structure table of FSM;
- construction of systems Φ and Y on the base of the structure table;
- implementation of FSM logic circuit using some logic elements.

Let us discuss some examples of FSM synthesis using GSA Γ_1 to represent the control algorithm to be interpreted.

In case of Mealy FSM, marked GSA is constructed in the following way [9]:

- the output of the initial vertex b_0 and the input of the final vertex b_E are marked by an initial state a_1 (it is a final state, too);
- inputs of vertices $b_t \in B$, connected with outputs of operator vertices, are marked by unique states $a_2,\ldots,a_M$;
- any input can be marked only once.

Application of this procedure to GSA Γ_1 leads to the marked GSA Γ_1 (Fig. 1.5a), corresponding to the graph of Mealy FSM S_1 (Fig. 1.5b). The vertices of this graph correspond to the states of Mealy FSM S_1, whereas its arcs correspond to the transitions among the states. Each arc is marked by a pair $\langle$input signal, output signal$\rangle$. Input signal $X_h (h = 1,\ldots,H)$ corresponds to conjunction of some variables from the set X (or their complements). Output signal $Y_h \subseteq Y$ corresponds to some collection of microoperations $y_n \in Y$, written into an operator vertex, which belongs to the transition h of Mealy FSM $(h = 1,\ldots,H)$. Thus, Mealy FSM S_1 is described by sets $X = \{x_1,x_2\}$, $Y = \{y_1,\ldots,y_4\}$, $A = \{a_1,a_2,a_3\}$ and has $H = 5$ transitions among its states.

There are many methods of state assignment [1,4,7,12,13,17,18,20,21,23,25, 28–34,36,37,39,40,42,44,46,47,51,53,54,56,57,59,61–63,65–67,70], targeted on optimization of hardware amount of the combinational circuit CC. These methods depend strongly on logic elements in use. Let us use a trivial encoding of states first, using minimum possible amount of state variable to encode the states. In case of the Mealy FSM S_1, we have $M = 3$, $R = 2$, $T = \{T_1,T_2\}$. Let the states are encoded

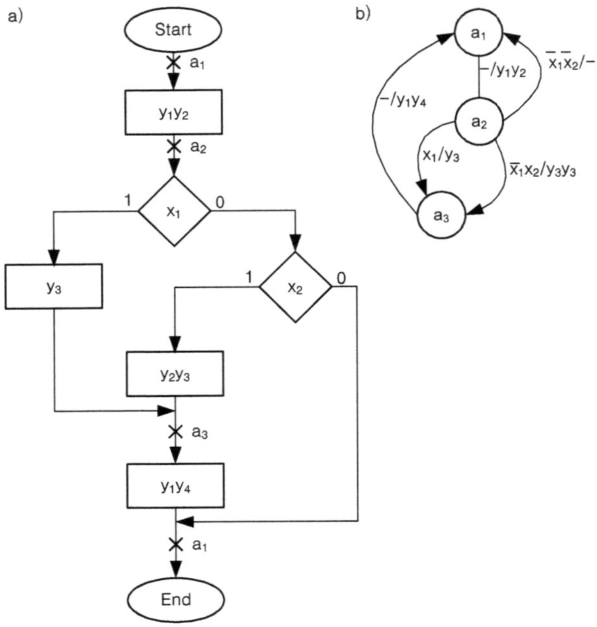

Fig. 1.5 Marked GSA $\Gamma_1\,(a)$ and graph of Mealy FSM $S_1\,(b)$

using the following codes:$K(a_1) = 00$, $K(a_2) = 01$ and $K(a_3) = 10$. Let us point out that the code $K(a_1)$ should include only zeros to simplify the circuit of setting FSM into the initial state $a_1 \in A$.

An FSM structure table (ST) can be viewed as the FSM graph represented by a list of interstate transitions. This table includes a column Φ_h with input memory functions, which are equal to 1 in order to change the states of particular FSM memory flip-flops. This table includes the following columns [9]: a_m is the current state of FSM; $K(a_m)$ is the code of the state $a_m \in A$; a_s is the state of transition (next state of FSM); $K(a_s)$ is the code of this state; X_h is the input signal determined the transition $\langle a_m, a_s \rangle$; Y_h is the output signal produced during the transition$\langle a_m, a_s \rangle$; Φ_h is the collection of input memory functions, which are equal to 1 to change the register content from $K(a_m)$ into $K(a_s)$; $h = \overline{1,H}$ is the number of transition.

Structure table is constructed in a trivial way using the automaton graph. In case of Mealy FSM S_1 this table contains $H = 5$ lines (Table 1.1).

Functions (1.3) – (1.4) are derived from the FSM structure table as the sums of products (SOP) depending on the following product terms

$$F_h = A_m X_h \quad (h = 1, \ldots, H). \tag{1.6}$$

In this formula, term A_m is the conjunction of state variables $T_r \in T$ corresponding to the code of state $a_m \in A$ from the line h of the structure table:

Table 1.1 Structure table of Mealy FSM S_1

a_m	$K(a_m)$	a_s	$K(a_s)$	X_h	Y_h	Φ_h	h
a_1	00	a_2	01	1	y_1y_2	D_2	1
a_2	01	a_3	10	x_1	y_3	D_1	2
		a_3	10	$\bar{x}_1x_2$	y_2y_3	D_1	3
		a_1	00	$\bar{x}_1\bar{x}_2$	y_1D_1	–	4
a_3	10	a_1	00	1	–	–	5

$$A_m = T_1^{l_{m_1}} \cdots T_R^{l_{m_R}}. \tag{1.7}$$

In this formula, variable $l_{m_r} \in \{0,1\}$ is the value of bit r of the code $K(a_m)$, and $T_r^0 = \bar{T}_r, T_r^1 = T_r (r = 1,\dots,R; m = 1,\dots,M)$. Systems (1.3) - (1.4) are represented as the following SOPs:

$$\phi_r = \bigvee_{h=1}^{H} C_{rh}F_h \quad (r = 1,\dots,R); \tag{1.8}$$

$$y_n = \bigvee_{h=1}^{H} C_{nh}F_h \quad (n = 1,\dots,N). \tag{1.9}$$

In these expressions, $C_{rh}(C_{nh})$ is a Boolean variable equal to 1, if and only if (iff) the line of the ST includes the variable $\phi_r(y_n)$.

For example, from Table 1.1 we get the following equations: $F_1 = \bar{T}_1\bar{T}_2$; $F_2 = \bar{T}_1T_2x_1$; $F_3 = \bar{T}_1T_2\bar{x}_1x_2$; $F_4 = \bar{T}_1T_2\bar{x}_1\bar{x}_2$, $F_5 = \bar{T}_1\bar{T}_2$; $y_1 = F_1 \vee F_4$; $y_2 = F_1 \vee F_3$; $y_3 = F_2 \vee F_3$; $y_4 = F_4$; $D_1 = F_2 \vee F_3$; $D_2 = F_1$.

Implementation of FSM circuit depends strongly on particular properties of logic elements in use. This step will be discussed a bit later.

The marked GSA of Moore FSM is constructed using the following procedure [9]:

- the vertices b_0 and b_E are marked by the initial state a_1;
- the operator vertices $b_t \in B_1$ are marked by unique states $a_2,\dots,a_M$.

Application of this procedure to the GSA Γ_1 leads to the marked GSA Γ_1 (Fig. 1.6a), corresponding to the automaton graph of Moore FSM S_2(Fig. 1.6b).

The vertices of Moore automaton graph are marked by output signals $y_n \in Y$, because of it its arcs are marked only b_y input signals determining the transitions among the states. Thus, the Moore FSM S_2 is represented by the sets $X = \{x_1, x_2\}$, $Y = \{y_1,\dots,y_4\}$, $A = \{a_1,\dots,a_5\}$, and it has $H = 7$ transitions.

In case of the Moore FSM S_2, we have $R = 3$, $T = \{T_1, T_2, T_3\}$. Let us encode its states in the following manner:$K(a_1) = 000,\dots,K(a_5) = 100$. The structure table of Moore FSM is constructed using the automaton graph (or the marked GSA). This table has the following columns: a_m, $K(a_m)$, a_s, $K(a_s)$, X_h, Φ_h, h. Information about output signals to be produced is placed into the column a_m [9]. In case of the Moore FSM S_2, the structure table contains $H = 7$ lines (Table 1.2).

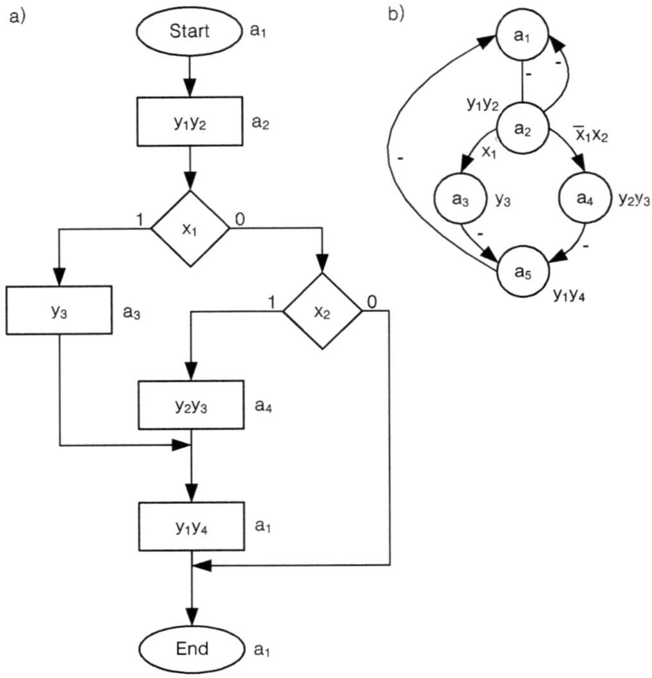

Fig. 1.6 Marked GSA $\Gamma_1(a)$ and automaton graph of Moore FSM $S_2(b)$

Boolean systems (1.3) and (1.5) are derived from the structure table; let us point out that system (1.3) depends on terms (1.6) and its SOP is similar to system (1.8). Functions (1.5) are represented in the form

$$y_n = \bigvee_{m=1}^{M} C_{nm}A_m \quad (n = 1, \ldots, N), \tag{1.10}$$

where $C_{nm}$is a Boolean variable equal to 1 iff microoperations $y_n \in Y$ are executed, when FSM is in the state $a_m \in A$.

For Moore FSM S_2, we get from Table 1.2: $F_1 = \bar{T}_1\bar{T}_2\bar{T}_3$, $F_1 = \bar{T}_1\bar{T}_2\bar{T}_3x_1, \ldots, F_7 = T_1\bar{T}_2\bar{T}_3$; $D_1 = F_5 \vee F_6$; $D_2 = F_2 \vee F_3$; $D_3 = F_1 \vee F_3$; $A_1 = \bar{T}_1\bar{T}_2\bar{T}_3$; $\ldots, A_5 = T_1\bar{T}_2\bar{T}_3$; $y_1 = A_2 \vee A_5$; $y_2 = A_2 \vee A_4$; $y_3 = A_3 \vee A_4$; $y_4 = A_5$.

Automata S_1 and S_2 are equivalent in the sense that they interpret the same GSA Γ_1. Comparison of automata S_1 and S_2 leads to the following conclusions satisfied for all equivalent Mealy and Moore automata:

- Moore FSM has, as a rule, more states and transitions than the equivalent Mealy FSM;
- system of output signals of Moore FSM has regular form, because it depends only on the states of FSM.

Table 1.2 Structure table of Moore FSM S_2

a_m	$K(a_m)$	a_s	$K(a_s)$	X_h	Φ_h	h
$a_1(-)$	000	a_2	001	1	D_3	1
$a_2(y_1y_2)$	001	a_3	010	x_1	D_2	2
		a_4	011	$\bar{x}_1x_2$	D_2	3
		a_1	000	$\bar{x}_1\bar{x}_2$	-	4
$a_3y_3)$	010	a_5	100	1	D_1	5
$a_4(y_2y_3)$	011	a_5	100	1	D_1	6
$a_5(y_1y_4)$	100	a_1	000	1	-	7

Let us point out that model of Moore FSM is used more often in practical design [48] because it offers more stable control than the control units based on the Mealy FSM model. Moreover, system (1.5) is regular, what means that it is specified for more than 50% of all possible input assignments. This regularity makes possible implementation of this system using either read-only memory (ROM) chips or random-access memory (RAM) blocks [11].

The number of product terms in the input memory functions system can be reduced due to existence of the pseudoequivalent states of Moore FSM [14]. The states $a_m, a_s \in A$ are called pseudoequivalent states of Moore FSM, if there exist the arcs $\langle b_i, b_t \rangle, \langle b_j, b_t \rangle \in E$, where vertex $b_i \in B_1$ is marked by state $a_m \in A$ and vertex $b_j \in B_1$ by state $a_s \in A$. Thus, the states a_3 and a_4 of the Moore FSM S_2 are pseudoequivalent states. They cannot be treated as equivalent states [9] because of different output signals generated for these states. As follows from Table 1.2, the columns $a_s - -\Phi_h$ of structure table for the states a_3 and a_4 contain the same information. Let $\Pi_A = \{B_1, \ldots, B_I\}$ be a partition of set A into the classes of pseudoequivalent states . For example, in the case of Moore FSM S_2 we have $\Pi_A = \{B_1, \ldots, B_4\}$, with $B_1 = \{a_1\}$, $B_2 = \{a_2\}$, $B_3 = \{a_3, a_4\}$, $B_4 = \{a_5\}$. The number of terms in system Φ can be reduced due to optimal state encoding [14], when the codes of pseudoequivalent states from some class $B_i \in \Pi_A$ belong to a single generalized interval of an R-dimensional Boolean space . For example, the well-known algorithms NOVA, ASYL or ESPRESSO [47] can be used for the state encoding mentioned above .

The optimal state encoding for the Moore FSM S_2 is shown in the Karnaugh map (Fig. 1.7).

As follows from Fig. 1.7, the class B_1 corresponds to the interval $K(B_1) = 000$, $B_2 \to K(B_2) = *01$, $B_3 \to K(B_3) = *1*$, $B_4 \to K(B_4) = 1**$, where sign "*"

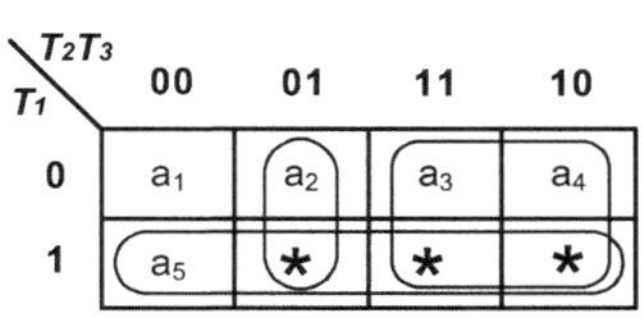

Fig. 1.7 Optimal state codes for Moore FSM S_2

determines "don't care" value of state variable $T_r \in T$. These intervals can be considered as the codes of classes $B_i \in \Pi_A$. Let us construct a transformed structure table of Moore FSM with the following columns:B_i, $K(B_i)$, a_s, $K(a_s)$, X_h, Φ_h, h. To do this we replace the column a_m by the column B_i, and the column $K(a_m)$ by the column $K(B_i)$. If the structure table transformed in this way contains equal lines, only one of them should find place in the final transformed table. For example, the transformed structure table of Moore FSM S_2 (Table 1.3) contains $H = 6$ lines.

The transformed structure table serves as the base to form product terms (1.6), but now these terms include variables $l_{mr} \in \{0, 1, *\}$, where $T_r^0 = \bar{T}_r$, $T_r^1 = T_r$, $T_r^* = 1 (m = 1, \ldots, M; r = 1, \ldots, R)$. Presence of "don't care" input assignments makes possible to reduce the number of product terms in system (1.8), which has only H_0 terms. Thus, in case of the Moore FSM S_2 we have: $F_1 = \bar{T}_1\bar{T}_2\bar{T}_3$; $F_1 = \bar{T}_2T_3x_1$; $F_3 = \bar{T}_2T_3\bar{x}_1x_2$; $F_4 = \bar{T}_2T_3\bar{x}_1\bar{x}_2$; $F_5 = T_2$; $F_6 = T_1$. We find that terms F_4 and F_6 are not the parts of SOP (1.8).

Table 1.3 Transformed structure table of Moore FSM S_2

B_i	$K(B_i)$	a_s	$K(a_s)$	X_h	Φ_h	h
B_1	000	a_2	001	1	D_3	1
B_2	*01	a_3	011	x_1	D_2D_3	2
		a_4	010	$\bar{x}_1x_2$	D_2	3
		a_1	000	$\bar{x}_1\bar{x}_2$	–	4
B_3	*1*	a_5	100	1	D_1	5
B_4	1**	a_1	000	1	–	6

It was shown in [14] that optimal state encoding permits to compress the transformed structure table of Moore FSM up to corresponding size of the equivalent Mealy FSM structure table.

As a rule, models of FSM are used for implementation of fast operational units [1]. If system performance is not important for a project, the control unit can be implemented as a microprogram control unit (MCU).

1.3 Control Algorithm Interpretation with Microprogram Control Units

Microprogram control units are based on the operational - address principle for presentation of control words (microinstructions) kept in a special control memory [1]. The typical method of MCU design includes the following steps [1]:

- transformation of initial graph-scheme of algorithm;
- generation of microinstructions with given format;
- microinstruction addressing;

- encoding of operational and address parts of microinstructions;
- construction of control memory content;
- synthesis of logic circuit of MCU using given logic elements.

The mode of microinstruction addressing affected tremendously the method of MCU synthesis [2, 3, 5]. Three particular addressing modes are used most often nowadays:

- compulsory addressing of microinstructions ;
- natural addressing of microinstructions; ;
- combined addressing of microinstructions.;

As a rule, microinstruction formats include the following fields:FY, FX, FA_0 and FA_1. The field FY, operational part of the microinstruction, contains information about microoperations $y_n \in Y$ $(t = 0, 1, \ldots)$, which are executed in cycle t of control unit operation. The field FX contains information about logical condition $x_l^t \in X$, which is checked at time $t (t = 0, 1, \ldots)$. The field FA_0 contains next microinstruction address A^{t+1} (transition address), either in case of unconditional transition (go to type), or if $x_l^t = 0$. The field FA_1 contains next microinstruction address for the case when $x_l^t = 1$. The fields FX, FA_0 andFA_1 form the address part of microinstruction.

Consider an example of MCU design with compulsory microinstruction addressing S_3 interpreting GSA Γ_1 (Fig. 1.3). The microinstruction format is shown in Fig. 1.8.

Fig. 1.8 Format of microinstructions with compulsory addressing

FY	FX	FA_0	FA_1

The address of next microinstruction A^{t+1} is determined by contents of the fields $[FX]^t$, $[FA_0]^t$ and $[FA_1]^t (t = 0, 1, \ldots)$ using the following rules:

$$A^{t+1} = \begin{cases} [FA_0]^t, \text{ if } [FX]^t = \emptyset; \\ [FA_0], \text{ if } x_l^t = 0; \\ [FA_1], \text{ if } x_l^t = 1. \end{cases} \tag{1.11}$$

First line of expression 1.11 determines the address of transition in case of unconditional jump, whereas the second and third lines determine this address for the conditional jump.

Structural diagram of MCU with compulsory microinstruction addressing (Fig. 1.9) includes the following blocks [13]:

- sequencer CFA, calculating transition address from (1.11);
- register of microinstruction address RAMI, keeping address A^t;
- control memory CM, keeping microinstructions;
- block of microoperation generation CMO;
- fetch flip-flop TF used to organize the stop mode of the MCU.

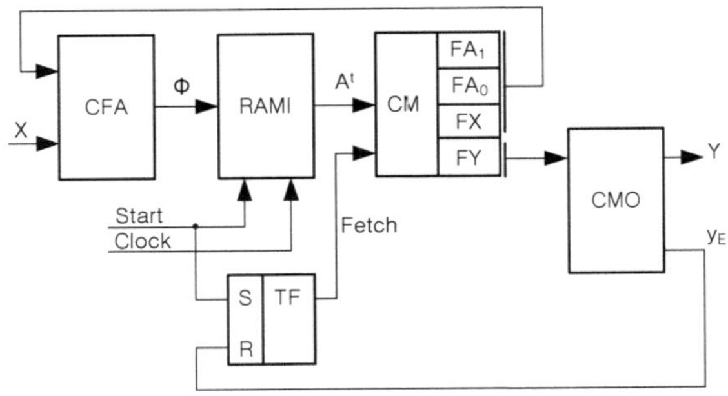

Fig. 1.9 Structural diagram of MCU with compulsory addressing

The control unit S_3 operates as follows. The pulse "Start" is used to load the address of first microinstruction to be executed (start address) into RAMI. At the same time the flip-flop TF is set up; signal Fetch=1 initiates reading of a microinstruction from the control memory. Let some address A^t be located in the register RAMI at time t $(t = 0, 1, \ldots)$. Corresponding microinstruction is then fetched from the memory CM. The operational part of this microinstruction is next transformed by the block CMO into microoperations $y_n \in Y$, which are directed to a system data-path. The sequencer CFA processes both the microinstruction address part and logical conditions X to produce the functions Φ, which form a transition address A^{t+1} sent into register RAMI. This address is loaded into RAMI by synchronization pulse "Clock". If the end of microprogram is reached, then special signal y_E is generated to clear the flip-flop TF. It causes termination of microinstruction fetching from memory CM, which means the end of MCU operation.

The transformation of initial GSA Γ is executed using the following rules [10]:

- if there is an arc $\langle b_q, b_E \rangle \in E$, such that $b_q \in B_1$, the variable y_E is assigned to the vertex b_q;
- if there is an arc $\langle b_q, b_E \rangle \in E$, such that $b_q \in B_2$, an additional operator vertex b_{Q+1} $(Q = |B| - 2)$ with the variable y_E is inserted into GSA Γ, and the arc $\langle b_q, b_E \rangle$ is replaced by arcs $\langle b_q, b_{Q+1} \rangle$ and $\langle b_{Q+1}, b_E \rangle$.

Therefore, the transformation of GSA for MCU with compulsory addressing of microinstructions is necessary to organize the ending mode of the MCU. Thus, transformation of the GSA Γ_1 is reduced to inserting the variable y_E into the vertex $b_6 \in B_1$ and adding the vertex b_7. The transformed GSA $\Gamma_1(S_3)$ thus obtained is shown in Fig. 1.10.

Generation of microinstructions with compulsory addressing is reduced to successive analysis of pairs of vertices$\langle b_q, b_t \rangle \in E$. All possible vertices pair configurations are shown in Fig. 1.11

There are four possible configurations:

- $b_q, b_t \in B_1$ (Fig. 1.11a). In this case the vertex $b_q \in B$ corresponds to microinstruction with empty fields FX and FA_1, whereas its field FY contains the set of microoperations Y_q and field FA_0 contains the microinstruction address, corresponding to vertex $b_t \in B_1$. The analysis should be continued for the vertex

$b_t \in B_1$. If, in such a pair, second vertex is the final vertex of GSA ($b_t = b_E$), then the vertex b_q corresponds to microinstruction with empty fields FX, FA_0 and FA_1;

- $b_q \in B_1$, $b_t \in B_2$ (Fig. 1.11b). In this case, the pair of vertices corresponds to one microinstruction with all fields containing useful information;
- $b_q, b_t \in B_2$ (Fig. 1.11c). In this case the vertex $b_t \in B_2$ corresponds to microinstruction with empty field FY, and the analysis should be continued for the vertex $b_q \in B_2$;
- $b_q \in B_2$, $b_t \in B_1$ (Fig. 1.11d). In this case the analysis should be continued for the both vertices of the pair.

Let us denote microinstructions by symbols $O_m (m = 1, \dots, M)$; now the following microinstructions can be generated using the transformed GSA $\Gamma_1(S_2)$: $O_1 = \langle b_1, b_2 \rangle$, $O_2 = \langle b_3, \emptyset \rangle$, $O_3 = \langle \emptyset, b_4 \rangle$, $O_4 = \langle b_5, \emptyset \rangle$, $O_5 = \langle b_6, \emptyset \rangle$, $O_6 = \langle b_7, \emptyset \rangle$.

Addresses of microinstructions with compulsory addressing can be appointed in the following manner. Each microinstruction O_m corresponds (one-to-one) to a binary code A_m with $R = \lceil \log_2 M \rceil$ bits $(m = 1, \dots, M)$. A microinstruction with start address is determined by the arc $\langle b_0, b_q \rangle \in E$. In the case under consideration there is the arc $\langle b_0, b_1 \rangle \in E$ (Fig. 1.10), and therefore the start address belongs to the microinstruction O_1, corresponding to the pair with vertex $b_1 \in B_1$. All other microinstructions are addressed in arbitrary manner.

The microprogram of MCU $S_3(\Gamma_1)$ includes $M = 6$ microinstructions, thus $R = 3$; it is clear that $A_1 = 000$. Let $A_2 = 001, \dots, A_6 = 101$.

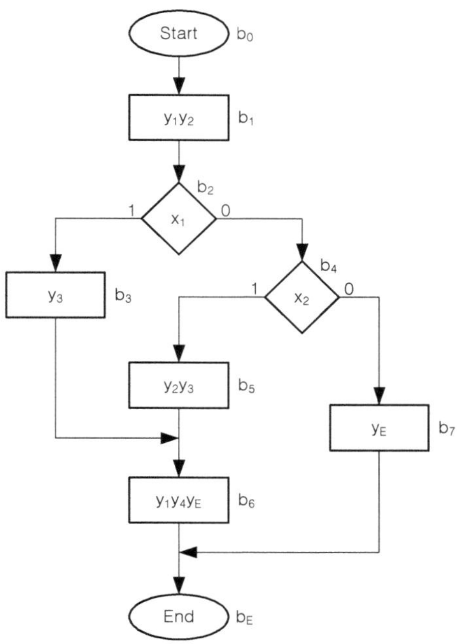

Fig. 1.10 Transformed GSA $\Gamma_1(S_3)$

Because the control memory can keep only some bit strings, the encoding of operational and address parts of microinstructions is necessary to load microinstructions into the control memory. Addressing of microinstructions gives information, which should be written into the fields FA_0 and FA_1.There are many methods to encode operational part of microinstructions [11]. Let us choose the one-hot encoding of microoperations to design the control memory of MCU $S_3(\Gamma_1)$, where $S_3(\Gamma_1)$ means that the GSA Γ_1 is interpreted by MCU with compulsory addressing of microinstructions. In case of one-hot encoding the length (bit capacity) n_1 of the field FY is determined by the following formula:

$$n_1 = N + 1. \tag{1.12}$$

For MCU $S_3(\Gamma_1)$ this formula gives the value $n_1 = 5$.

Let us encode logical conditions $x_l \in X$ using binary codes with minimum length (called sometimes minimal-length codes)

$$n_2 = \lceil \log_2(L+1) \rceil . \tag{1.13}$$

The value 1 is added into (1.13) in order to take into account the code for unconditional jump, when $[FX] = \emptyset$. For MCU $S_3(\Gamma_1)$ this formula gives the value $n_2 = 2$. Let $K(\emptyset) = 00$; $K(x_1) = 01$; $K(x_2) = 10$.

Construction of the control memory content results in construction of a table with lines keeping microinstruction addresses and binary codes of particular microinstructions. Control memory of MCU S_3 keeps M microinstructions with

$$n_3 = n_1 + n_2 + 2R \tag{1.14}$$

bits. In case of MCU $S_3(\Gamma_1)$ this formula gives the value $n_3 = 13$. The control memory content for MCU $S_3(\Gamma_1)$ is shown in Table 1.4.

In this table, microinstruction addresses are represented by variables from the set $A = \{a_1, a_2, a_3\}$, whereas the codes of microinstructions are represented by variables

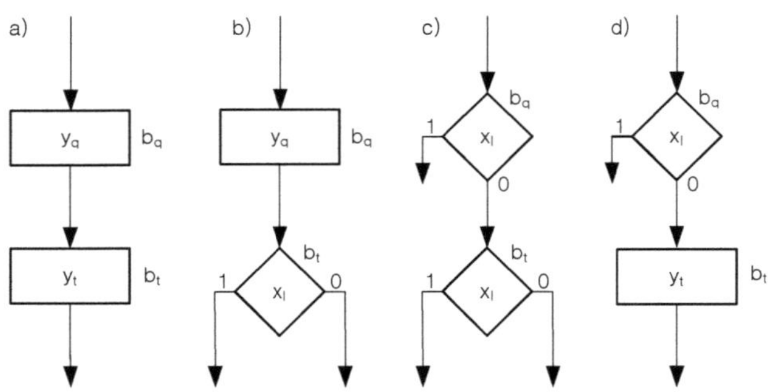

Fig. 1.11 Possible configurations for pairs of GSA vertices

Table 1.4 Control memory content for MCU $S_3(\Gamma_1)$

Address $a_1a_2a_3$	FY $v_1v_2v_3v_4v_5$	FX v_6v_7	FA_0 $v_8v_9v_{10}$	FA_1 $v_{11}v_{12}v_{13}$	Formula of transition
000	11000	01	010	001	$O_1 \to \bar{x}_1O_3 \vee x_3O_2$
001	00100	00	100	000	$O_2 \to O_5$
010	00000	10	101	011	$O_3 \to \bar{x}_2O_6 \vee x_2O_4$
011	01100	00	100	000	$O_4 \to O_5$
100	10011	00	000	000	$O_5 \to End$
101	00001	00	000	000	$O_6 \to End$

from the set $V = \{v_1, \ldots, v_{13}\}$, where $|A| = R$, $|V| = n_3$. The last column of the table contains formula of transitions for microinstructions, which are direct analogues of the formulae of transitions for operators of GSA [9].

Analysis of this table shows the main drawbacks of MCU with compulsory addressing of microinstructions, such as:

- an empty field FY for microinstructions, corresponding to the pairs $\langle \emptyset, b_t \rangle$, where $b_t \in B_2$;
- empty fields FX and FA_1 for microinstructions, corresponding to the pairs $\langle b_t, \emptyset \rangle$, where $b_t \in B_1$.

It results in the inefficient use of control memory volume, but a positive feature of compulsory addressing is the minimum number of microinstructions for the particular GSA, in comparison with MCU with other modes of microinstruction addressing [1]. Synthesis of the logic circuit of MCU S_3 is reduced to the implementation of block CFA using standard multiplexers and control memory using standard memory blocks, such as PROM or RAM chips [1]. Let us point out that some logic elements should be used to implement the block CMO [1].

Assume that content of the field FA_1 is loaded into register RAMI if $z_1 = 1$, otherwise (if $z_1 = 0$) RAMI is loaded from the field FA_0 of current microinstruction. Thus, expression (1.11) can be represented as

$$A^{t+1} = \bar{z}_1[FA_0] \vee z_1[FA_1]. \tag{1.15}$$

The variable $z_1 = 1$, if a logical condition to be checked is equal to 1; it means that

$$z_1 = \bigvee_{l=1}^{L} V_l x_l, \tag{1.16}$$

where V_l is a conjunction of variables $v_r \in V$, corresponding to the code $K(x_l)$ $(l = 1, \ldots, L)$.

In case of the MCU $S_3(\Gamma_1)$ expression (1.15) is represented as

$$\begin{aligned} a_1 &= \bar{z}_1 v_8 \vee z_1 v_{11}; \\ a_2 &= \bar{z}_1 v_9 \vee z_1 v_{12}; \end{aligned} \tag{1.17}$$

$$a_3 = \bar{z}_1 v_{10} \vee z_1 v_{13},$$

and expression (1.16) has now the form

$$z_1 = \bar{v}_1 \bar{v}_2 0 \vee \bar{v}_1 v_2 x_1 \vee v_1 \bar{v}_2 x_2. \quad (1.18)$$

This formula specifies standard multiplexer with two control inputs and three data inputs in use. The first term of expression (1.18) corresponds to unconditional jump. Symbol "0" represents the fact that logic 1 should be connected with informational input of the multiplexer corresponding to code 00; variables a_r from (1.17) coincide with variables $D_r (r = 1, \ldots, R)$. Expressions (1.17) – (1.18) determine the logic circuit of sequencer CFA of the MCU $S_3(\Gamma_1)$, shown in Fig. 1.12. Operation of this circuit can be easily deduced from Fig. 1.12.

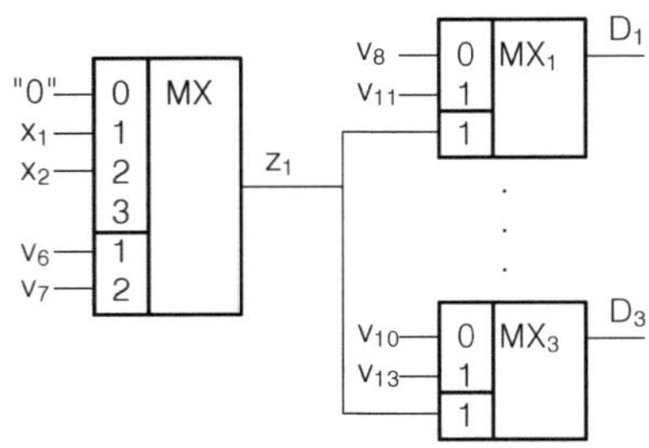

Fig. 1.12 Logic circuit of CFA of MCU $S_3(\Gamma_1)$

There are two microinstruction formats in case of natural microinstruction addressing [1, 12]: operational microinstructions corresponding to operator vertices of GSA Γ and control microinstructions corresponding to conditional vertices of GSA Γ (Fig. 1.13).

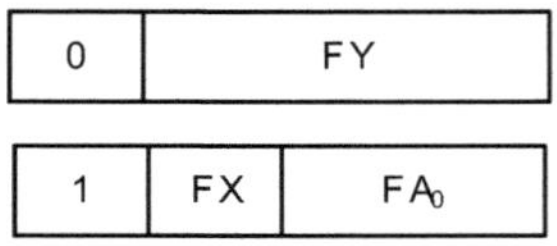

Fig. 1.13 Microinstruction formats for MCU with natural addressing of microinstructions

First bit of each format represents field FA, used to recognize the type of microinstruction. Let $FA = 0$ correspond to operational microinstruction and $FA = 1$ to control microinstruction. As follows from Fig. 1.13, next address is not included into operational microinstructions. The same is true for the case, when a logical condition to be checked is equal to 1. In both cases mentioned above current address A^t is used to calculate next address:

$$A^{t+1} = A^t + 1. \quad (1.19)$$

Hence the following rule is used for next address calculation:

$$A^{t+1} = \begin{cases} A^t + 1, \text{ if } [FA]^t = 0; \\ A^t + 1, \text{ if } (x_l^t) \wedge ([FA]^t = 1), \\ [FA_0]^t, \text{ if } (x_l^t = 0) \wedge ([FA]^t = 1); \\ [FA_0]^t, \text{ if } ([FX]^t = \emptyset) \wedge ([FA]^t = 1). \end{cases} \quad (1.20)$$

Analysis of (1.20) shows that MCU with natural addressing of microinstructions should include a counter CAMI. Corresponding structure is shown in Fig. 1.14.

Fig. 1.14 Structural diagram of MCU with natural addressing of microinstructions

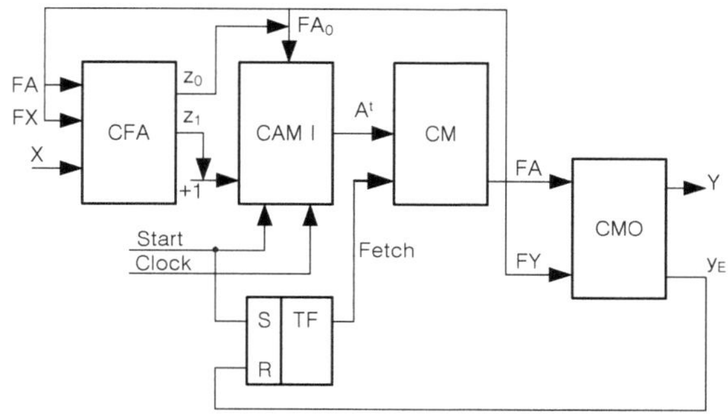

This MCU operates in the following manner. The pulse "Start" initiates loading of start address into CAMI. At the same time flip-flop TF is set up. Let an address A^t be located in CAMI at time t $(t = 0, 1, \ldots,)$. If this address determines an operational microinstruction, the block CMO generates microoperations $y_n \in Y$, and the sequencer CFA produces signal z_1. If this address determines a control microinstruction, microoperations are not generated, and the sequencer produces either signal z_0 (corresponding to an address loaded from the field FA_0), or signal z_1 (it corresponds to adding 1 to the content of CAMI). The content of counter CAMI can be changed by pulse "Clock". If variable y_E is generated by CMO, then the flip-flop TF is cleared and operation of MCU terminated.

Let symbol S_4 stand for this kind of MCU. Now we use an example of MCU $S_4(\Gamma_1)$ to discuss some particular problems of such a design.

The transformation of initial GSA is executed in two consecutive steps. First step involves the same transformations as in case of MCU S_3. Addressing conflicts between microinstructions [1,12] are eliminated during the second step. Let us point out that in case of MCU S_4 operational microinstructions correspond to operator vertices $b_q \in B_1$ and control microinstructions correspond to conditional vertices $b_q \in B_2$. Nature of addressing conflicts is the consequence of implicit transition addresses, as expressed by (1.19).

Let some GSA include two arcs $\langle b_i, b_q \rangle, \langle b_j, b_q \rangle \in E$, where $b_i, b_j \in B_1$ (Fig. 1.15a). Let indexes of vertices, corresponding microinstructions and microinstruction addresses be the same and take $A_i = 100$. According to (1.19) we find that

$A_q = 101$, which means that the address A_j should be equal to 100. Thus, microinstructions O_i and O_j should have the same address. We call this situation addressing conflict. Some conditional vertex b_t with logical condition x_0 should be inserted in the initial GSA to eliminate this conflict. This condition corresponds to the unconditional jump, when $FX = \emptyset$ (Fig. 1.16a).

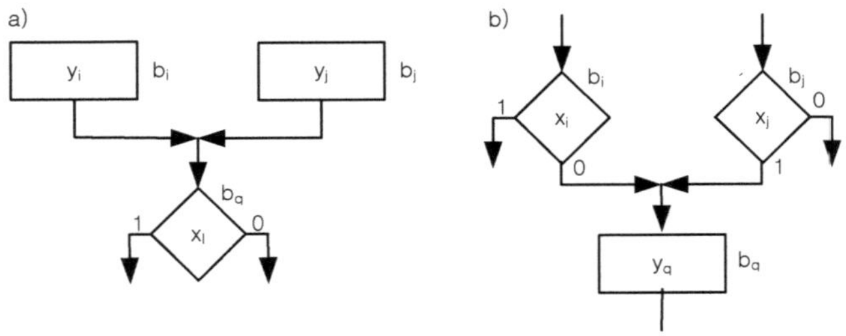

Fig. 1.15 Addressing conflicts in MCU S_4

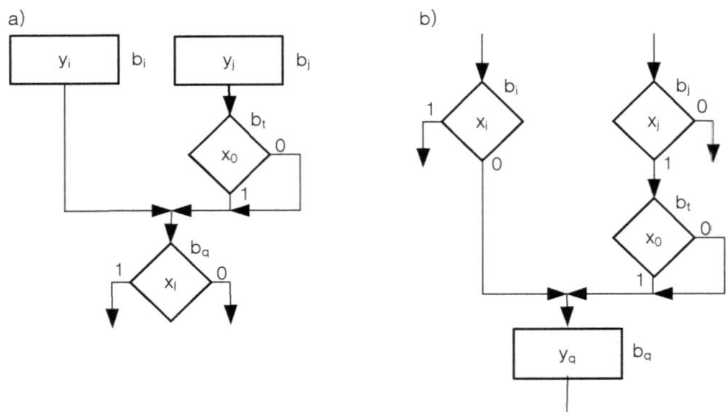

Fig. 1.16 Elimination of addressing conflicts

Now, if $A_i = 100$, we have $A_q = 101$, and the field FA_0 of microinstruction O_t contains address $A_q = 101$. Addressing conflict is possible also between control microinstructions (Fig. 1.15b), and its elimination requires inserting of some additional vertex (Fig. 1.16b).

Let us point out that GSA subgraphs, similar to ones shown in Fig. 1.15, can have arbitrary number of vertices. Addressing conflicts can arise also among operational microinstructions and control microinstructions [11].

The transformed GSA $\Gamma_1(S_4)$ contains $M = 8$ vertices (Fig. 1.17). As the result of transformation, variable y_E is inserted into vertex b_6, vertex b_7 with y_E is added, and vertex b_8 is also added, to eliminate addressing conflict between microinstructions corresponding to vertices b_3 and b_5.

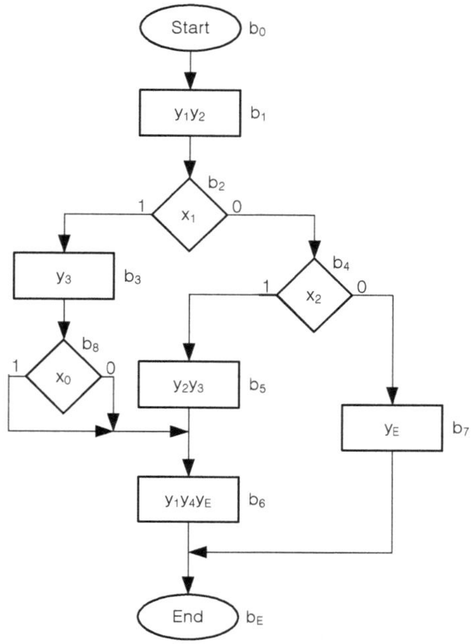

Fig. 1.17 Transformed GSA $\Gamma_1(S_4)$

As it was mentioned above, each operator vertex corresponds to an operational microinstruction and each conditional vertex corresponds to a control microinstruction. It means that microinstructions are generated in a very simple way. For example, the microprogram of MCU $S_4(\Gamma_1)$ includes $M = 8$ microinstructions.

Generation of special microinstruction sequences is needed in case of natural addressing of microinstructions. These sequences are created as follows. Let us build a set $I(\Gamma)$, elements of which are inputs of the sequences. Vertex $b_q \in B_1 \cup B_2$ is the input of a sequence, if the input of this vertex is connected either with the output of vertex b_0 or with the output of conditional vertex, marked as "0".

In case of the MCU $S_4(\Gamma_1)$, the set $I(\Gamma) = \{b_1, b_4, b_7\}$ and microinstruction sequences are started by corresponding microinstructions. End points of these sequences are microinstructions corresponding to vertices connected with final vertex b_E, or conditional vertex with x_0. Let α_g denote a microinstruction sequence. There are three such sequences in case of MCU $S_4(\Gamma_1)$, namely $\alpha_1 = \langle O_1, O_2, O_3, O_8 \rangle$, $\alpha_2 = \langle O_4, O_5, O_6 \rangle$, $\alpha_3 = \langle O_7 \rangle$. The zero address is assigned to the microinstruction corresponding to vertex b_t, where $\langle b_0, b_t \rangle \in E$. Addresses of next microinstructions belonging to this sequence are calculated according to (1.19). The address of current sequence input is calculated by adding 1 to the address of last microinstruction from previous sequence, and so on. Application of this procedure to the case of MCU $S_4(\Gamma_1)$, when $R = 3$, results in the microinstruction addresses shown in Table 1.5.

Encoding of operational and address parts of microinstructions is executed in the same manner as in case of MCU S_3. Let us take the case of MCU $S_4(\Gamma_1)$, and

Table 1.5 Microinstruction addresses of MCU $S_4(\Gamma_1)$

O_m	A_m	O_m	A_m	O_m	A_m	O_m	A_m
O_1	000	O_3	010	O_4	100	O_6	110
O_2	001	O_8	011	O_5	101	O_7	111

use one-hot codes for microoperations ($n_1 = 5$), as well as minimal-length codes for logical conditions ($n_2 = 2$). Let corresponding codes for both MCU $S_3(\Gamma_1)$ and $S_4(\Gamma_1)$ be the same.

Construction of the control memory content is executed due to the fact that the usage of microinstruction bits depends on microinstruction type. The control memory of MCU S_4 contains M microinstructions with

$$n_4 = \max(n_1 + 1, n_2 + R + 1) \tag{1.21}$$

bits; for example, for MCU $S_4(\Gamma_1)$ it can be found that $n_4 = 6$. The control memory content of MCU $S_4(\Gamma_1)$ is shown in Table 1.6.

Table 1.6 Microinstruction addresses of MCU $S_4(\Gamma_1)$

Address	FA	FX FA_0 FY	Formula of transitions
$a_1a_2a_3$	v_1	$v_2v_3v_4v_5v_6$	
000	0	11000	$O_1 \to O_2$
001	1	01100	$O_2 \to \bar{x}_1 O_4 \vee x_1 O_3$
010	0	00100	$O_3 \to O_8$
011	1	00110	$O_8 \to \bar{x}_0 O_6 \vee x_0 O_6$
100	1	10111	$O_4 \to \bar{x}_2 O_7 \vee x_2 O_5$
101	0	01100	$O_5 \to O_6$
110	0	10011	$O_6 \to End$
111	0	00001	$O_7 \to End$

Let us discuss now the design of sequencer CFA for MCU S_4. The variable $z_1 = 1$ should be generated either if $x_l^t = 1$ or when an operational microinstruction is executed at time t. Thus, the logical expression for calculation of z_1 can be obtained by the following transformation of expression (1.16):

$$z_1 = (\bigvee_{l=1}^{L} V_l x_l) \vee \bar{v}_1, \tag{1.22}$$

where $v_1 = 0$ corresponds to FA=0. It is clear that $z_0 = \bar{z}_1$.

Let the counter CAMI have input C_1, used to increment the counter content and input C_2 to load the input parallel code into the counter under the influence of pulse "Clock". The corresponding Boolean expressions for C_1 andC_2 have the form:

$$C_1 = z_1 \cdot Clock, \\ C_2 = \bar{z}_1 \cdot Clock. \quad (1.23)$$

Expressions (1.23) serve to design the logic circuit of sequencer CFA for MCU $S_4(\Gamma_1)$ shown in Fig. 1.18.

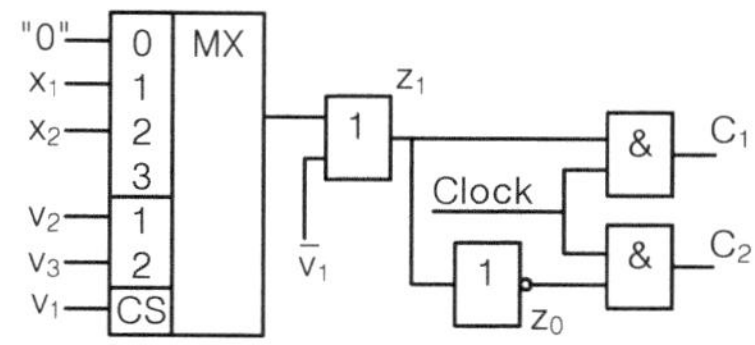

Fig. 1.18 Implementation of the block CFA for MCU $S_4(\Gamma_1)$

In this circuit multiplexer MX is active if $v_1 = 1$ is applied to the "enable" input CS of the chip. It corresponds to a control microinstruction. Remaining elements of this circuit follow directly from expressions (1.22) and (1.23). The methods of control memory implementation will be discussed later.

The comparative analysis of Tables 1.4 and 1.6 shows that MCU S_4 is characterized by longer microprogram, than the equivalent MCU S_3. In case of MCU S_4, control algorithm execution requires more time, than in case of the equivalent MCU S_3. A positive feature of MCU S_4 is smaller microinstruction length. In case of our example we find that $n_3 = 2,17n_4$.

Microprogram control units with combined microinstruction addressing (Fig. 1.19) represent a compromise settlement with average number of microinstructions, of bit capacity and of control algorithm execution time.

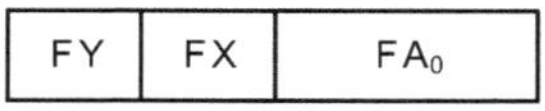

Fig. 1.19 Microinstruction format with combined addressing

In this case, transition address is described by the expression:

$$A^{t+1} = \begin{cases} [FA_0], & \text{if } |FX|^t = \emptyset; \\ [FA_0]^t, & \text{if } x_l^t = 0; \\ [FA_1]^t, & \text{if } x_l^t = 1. \end{cases} \quad (1.24)$$

It follows from (1.24), that addressing conflicts are possible only between microinstructions with $[FX] \neq \emptyset$. A design method, which can be used for MCU with combined microinstruction addressing, can be found in [1, 11].

Microprogram control units were very popular in the past [1, 12, 13, 16, 19, 24, 27, 38, 41, 43, 52, 55, 58, 60, 64], but they have one serious disadvantage, namely inferior performance in comparison with the equivalent finite state machines. As a rule, only single logical condition is checked during one cycle of MCU operation. Thus, multidirectional transitions depending on $k > 1$ logical conditions need $k > 1$ cycles for its execution, and the controlled data-path will have $k-1$ idle cycles, when its resources are not in use. Positive feature of MCU is the use of regular control memory to implement the microinstruction system, because MCU are Moore FSM [1]. Besides, the sequencer CFA is very simple and can be implemented using standard multiplexers. As a rule, any change in the control algorithm leads to the redesign of corresponding FSM, but only small modifications of the control memory content in the equivalent MCU are needed.

1.4 Organization of Compositional Microprogram Control Units

The properties of the interpreted control algorithm have great influence on the hardware amount of corresponding control unit [11]. One of such properties is the existence of operational linear chains corresponding to the paths of GSA, which include operator vertices only. Let us call a GSA Γ the linear GSA (LGSA), if the number of its operator vertices exceeds 75% of the total number of vertices. Existence of operational linear chains allows simplification of input memory functions and reduction of hardware amount in the logic circuit of control unit. In this case either shift register [50] or up counter [4,49] is used to keep state codes.

One of the approaches for linear GSA interpretation is the use of compositional microprogram control units (CMCU), which can be viewed as a composition of the finite state machine and microprogram control unit [15]. These units have several particularities, distinguishing them from other control units:

1. Microinstruction format includes the operational part only. It permits to minimize the control word bit capacity. Thus control words kept in control memory have minimum possible length in comparison with all organizations of MCU mentioned above.
2. Microprograms have minimum possible length (the number of microinstructions), because the CMCU control memory is free from control microinstructions.
3. Multidirectional transitions are executed in one cycle of CMCU operation. It provides minimum time of control algorithm interpretation. Thus, such control units have similar performance, as compared with the equivalent FSM.

Let us introduce some definitions helping to understand the features of CMCU.

Definition 1.1. An operational linear chain (OLC) of GSA Γ is a finite vector of operator vertices $\alpha_g = \langle b_{g1}, \ldots, b_{gF_g} \rangle$, such that an arc $\langle b_{g_i}, b_{g_{i+1}} \rangle \in E$ corresponds to each pair of adjacent vertices $b_{g_i}, b_{g_{i+1}}$, where i is the component number of vector α_g.

Let D^g be a set of operator vertices, which are components of OLC α_g.

Definition 1.2. An operator vertex $b_q \in D^g$ is called an input of OLC α_g, if there is an arc $\langle b_t, b_q \rangle \in E$, such that $b_t \notin D^g$.

Definition 1.3. An input $b_q \in D^g$ is called a main input of OLC α_g, if GSA Γ does not include an arc $\langle b_t, b_q \rangle \in E$ such that $b_t \in B_1$.

Definition 1.4. An operator vertex $b_q \in D^g$is called an output of OLC α_g, if there is an arc $\langle b_q, b_t \rangle \in E$, where $b_t \notin D^g$.

It follows from the basic properties of GSA [9] that each OLC α_g corresponding to definitions given above should have at least one input and exactly one output. Let I_g^j stand for input j of OLC α_g and O_g for its output. Let inputs of OLC α_g form a set $I(\alpha_g)$.

For GSA Γ we have the following sets:

1. A set of OLC $C = \{\alpha_1, \ldots, \alpha_G\}$, satisfying the following condition

$$\begin{aligned} &D^1 \cup \ldots \cup D^G = B_1; \\ &|D^i \cap D^j| = 0 \quad (i \neq j; i, j \in \{1, \ldots, G\}); \\ &G \to \min. \end{aligned} \tag{1.25}$$

2. A set of inputs $I(\Gamma)$ of the operational linear chains of GSA Γ:

$$I(\Gamma) = \bigcup_{g=1}^{G} I(\alpha_g). \tag{1.26}$$

3. A set of outputs $O(\Gamma)$ of the operational linear chains of GSA Γ:

$$O(\Gamma) = \{O_1, \ldots, O_G\}. \tag{1.27}$$

Let the natural microinstruction addressing be executed for microinstructions corresponding to the adjacent components of each OLC $\alpha_g \in C$:

$$A(b_{g_{i+1}}) = A(b_{g_i}) + 1 \quad (i = 1, \ldots, F_g - 1). \tag{1.28}$$

In expression (1.28) symbol $A(b_{g_i})$ stands for the address of microinstruction corresponding to component i of vector $\alpha_g \in C$, where $i = 1, \ldots, F_g - 1$.

In this case GSA Γ can be interpreted by compositional microprogram control unit with basic structure of Fig. 1.20 [15]. Let us denote it as unit U_1.

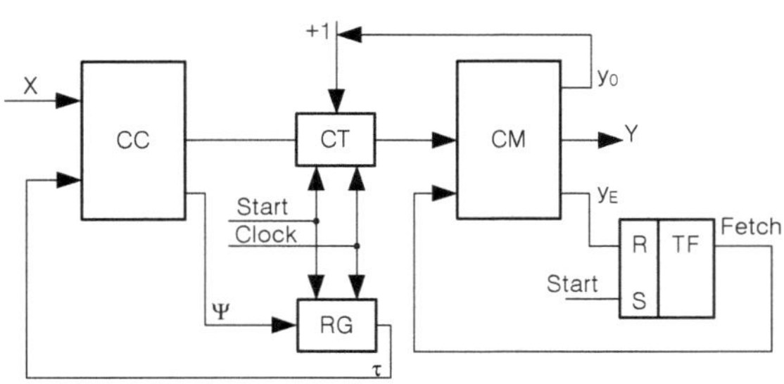

Fig. 1.20 Structural diagram of compositional microprogram control unit with basic structure

In the unit U_1, combinational circuit CC and register RG form a finite state machine S_1, which will be called microinstruction addressing unit or FSM S_1. Counter CT, control memory CM and flip-flop TF form microprogram control unit S_2 with natural microinstruction addressing. The unit U_1 operates in the following manner.

The pulse "Start" initializes following actions: the zero code of FSM S_1 initial state is loaded into register RG; start address of microprogram is loaded into counter CT; flip-flop TF is set up (Fetch=1). If Fetch=1, microinstructions can be fetched out of the control memory. Let at time t ($t = 0, 1, 2, \ldots$) the code of state $a_m \in A_1$, where A_1 is a set of FSM S_1 states, be loaded into the register RG and address $A(I_g^j)$ of the input j of OLC $\alpha_g \in C$ be loaded into the counter CT. Current microinstruction is read out of CM and its microoperations $y_n \in Y$ initialize some actions of the datapath. If this input is not the output of current OLC $\alpha_g \in C$ ($I_g^j \neq O_g$), additional variable $y_0 = 1$ is generated by MCU S_2. If $y_0 = 1$, the content of register RG is unchangeable and 1 is added to the content of counter CT. It corresponds to a transition between adjacent components of OLC $\alpha_g \in C$. If the output O_g is reached, then $y_0 = 0$. In this case circuit CC generates Boolean functions:

$$\Phi = \Phi(\tau, X), \tag{1.29}$$

$$\Psi = \Psi(\tau, X), \tag{1.30}$$

where $\tau = \{\tau_1, \ldots, \tau_{R_1}\}$ is a set of state variables encoding states $a_m \in A_1$. The minimum number of these variables is determined as

$$R_1 = \lceil \log_2 M_1 \rceil, \tag{1.31}$$

where $M_1 = |A_1|$. If there is a transition from output $O_g$to some input under influence of some values of logical conditions, functions (1.29) determine the address of this input $I_i^j \in I(\Gamma)$ which is to be loaded into the counter. Functions (1.30) calculate the code of next state $a_s \in A_1$ to be loaded into RG. Content of both CT and RG is changed by the pulse "Clock". Outputs of the CT, $T = \{T_1, \ldots, T_{R_2}\}$ determine next microinstruction address. This set includes

$$R_2 = \lceil \log_2 M_2 \rceil \tag{1.32}$$

variables, where $M_2 = |B_2|$. If CT contains the address of microinstruction corresponding to vertex $b_q \in B_1$ such that $\langle b_q, b_E \rangle \in E$, some additional variable $y_E = 1$ is generated. If $y_E = 1$, the flip-flop TF is cleared. Thus Fetch=0 and microinstruction fetching from the control memory is terminated.

As follows from (1.29), FSM S_1 of unit U_1 implements any multidirectional microprogram transition between output $O_g \in O(\Gamma)$ and input $I_i^j \in I(\Gamma)$ in one cycle of CMCU operation. At the same time MCU S_2 implements addressing rule (1.28), used to organize transitions between microinstructions corresponding to adjacent components of OLC $\alpha_g \in C$. Therefore, control memory CM should only keep microoperations $y_n \in Y$ and additional variables y_0, y_E. In other words, an address part is absent in the microinstruction format in case of CMCU U_1. The main disadvantage of CMCU U_1 is the loss of universality, because changes in the interpreted microprogram lead to the redesign of circuit CC. Fortunately, as it will be shown below, current achievements in semiconductor technology permit to eliminate this drawback.

There are two main methods used to decrease the number of outputs of the block CC of CMCU U_1 [11]:

1. The counter CT is used to represent both address of microinstruction and code of OLC. It results in CMCU with common memory [11].
2. Microinstruction address can be represented as concatenation:

$$A(b_t) = K(\alpha_g) * K(b_t), \tag{1.33}$$

where $A(b_t)$ is an address of microinstruction corresponding to the vertex $b_t \in B_1$, $K(\alpha_g)$ is a code of OLC, including the vertex b_t, $K(b_t)$ is a code of the vertex $b_t \in B_1$, as a component of OLC. It leads to CMCU with code sharing [11].

In our book, the methods of synthesis are discussed, which target on application-specific integrated circuits (ASIC) and standard programmable logic devices (PLD). These methods depend strongly on logic elements in use. Because of it, we should analyze main features of ASIC and PLD. In our book we use the terms *control unit*, *control automata* and *FSM* as synonyms.

References

1. Adamski, M., Barkalov, A.: Architectural and Sequential Synthesis of Digital Devices. University of Zielona Góra Press, Zielona Góra (2006)
2. Agerwala, T.: Microprogram optimization: A survey. IEEE Transactions of Computers (10), 962–973 (1976)
3. Agrawala, A., Rauscher, T.: Foundations of Microprogramming. Academic Press, New York (1976)
4. Amann, R., Baitinger, U.: Optimal state chains and states codes in finite state machines. IEEE Transactions on Computer-Aided Design 8(2), 153–170 (1989)
5. Anceau, F.: The Architecture of Microprocessors. Addison-Wesley, Workingham (1986)
6. Asahar, P., Devidas, S., Newton, A.: Sequential Logic Synthesis. Kluwer Academic Publishers, Boston (1992)
7. Bacchetta, P., Daldos, L., Sciuto, D., Silvano, C.: Low-power state assignment techniques for finite state machines. In: Proc. of the IEEE Inter. Symp. on Circuits and Systems (ISCAS 2000), vol. 2, pp. 641–644 (2000)
8. Baranov, S.: Logic and System Design of Digital Systems. TUT Press, Tallinn (2008)
9. Baranov, S.I.: Logic Synthesis of Control Automata. Kluwer Academic Publishers, Dordrecht (1994)
10. Barkalov, A., Salomatin, V., Starodubov, K., Das, K.: Optimization of mealy automaton logic using programmable logic arrays. Cybernetics and system analysis 27(5), 789–793 (1991)
11. Barkalov, A., Titarenko, L.: Logic Synthesis for Compositional Microprogram Control Units. Springer, Berlin (2008)
12. Barkalov, A., Titarenko, L.: Synthesis of Operational and Control Automata. UNITECH, Donetsk (2009)

13. Barkalov, A., Węgrzyn, M.: Design of Control Units With Programmable Logic. University of Zielona Góra Press (2006)
14. Barkalov, A.A.: Principles of optimization of logic circuit of Moore FSM. Cybernetics and System Analysis (1), 65–72 (1998) (in Russian)
15. Barkalov, A.A.: Microprogram control unit as composition of automate with programmable and hardwired logic. Automatics and computer technique (4), 36–41 (1983) (in Russian)
16. Bomar, B.W.: Implementation of microprogrammed control in FPGAs. IEEE Transactions on Industrial Electronics 49(2), 415–422 (2002)
17. Brayton, R., Hatchel, G., McMullen, C., Sangiovanni-Vincentelli, A.: Logic Minimization Algorithms for VLSI Synthesis. Kluwer Academic Publishers, Boston (1984)
18. Brayton, R., Rudell, R., Sangiovanni-Vincentelli, A., Wang, A.: MIS: a multi- level logic optimization system. IEEE Transactions on Computer-Aided Design 6, 1062–1081 (1987)
19. Webb, C., Liptay, J.: A high-frequency custom cmos s/390 microprocessor. IBM Journal of research and Development 41(4/5), 463–473 (1997)
20. Chattopadhyay, S.: Area conscious state assignment with flip-flop and output polarity selection for finite state machines synthesis – a genetic algorithm. The Computer Journal 48(4), 443–450 (2005)
21. Chattopadhyay, S., Chaudhuri, P.: Genetic algorithm based approach for integrated state assignment and flipflop selection in finite state machines synthesis. In: Proc. of the IEEE Inter. Conf. on VLSI Design, pp. 522–527. IEEE Computer Society, Los Alamitos (1998)
22. Chu, Y.C.: Computer Organization and Microprogramming. Prentice Hall, Englewood Cliffs (1972)
23. Ciesielski, M., Jang, S.: PLADE: A two-stage PLA decomposition. IEEE Transactions on Computer-Aided Design 11(8) (1992)
24. Clements, A.: The Principles of Computer Hardware. Oxford University Press, Inc., New York (2000)
25. Czerwinski, R., Kania, D.: State assignment method for high speed FSM. In: Proc. of Programmable Devices and Systems, pp. 216–221 (2004)
26. Czerwinski, R., Kania, D.: State assignment for PAL-based CPLDs. In: Proc. of 8th Euromicro Sym. on Digital System Design, pp. 127–134 (2005)
27. Dasgupta, S.: The organization of microprogram stores. ACM Computing Survey (24), 101–176 (1979)
28. Sasao, T., Debnath, D.: Doutput phase optimization for and-or-exor plas with decoders and its application to design of adders. IFICE Transactions on Information and Systems E88-D (7), 1492–1500 (2005)
29. Deniziak, S., Sapiecha, K.: An efficient algorithm of perfect state encoding for CPLD based systems. In: Proceedings of IEEE Workshop on Design and Diagnostic of Electronic Circuits and Systems (DDECS 1998), pp. 47–53 (1998)
30. Devadas, S., Ma, H.: Easily testable PLA-based finite state machines. IEEE Transactions on Computer-Aided Design of Integrated Circuits and Systems 9(6), 604–611 (1990)
31. Devadas, S., Ma, H., Newton, A., Sangiovanni-Vincentelli, A.: MUSTANG: State assignment of finite state machines targeting multilevel logic implementation. IEEE Transactions on Computer-Aided Design 7(12), 1290–1300 (1988)
32. Devadas, S., Newton, A.: Exact algorithms for output encoding, state assignment, and four-level boolean minimization. IEEE Transactions on Computer-Aided Design 10(1), 143–154 (1991)
33. Du, X., Hachtel, G., Lin, B., Newton, A.: MUSE: a multilevel symbolic encoding algorithm for state assignment. IEEE Transactions on Computer-Aided Design of Integrated Circuits and Systems 10(1), 28–38 (1991)

34. Escherman, B.: State assignment for hardwired vlsi control units. ACM Computing Surveys 25(4), 415–436 (1993)
35. Gajski, D.: Principles of Digital Design. Prentice Hall, New York (1997)
36. Goren, S., Ferguson, F.: CHESMIN: a heuristic for state reduction of incompletely specified finite state machines. In: Proc. of the Design, Automation and Test in Europe Conf. and Exhibition (DATE 2002), pp. 248–254 (2002)
37. Gupta, B., Narayanan, H., Desai, M.: A state assignment scheme targeting performance and area. In: Proc. of 12th Inter. Conf. on VLSI Design, pp. 378–383 (1999)
38. Habib, S.: Microprogramming and Firmware Engineering Methods. John Wiley and Sons, New York (1988)
39. Hu, H., Xue, H., Bian, J.: A heuristic state assignment algorithm targeting area. In: Proc. of 5th Inter. Conf. on ASIC, vol. 1, pp. 93–96 (2003)
40. Huang, J., Jou, J., Shen, W.: ALTO: An iterative area/performance algorithms for LUT-based FPGA technology mapping. IEEE Transactions on VLSI Systems 18(4), 392–400 (2000)
41. Husson, S.: Microprogramming: Principles and Practices. Prentice Hall, Englewood Cliffs (1970)
42. Iranli, A., Rezvani, P., Pedram, M.: Low power synthesis of finite state machines with mixed D and T flip-flops. In: Proc. of the Asia and South Pacific– DAC, pp. 803–808 (2003)
43. Flynn, M.J., Rosin, R.F.: Microprogramming: An introduction and a viewpoint. IEEE transactions on Computers C–20(7), 727–731 (1971)
44. Kubatova, H.: Finie State Machine Implementation in FPGAs. In: Software Frameworks and Embedded Control Systems, pp. 177–187. Springer, New York (2005)
45. Maxfield, C.: The Design Warrior's Guide to FPGAs. Academic Press, Inc., Orlando (2004)
46. De Micheli, G.: Symbolic design of combinational and sequential logic implemented by two–level macros. IEEE Transactions on Computer-Aided Design 5(9), 597–616 (1986)
47. De Micheli, G.: Synthesis and Optimization of Digital Circuits. McGraw-Hill, New York (1994)
48. Minns, P., Elliot, I.: FSM-based digital design using Verilog HDL. John Wiley and Sons, Chichester (2008)
49. Papachristou, C.: Hardware microcontrol schemes using PLAs. In: Proceeding of 14th Microprogramming Workshop, vol. 2, pp. 3–15 (1981)
50. Papachristou, C., Gambhir, S.: A microsequencer architecture with firmware support for modular microprogramming. ACM SIGMICRO Newsletters 13(4) (1982)
51. Park, S., Yang, S., Cho, S.: Optimal state assignment technique for partial scan designs. Electronic Letters 36(18), 1527–1529 (2000)
52. Patterson, D., Henessy, J.: Computer Organization and Design: The Hardware/Software Interface. Morgan Kaufmann, San Moteo (1998)
53. Pedram, C., Despain, A.: Low-power state assignment targeting two- and multilevel logic implementations. IEEE Transactions on Computer-Aided Design of Integrated Circuits and Systems 17(12), 1281–1291 (1998)
54. Pomerancz, I., Cheng, K.: STOIC: state assignment based on output/input functions. IEEE Transactions on Computer-Aided Design of Integrated Circuits and System 12(8), 1123–1131 (1993)
55. Pugh, E., Johnson, L., Palmer, J.: IBM's 360 and Early 370 Systems. MIT Press, Cambridge (1991)
56. Rho, J., Hatchel, F., Somenzi, R., Jacoby, R.: Exact and heuristic algorithms for the minimization of incompletely specified state machines. IEEE Transactions on Computer-Aided Design 13(2), 167–177 (1994)

57. Rudell, R., Sangiovanni-Vincentelli, A.: Multiple-valued minimization for pla optimization. IEEE Transactions on Computer-Aided Design 6(5), 727–750 (1987)
58. Salisbury, A.: Microprogrammable Computer Architectures. Am Elstein, New York (1976)
59. Sasao, T.: Input variable assignment and output phase optimization of pla optimization. IEEE Transactions on Computers 33(10), 879–894 (1984)
60. Smith, M.: Application-Specific Integrated Circuits. Addison-Wesley, Boston (1997)
61. Solovjev, V., Czyzy, M.: Refined CPLD macrocells architecture for effective FSM implementation. In: Proc. of the 25th EUROMICRO Conference, Milan, Italy, vol. 1, pp. 102–109 (1999)
62. Solovjev, V., Czyzy, M.: The universal algorithm for fitting targeted unit to complex programmable logic devices. In: Proc. of the 25th EUROMICRO Conference, Milan, Italy, vol. 1, pp. 286–289 (1999)
63. Solovjev, V.V.: Design of Digital Systems Using the Programmable Logic Integrated Circuits. In: Hot line – Telecom, Moscow (2001) (in Russian)
64. Tucker, S.: Microprogram control for system/360. IBM System Journal 6(4), 222–241 (1967)
65. Venkatamaran, G., Reddy, S., Pomerancz, I.: GALLOP: genetic algorithm based low power fsm synthesis by simultaneous partitioning and state assignment. In: Proc. of 16th Inter. Conf. on VLSI Design, pp. 533–538 (2003)
66. Villa, T., Saldachna, T., Brayton, R., Sangiovanni-Vincentelli, A.: Symbolic two-level minimization. IEEE Transactions on Computer-Aided Design 16(7), 692–708 (1997)
67. Villa, T., Sangiovanni-Vincentelli, A.: NOVA: State assignment of finite state machines for optimal two-level logic implementation. IEEE Transactions on Computer-Aided Design 9(9), 905–924 (1990)
68. Wilkes, M.: The best way to design an automatic calculating machine. In: Proc. of Manchester University Computer Inaugural Conference (1951)
69. Wilkes, M., Stringer, J.: Microprogramming and the design of the control circuits in an electronic digital computer. In: Proc. of Cambridge Philosophical Society, vol. 49, pp. 230–238 (1953)
70. Xia, Y., Almani, A.: Genetic algorithm based state assignment for power and area optimization. In: IEEE Proc. on Computers and Digital Techniques, vol. 149, pp. 128–133 (2002)

Chapter 2
Matrix Realization of Control Units

Abstract. The chapter discusses some problems, connected with logic synthesis and optimization of FSM implemented with custom matrix integrated circuits. The primitive matrix implementation of FSM circuit is analyzed first. It is reduced to direct interpretation of FSM structure table and is characterized by considerable redundancy. Next, the methods of logical condition replacement and encoding of collections of microoperations are considered. These methods allow decrease for circuit redundancy due increase of the number of FSM model levels. Next, it is shown that the model of Moore FSM offers an additional possibility for its circuit optimization due to existence of the classes of pseudoequivalent states. Each such class corresponds to one state of the equivalent Mealy FSM. Optimization methods are introduced based on different approaches for state encoding, as well as on transformation of state codes into class codes. The last part of the chapter is devoted to optimization of the block generating microoperations.

2.1 Primitive Matrix Realization of FSM

In case of ASIC, two-level circuits of the FSM type are often realized in a form of *custom matrices* [2,3]. The conception of *distributed logic* [10] is used in custom matrices; this conception can be explained as the following one.

Consider implementation of the following system of Boolean functions:

$$\begin{aligned} y_1 &= abc \vee a\bar{b}\bar{c} \vee \bar{a}b\bar{c} = F_1 \vee F_2 \vee F_3, \\ y_2 &= a\bar{b}\bar{c} \vee \bar{a}bc = F_2 \vee F_4 \end{aligned} \tag{2.1}$$

System (2.1) is characterized by the following parameters: the number of inputs $L = 3$, the number of functions $N = 2$, the number of product terms (conjunctions of arguments) $H = 4$. This system can be implemented as a two-level matrix circuit (Fig. 2.1).

The logic circuit shown in Fig. 2.1 can be viewed as a matrix M_1, which includes H AND gates (AND-plane), and a matrix M_2, which includes N OR gates (OR-plane). Each element of the matrix M_1 has up to L inputs (the number of inputs

A. Barkalov and L. Titarenko: Logic Synthesis for FSM-Based Control Units, LNEE 53, pp. 29–52.
springerlink.com

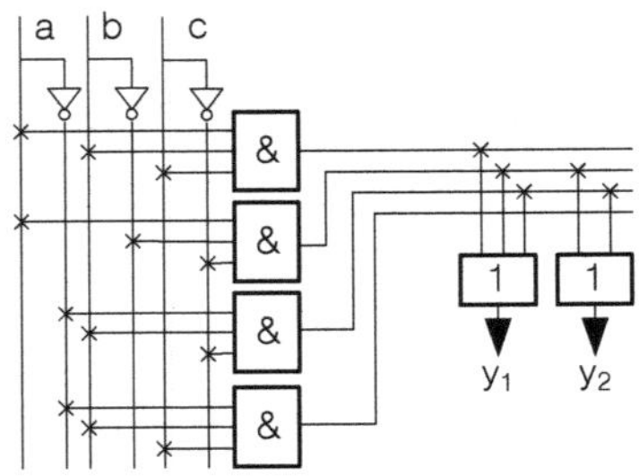

Fig. 2.1 Matrix implementation of system (2.1)

can be less than L, if an implemented Boolean system can be minimized). Each element of the matrix M_2 has up to H inputs. Because realization of AND and OR gates in modern technologies generally uses more transistors (and hence more delays and chip area), then both matrices M_1 and M_2 are implemented using NAND or NOR gates [10]. For the sake of simplification, let symbol AND stand for M_1 and OR for M_2.

Practical hardware implementation of such a circuit is very difficult because of problems with routing wires and building of multi input gates. For example, if a circuit has $L = 16$ inputs, then it should require up to 64 000 gates in the AND-plane. Moreover, each input could be connected with each gate of the matrix M_1 (up to 64000 connections), whereas each gate of OR-plane should have up to 64K inputs. Such an implementation is very space and time consuming because of large gates and long lines of connections among them. To eliminate these drawbacks, the gates are distributed along the rows and columns of the matrices [10].

The distributed NOR gate implementing the minterm F_1 is shown by the circuit in Fig. 2.2a, whereas Fig. 2.2b depicts the distributed implementation of the function y_1 from system (2.1).

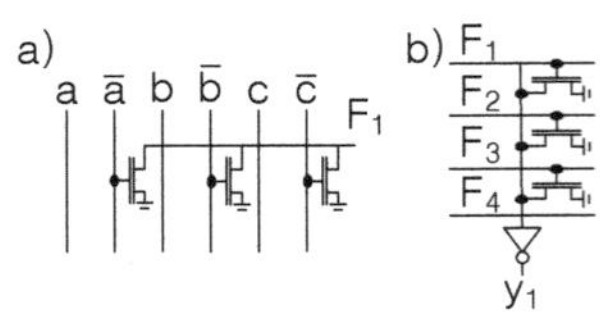

Fig. 2.2 Distributed implementation of terms and functions

CMOS transistors shown in Fig. 2.2 are used for interconnections of rows and columns of AND- and OR-planes (matrices M_1 and M_2). Let us point out that all ASIC are implemented using CMOS transistors [11]. We should add that technological aspects of logic circuits implementation are out the scope of this book.

Let us consider the matrix implementation of Mealy FSM represented by systems (1.8) – (1.9), which are derived from FSM structure table (Fig. 2.3).

A matrix M_1 has $2(L+R)$ inputs and implements H conjunctive terms $F_h \in F = \{F_1, \ldots, F_H\}$, which are members of functions Y and Φ. A matrix M_2 has H inputs and implements $N+R$ functions, depended on terms (1.6). The complexity of matrix

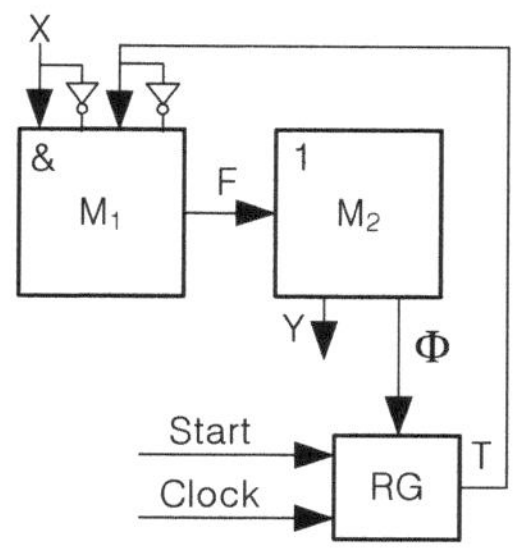

Fig. 2.3 Primitive matrix realization of Mealy FSM

realization can be estimated as a total area of matrices M_1 and M_2 [3]. Let $S(M_1)$, $S(M_2)$ and $S(M_T)$ denote respectively the areas of matrices M_1, M_2 and total area of the circuit shown in Fig. 2.3. These areas can be determined in the following way:

$$S(M_1) = 2(L+R)H; \tag{2.2}$$
$$S(M_2) = H(R+N); \tag{2.3}$$
$$S(M_T)_1 = (3R+2L+N)H. \tag{2.4}$$

Assessments (2.2) – (2.4) are rather theoretical, because they do not include technological coefficients to give real sizes of transistors, wires and spaces among these components of the circuit. Let us analyze the parameters of matrix implementation for the Mealy FSM S_5, represented by its structure table (Table 2.1).

In this case there is the set of terms $F = \{F_1,\dots,F_8\}$, where $F_1 = \bar{T}_1\bar{T}_2$, $F_2 = \bar{T}_1T_2x_1,\dots,F_8 = T_1T_2\bar{x}_4$. For the Mealy FSM S_5, functions $y_n \in Y$ and $\phi_r \in \Phi$ are represented as the following equations:

$$\begin{aligned} y_1 &= F_1 \vee F_4 \vee F_5 \vee F_8; & D_1 &= F_2 \vee F_3 \vee F_4 \vee F_6 \vee F_8;\\ y_2 &= F_1 \vee F_3 \vee F_5; & D_2 &= F_1 \vee F_4 \vee F_8.\\ y_3 &= F_2 \vee F_4 \vee F_8; && \\ y_4 &= F_3 \vee F_6; && \\ y_5 &= F_4 \vee F_7 \vee F_8; && \end{aligned} \tag{2.5}$$

The primitive matrix realization of FSM S_5 is shown in Fig. 2.4, where CMOS transistors are replaced by the sign "•".

The following values can be found for FSM S_5: $S(M_1) = 2(4+2)8 = 96$, $S(M_2) = 8(5+2) = 56$ and $S(M_T) = 152$. The circuit shown in Fig. 2.4 includes two levels of matrices on the path from inputs X to outputs Y. Let t_M and t_{RG} stand for propagation delay of a combinational circuit and a register respectively, then the cycle time of such a circuit is determined as

$$t(T) = 2t_M + t_{RG}. \tag{2.6}$$

In expression (2.6), the symbol T stands for the primitive matrix realization of FSM circuit.

Table 2.1 Structure table of Mealy FSM S_5

a_m	$K(a_m)$	a_s	$K(a_s)$	X_h	Y_h	Φ_h	h
a_1	00	a_2	01	1	y_1y_2	D_2	1
a_2	01	a_3	10	x_1	y_3	D_1	2
		a_3	10	$\bar{x}_1x_2$	y_2D_1	D_1	3
		a_4	11	$\bar{x}_1\bar{x}_2$	$y_1y_3y_5$	D_1D_2	4
a_3	10	a_1	00	x_3	y_1y_2	–	5
		a_3	10	$\bar{x_3}$	D_1	D_1	6
a_4	11	a_1	00	x_4	y_5	–	7
		a_4	11	$\bar{x}_4$	$y_1y_3y_5$	D_1D_2	8

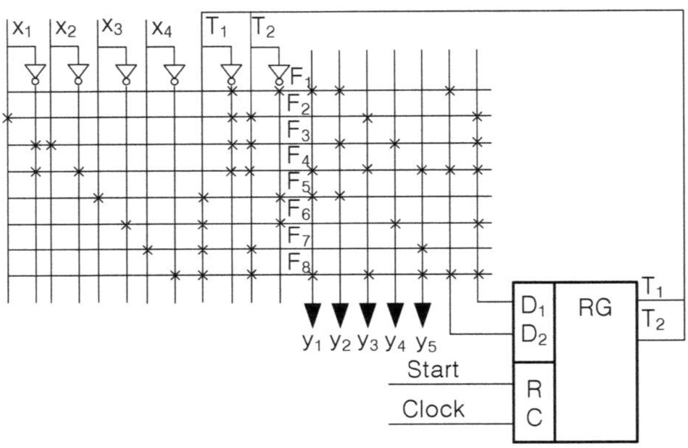

Fig. 2.4 Primitive matrix realization of FSM S_5

The primitive realization leads to logic circuits with maximal possible performance (minimal cycle time) among all possible matrix implementations of Mealy FSM. But such an approach leads to very redundant logic circuits. For example, an FSM with average complexity is characterized by the following values of parameters [3]:$H \approx 2000$, $R \approx 8$, $N \approx 5$, $L \approx 30$. It follows from (2.4), that such an FSM has $S(M_T) = 268000$. This parameter determines the number of possible interconnections in both matrices M_1 and M_2. Obviously, only some part of possible interconnections is used for a particular FSM. Let a term F_h include L_h letters from the input alphabet X, and let N_h microoperations and R_h input memory functions be produced for each FSM transition. It means that the number of interconnections for matrix M_i are determined as $S(M_i)_R$, where $i = 1,2,T$:

$$\begin{aligned} S(M_1)_R &= (L_h + R)H; \\ S(M_2)_R &= (N_h + R_h)H; \\ S(M_T)_R &= (L_h + N_h + R_h + R)H. \end{aligned} \tag{2.7}$$

Let the symbol E_T stand for efficiency of use of matrix areas for matrices M_1 and M_2 in the case of FSM primitive realization. For Mealy FSM, it is determined as:

$$E_T = S(M_T)_R / S(M_T). \tag{2.8}$$

Let an FSM of average complexity be characterized by the values $L_h = N_h = 6$ and $R_h = 4$, then $E_T = 16/134 \approx 0,12$. It means that near 88% of matrix area is wasted in case of the primitive realization. If a designer does not strive for ultimate performance, then the primitive Mealy FSM matrix realization is not used.

Outputs of Moore FSM depend only on its states $a_m \in A$, as follows from (1.10). Thus, functions $y_n \in Y$ are independent on terms F_h, represented as (1.6). The terms $A_m \in A_0$, corresponding to states $a_m \in A$, are used as the minterms of output functions $y_n \in Y$. Therefore, the primitive matrix realization of Moore FSM can be represented as it is shown in Fig. 2.5.

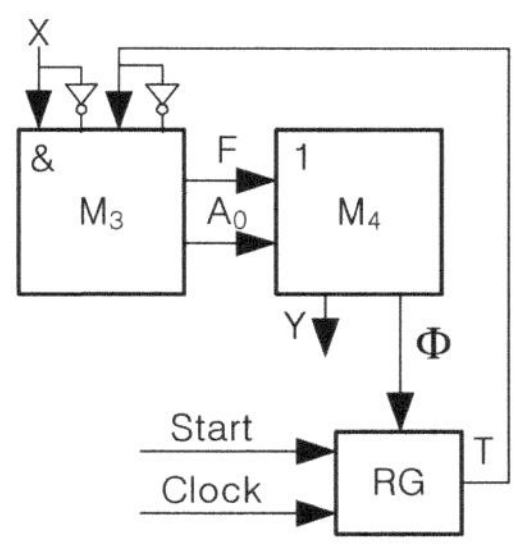

Fig. 2.5 Primitive matrix realization of Moore FSM

A conjunctive matrix M_3 has $2(L+R)$ inputs; it implements H terms $F_h \in F$ from system Φ, and M terms $A_m \in A_0$ from system Y. A disjunctive matrix M_4 has $H+M$ inputs and implements $N+R$ functions. Let us find the areas of matrices M_3, M_4 and the total area occupied by the FSM circuit. By analogy with (2.2) – (2.4), these areas can be found as the following:

$$S(M_3) = 2(L+R)(H+M); \tag{2.9}$$

$$S(M_4) = (H+M)(R+N); \tag{2.10}$$

$$S(M_T)_2 = (3R+2L+N)(H+M). \tag{2.11}$$

Consider an example of the primitive matrix realization for the Moore FSM S_6, represented by its structure table (Table 2.2).

For the FSM S_6, we have $L = 4$, $R = 3$, $H = 15$, $M = 5$, $N = 6$, therefore, $S(M_3) = 280$, $S(M_4) = 180$, $S(M_T)_2 = 460$. In the case of FSM S_6, as for any Moore FSM, there are two sets of terms, namely the set $F = \{F_1, \ldots, F_{15}\}$, where, for example, $F_3 = \bar{T}_1\bar{T}_2\bar{T}_3\bar{x}_1\bar{x}_2$, and the set $A_0 = \{A_1, \ldots, A_8\}$, where, for example, $A_3 = \bar{T}_1 T_2 \bar{T}_3$. The Boolean expressions for functions $y_n \in Y$ and $\phi_r \in \Phi$ are derived from Table 2.2.

Table 2.2 Structure table of Moore FSM S_6

a_m	$K(a_m)$	a_s	$K(a_s)$	X_h	Φ_h	h
$a_1(-)$	000	a_2	001	x_1	D_3	1
		a_3	010	$\bar{x}_1x_2$	D_2	2
		a_4	011	$\bar{x}_1\bar{x}_2$	D_2D_3	3
$a_2(y_1y_3D_1y_5y_6)$	001	a_2	001	x_1x_3	D_3	4
		a_1	000	$x_1\bar{x}_3$	–	5
		a_5	100	$\bar{x}_1$	D_1	6
$a_3(y_1y_2D_1y_5)$	010	a_2	001	x_1x_3	D_3	7
		a_1	000	$x_1\bar{x}_3$	–	8
		a_5	100	$\bar{x}_1$	D_1	9
$a_4(y_2y_6)$	011	a_1	000	x_3x_4	–	10
		a_4	011	$x_3\bar{x}_4$	D_2D_3	11
		a_4	011	$\bar{x}_3$	D_2D_3	12
$a_5(y_1y_3D_1y_6)$	100	a_1	000	x_3x_4	–	13
		a_4	011	$x_3\bar{x}_4$	D_2D_3	14
		a_4	011	$\bar{x}_3$	D_2D_3	15

In the case of FSM S_6, the matrix M_4 implements the following systems of Boolean functions $y_1 = A_2 \vee A_3 \vee A_5, \ldots, y_6 = A_2 \vee A_4 \vee A_5, D_1 = F_6 \vee F_9, \ldots, D_3 = F_1 \vee F_3 \vee F_4 \vee F_7 \vee F_{11} \vee F_{12} \vee F_{14} \vee F_{15}$. The primitive matrix realization of FSM S_6 is shown in Fig. 2.6.

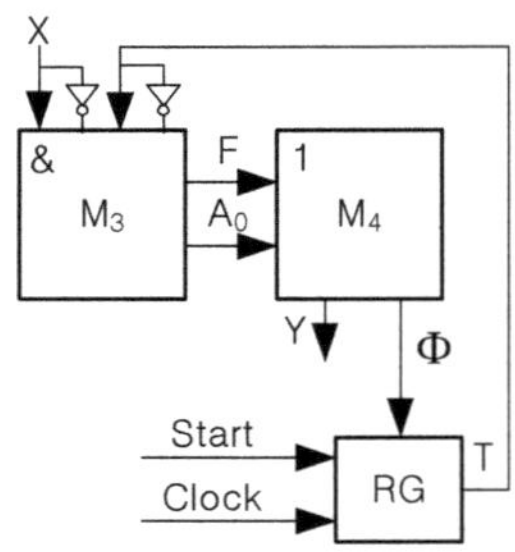

Fig. 2.6 Primitive matrix realization of Moore FSM S_6

As follows from analysis of Fig. 2.6, both matrices M_3 and M_4 are used very ineffectively. There are 280 possible interconnections for the matrix M_3, but only 85 of them are in use (less than 30%). As well, only 31 from 180 interconnections are used in the matrix M_4, it gives us near 17% of all possible interconnections. Thus, in whole only 116 from 460 possible interconnections are used, it means that 75% of mutual area of matrices M_3 and M_4 is free. So, the primitive matrix realization results in very redundant logic circuits of Moore FSM.

The real number of needed interconnections can be found from analysis of Fig. 2.6, namely:

$$\begin{aligned} S(M_3)_R &= (L_h + R)H + RM; \\ S(M_4)_R &= N_m M + R_h H; \\ S(M_T)_R &= (L_h + R_h + R)H + (R + N_m)M, \end{aligned} \tag{2.12}$$

where N_m is an average number of microoperations executed in one state of Moore FSM. Using both expression (2.8) and parameters of an FSM with average complexity, it could be found that $E_T \approx 0.13$ (if $H = 2000$, $L = 30$, $N = 50$, $L_h = 6$, $R_h = N_m = 4$, $M = 200$, $R = 8$). Thus, approximately 87% of a chip area occupied by logic circuit of Moore FSM is not used.

Thus, primitive matrix realizations of Moore and Mealy FSMs are redundant considering use of a chip area. In both cases, only 12–13% is used to implement the really needed interconnections. The value of parameter E_T can be increased due to the multilevel realization of FSM logic circuits [3].

2.2 Optimization of Mealy FSM Matrix Realization

Let $X(a_m)$ be a set of logical conditions determining transitions from the state $a_m \in A$, and let

$$G = \max(|X(a_1)|, \ldots, |X(a_M)|). \tag{2.13}$$

If condition

$$G \ll L \tag{2.14}$$

takes place, then the method of *logical condition replacement* [3] can be used to improve the quality of the FSM matrix realization. The main idea of the method is reduced to construction of some additional variables $p_g \in P$ used for replacement of logical conditions $x_l \in X$, where $|P| = G$.

From analysis of Table 2.1, we can find for the Mealy FSM S_5 the following sets: $X(a_1) = \emptyset$, $X(a_2) = \{x_1, x_2\}$, $X(a_3) = \{x_3\}$, $X(a_4) = \{x_4\}$; it means that $G = 2$. Therefore, the set X can be replaced by a set $P = \{p_1, p_2\}$. Let us construct the table of logical condition replacement, having G columns marked by variables $p_g \in P$, and M rows marked by states $a_m \in A$. If a variable p_g replaces in a state a_m a logical condition x_l, then the symbol x_l is written on the intersection of the row a_m and column p_g of the table. To minimize the hardware amount of logic circuit used for the logical condition replacement, the distribution of logical conditions is executed in such a manner that each variable $x_l \in X$ is always placed in the same column of the table for all states of FSM. In the case of Mealy FSM S_5, above mentioned distribution of logical conditions is shown in Table 2.3. There are formal methods for execution of required distribution in cases of very complex FSM which can be found in [3].

The following system of Boolean functions should be constructed to replace the logical conditions:

$$P = P(X, T). \tag{2.15}$$

In our particular case, this system is the following one: $p_1 = A_2x_1 \vee A_3x_3, p_2 = A_2x_2 \vee A_4x_4$. Generally, system (2.16) is represented as:

Table 2.3 Logical condition replacement for Mealy FSM S_5

a_m	p_1	p_2
a_1	–	–
a_2	x_1	x_2
a_3	x_3	–
a_4	–	x_4

$$p_g = \bigvee_{m=1}^{M} C_{ml} A_m x_l \quad (g = 1, \dots, G), \tag{2.16}$$

where C_{ml} is a Boolean variable, which is equal to 1, iff the variable p_g replaces the logical condition x_l for the state a_m. For the FSM S_5, system (2.16) can be implemented as a two-level matrix circuit shown in Fig. 2.7.

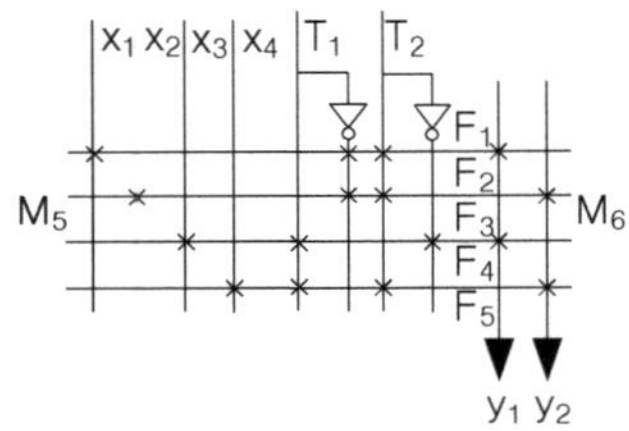

Fig. 2.7 Matrix realization of logical condition replacement for Mealy FSM S_5

In this circuit, a matrix M_5 implements terms of the system (2.16), making a set of terms $V = \{v_1, \dots, v_I\}$, whereas a matrix M_6 implements functions $p_g \in P$ as some disjunctions from terms $v_i \in V$. After these transformations, the matrix realization of Mealy FSM includes four matrices (Fig. 2.8).

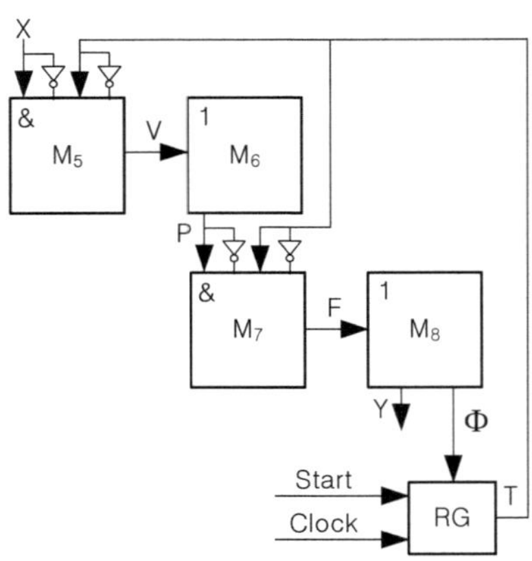

Fig. 2.8 Matrix realization of Mealy FSM with logical condition replacement

In Fig. 2.8, a matrix M_7 implements terms from the system

$$F = F(P,T), \tag{2.17}$$

used to form functions Y and Φ. To construct the system (2.17), the initial structure table of Mealy FSM should be transformed in such a way that its column X_h is replaced by a column P_h. The column P_h contains conjunctions of variables $p_g \in P$, replaced the corresponding conjunctions of logical conditions $x_l \in X$. As a rule [1], Mealy FSM shown in Fig. 2.3 are named P FSM, and the term MP FSM is used to denote Mealy FSM shown in Fig. 2.8. In case of MP FSM, a block BM, used to replace the logical conditions, is represented by matrices M_5 and M_6.

The transformed structure table of Mealy MP FSM S_5 (Table 2.4) is constructed using both Table 2.1 and Table 2.3.

Table 2.4 Transformed structure table of MP Mealy FSM S_5

a_m	$K(a_m)$	a_s	$K(a_s)$	P_h	Y_h	Φ_h	h
a_1	00	a_2	01	1	y_1y_2	D_2	1
a_2	01	a_3	10	p_1	y_3	D_1	2
		a_3	10	$\bar{p}_1p_2$	y_2D_1	D_1	3
		a_4	11	$\bar{p}_1\bar{p}_2$	$y_1y_3y_5$	D_1D_2	4
a_3	10	a_1	00	p_1	y_1y_2	–	5
		a_3	10	$\bar{p}_1$	D_1	D_1	6
a_4	11	a_1	00	p_2	y_5	–	7
		a_4	11	$\bar{p}_2$	$y_1y_3y_5$	D_1D_2	8

System (2.17) is derived from Table. 2.4, for example, $F_1 = \bar{T}_1\bar{T}_2$, $F_2 = \bar{T}_1T_2p_1, \ldots,$ $F_8 = T_1T_2\bar{p}_2$. The system of terms

$$V = V(X,T) \tag{2.18}$$

is implemented by the matrix M_5. For our example, this system is derived from Table 2.3 as the following one: $V_1 = \bar{T}_1T_2x_1$, $V_2 = \bar{T}_1T_2x_2$, $V_3 = T_1\bar{T}_2x_3$, $V_4 = T_1T_2x_4$. The matrix M_6 implements equations from the system (2.16), namely, $p_1 = v_1 \vee v_3$, $p_2 = v_2 \vee v_4$.

Complexity of MP Mealy FSM can be estimated adding the matrix areas:

$$\begin{aligned} S(M_5) &= (L+2R)I; \\ S(M_6) &= IG; \\ S(M_7) &= 2(G+R)H; \\ S(M_8) &= H(N+R). \end{aligned} \tag{2.19}$$

Analysis of system (2.19) shows that decrease for areas of both matrices M_5 and M_6 can be reached due to decrease of the value I. Obviously, its minimal value is equal to the number of logical conditions, L. In the considered example, this minimum was reached automatically, because there is no another variant of the logical condition replacement. In common case, this problem is solved using two approaches from [3].

Let $X(p_g)$ be a set of logical conditions written in the column p_g of replacement table. Let $A(x_l)$ be a set of states which can be extracted from terms (2.17) for logical condition $x_l \in X$. First, the logical conditions should be distributed among the table columns in such a manner, that the following condition takes place:

$$X(p_i) \cap X(p_j) = \emptyset \quad (i \neq j; i, j \in \{1, \ldots, G). \tag{2.20}$$

Let B_l be a disjunction of terms A_m, corresponding to states $a_m \in A(x_l)$. The second task is reduced to such a state assignment for states $a_m \in A$, that each disjunction B_l is represented by a single conjunctive term.

For example, let the following solution of the first problem be obtained for some FSM S:

$$\begin{aligned} p_1 &= (A_2 \vee A_{12} \vee A_{13})x_1 \vee (A_1 \vee A_8)x_4 \vee (A_7 \vee A_9 \vee A_{10})x_5; \\ p_2 &= (A_4 \vee A_5 \vee A_6 \vee A_7)x_2 \vee (A_3 \vee A_4 \vee A_5)x_3 \vee (A_6 \vee A_{11} \vee A_{12})x_6 \vee \\ &\quad (A_5 \vee A_8 \vee A_{13})x_7. \end{aligned}$$

Obviously, in this case value of parameter I can be any, from 21 (it is determined by the number of terms for logical condition replacement) to $L = 7$. The well-known algorithm ESPRESSO [9] can be used for such an encoding, in this case sets $A(x_l)$ are used as restrictions of the algorithm [12, 13]. For the FSM S, there are $M = 13$ states, and it is enough $R = 4$ state variables $T_r \in T$ for their encoding. One of the possible encoding variants is shown in Fig. 2.9, for this variant we can find the following values: $I = L = 7$.

$T_1T_2 \backslash T_3T_4$	00	01	11	10
00	a_1	a_4	a_3	a_2
01	a_8	a_5	*	a_{13}
11	a_{11}	a_6	*	a_{12}
10	a_9	a_7	a_{10}	*

Fig. 2.9 Codes for Mealy FSM S

Taking into account the "don't care" input assignments, the next system can be derived from the Karnaugh map (Fig. 2.9): $p_1 = T_3\bar{T}_4x_1 \vee T_1\bar{T}_3\bar{T}_4x_4 \vee T_1\bar{T}_2x_5$; $p_2 = \bar{T}_3T_4x_2 \vee \bar{T}_1T_4x_3 \vee T_1T_2x_6 \vee \bar{T}_1T_2x_7$. Thus, the appropriate state assignment allows three times decrease for the number of terms in system (2.18).

The logical condition replacement targets in reducing of the area occupied by the matrix M_1 (Fig. 2.3), which is replaced by matrices M_5, M_6, M_7 (Fig. 2.8). The method of *encoding of collections of microoperations* [1, 3] is used to decrease the area of matrix M_2 (Fig. 2.3); it leads to PY Mealy FSM [7]. Let the structure table of Mealy FSM include T_0 different collections of microoperations $Y_t \subseteq Y$. Encode each collection Y_t by a binary code $K(Y_t)$ having

$$R_3 = \lceil \log_2 T_o \rceil \tag{2.21}$$

bits. Let us use variables $z_r \in Z$, where $|Z| = R_3$, for such an encoding. Let $B(y_n)$ be a set of collections of microoperations containing the microoperation $y_n \in Y$, and $C(Y_t)$ be a conjunction of variables $z_r \in Z$, corresponding to the code $K(Y_t)$. Now, system $Y = Y(T,X)$ is transformed into the following systems:

$$Z = Z(T,X); \tag{2.22}$$

$$Y = Y(Z). \tag{2.23}$$

The matrix realization used for implementation of system (2.23) includes matrices M_9 and M_{10} (Fig. 2.10).

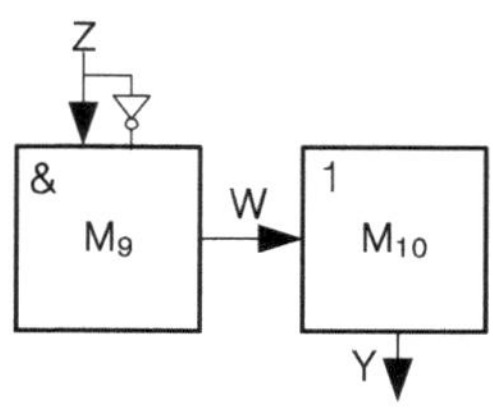

Fig. 2.10 Matrix realization of system of microoperations for PY Mealy FSM

In the circuit (Fig. 2.10), a matrix M_9 implements terms w_j from a set of terms W, and a matrix M_{10} implements functions $y_n \in Y$ as some disjunctions of the terms W_j $(j = 1, \dots, J)$. The complexity of matrix realization shown in Fig. 2.10 is determined as result of summation for areas of matrices M_9 and M_{10}:

$$S(M_9) = 2R_3 J, \tag{2.24}$$

$$S(M_{10}) = JN. \tag{2.25}$$

Values of variables R_3 and N are determined by initial GSA to be interpreted, whereas parameter J can be changed from T_0 to N (the value N corresponds to such a situation when each microoperation is represented as a single term). Let us point out that the matrix M_{10} is disappeared if $J = N$ [2, 3].

For the FSM S_5, we have $T_0 = 6$, because the structure fable (Table 2.4) includes the following collections of microoperations: $Y_1 = \{y_1, y_2\}$, $Y_2 = \{y_3\}$, $Y_3 = \{y_2, y_4\}$, $Y_4 = \{y_1, y_3, y_5\}$, $Y_5 = \{y_6\}$, $Y_6 = \{y_5\}$. Therefore, $R_3 = 3$ and $Z = \{z_1, z_2, z_3\}$. Encode the collections Y_t in a trivial way, namely: $K(Y_1) = 000, \dots, K(Y_6) = 101$. It can be derived from the column Y_h (Table 2.4) that: $y_1 = Y_1 \vee Y_4$, $y_2 = Y_1 \vee Y_3$, $y_3 = Y_2 \vee Y_4$, $y_4 = Y_3 \vee Y_5$, $y_5 = Y_4 \vee Y_6$. To get system (2.22), the initial structure table of Mealy FSM should be transformed by the replacement of the column Y_h by the column Z_h (Table 2.5).

Table 2.5 Transformed structure table of PY Mealy FSM S_5

a_m	$K(a_m)$	a_s	$K(a_s)$	X_h	Z_h	Φ_h	h
a_1	00	a_2	01	1	–	D_2	1
a_2	01	a_3	10	x_1	z_3	D_1	2
		a_3	10	$\bar{x}_1 x_2$	z_2	D_1	3
		$a_4 \bar{x}_4$	11	$\bar{x}_1 \bar{x}_2$	$z_2 z_3$	$D_1 D_2$	4
a_3	10	a_1	00	x_3	–	–	5
		a_3	10	$\bar{x}_3$	z_1	D_1	6
a_4	11	a_1	00	x_4	$z_1 z_3$	–	7
		a_4	11		$z_2 z_3$	$D_1 D_2$	8

An approach for filling of the column Z_h is the following one: if the row h of initial ST contains a collection Y_t, then the column Z_h should contain variables $z_r \in Z$, corresponding to 1 in the code $K(Y_t)$. The matrix realization of PY Mealy FSM is shown in Fig. 2.11.

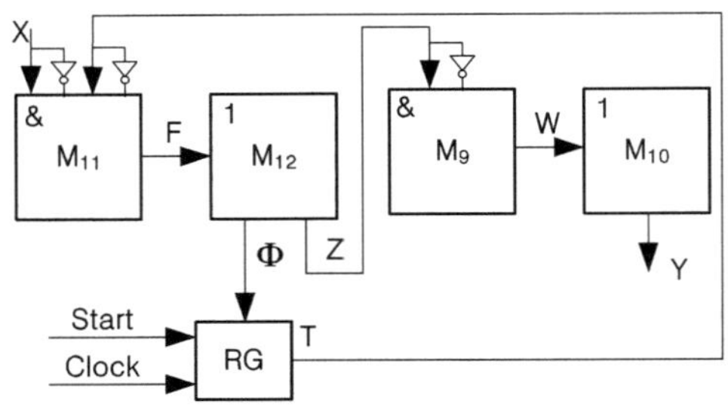

Fig. 2.11 Matrix realization of PY Mealy FSM

To minimize the areas of matrices M_9 and M_{10}, the problem of optimal *encoding of collections of microoperations* [2, 3] should be solved, when each expression (2.23) is represented by SOP with the minimal possible number of terms. The well-known algorithm ESPRESSO [9] can be applied to solve this problem. For the FSM S_5, optimal codes of collections of microoperations are represented by the Karnaugh map shown in Fig. 2.12.

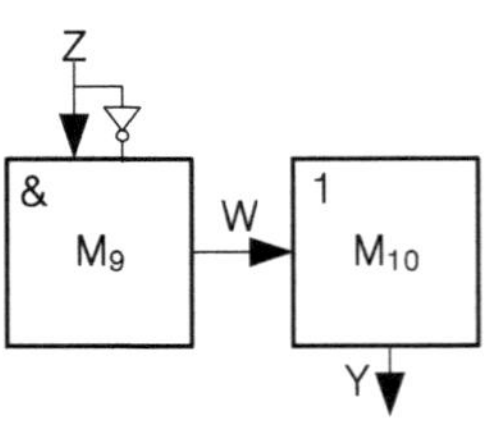

Fig. 2.12 Optimal codes of collections of microoperations for PY Mealy FSM S_5

Taking these codes into account, the system (2.23) is represented as the following one for FSM S_5: $y_1 = \bar{z}_1\bar{z}_2$, $y_2 = \bar{z}_1 z_3$, $y_3 = \bar{z}_1\bar{z}_3$, $y_4 = z_2 z_3$, $y_5 = \bar{z}_2\bar{z}_3$. This system is implemented by the matrix circuit shown in Fig. 2.13.

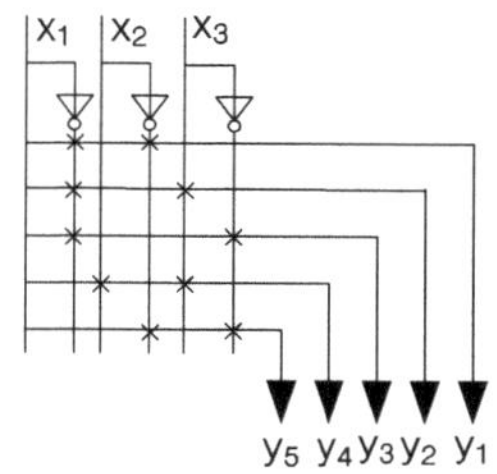

Fig. 2.13 Matrix realization of microoperations for PY Mealy FSM S_5

In the circuit from Fig. 2.13, the wire z_1 is not used, because of it the matrix area can be calculated as $5 \times 5 = 25$. Let us point out that the initial values obtained from (2.24) and (2.25) are equal to $S(M_9) = 2 \cdot 3 \cdot 6 = 36$ and $S(M_{10}) = 6 \cdot 5 = 30$. It gives the total area equal to 66. Thus, the application of optimal encoding results in the block BY having 2.6 times less hardware, than in case of a straightforward implementation of microoperations.

The joint application of logical condition replacement and encoding of collections of microoperations leads to MPY Mealy FSM (Fig. 2.14).

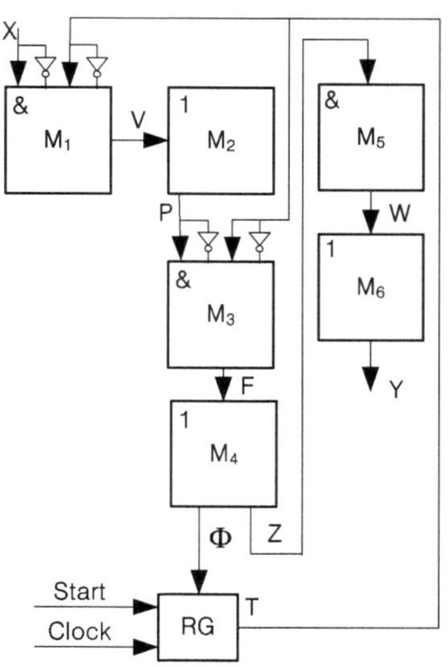

Fig. 2.14 Matrix realization of MPY Mealy FSM

Functions of all matrices from this matrix realization are clear from preceding information as well as formulae for their areas. The synthesis method of MPY Mealy FSM includes the following steps:

1. Construction of marked GSA.
2. Construction of transition table of Mealy FSM.
3. Construction of table of logical condition replacement.
4. State assignment targeted in reduction of hardware for the block BP.
5. Optimal encoding of collections of microoperations.
6. Construction of transformed structure table of Mealy FSM.
7. Construction of systems for realization of matrices $M_1 - M_6$.
8. Design of FSM logic circuit using functions obtained in step 7.

The transformed structure table of MPY Mealy FSM S_5 (Table 2.6) is constructed taking into account the codes of collections of microoperations from Fig. 2.12.

Table 2.6 Transformed structure table of MPY Mealy FSM S_5

a_m	$K(a_m)$	a_s	$K(a_s)$	P_h	Z_h	Φ_h	h
a_1	00	a_2	01	1	z_3	D_2	1
a_2	01	a_3	10	p_1	z_2	D_1	2
		a_3	10	$\bar{p}_1 p_2$	$z_2 z_3$	D_1	3
		a_4	11	$\bar{p}_1 \bar{p}_2$	–	$D_1 D_2$	4
a_3	10	a_1	00	p_1	z_3	–	5
		a_3	10	$\bar{p}_1$	$z_1 z_2 z_3$	D_1	6
a_4	11	a_1	00	p_2	z_1	–	7
		a_4	11	$\bar{p}_2$	–	$D_1 D_2$	8

It is really easy to design the logic circuit (matrix realization) of MPY Mealy FSM S_5, because all its components were already designed. Obviously, the discussed method can be adapted to design either Mealy MP– or PY FSM. The results of many researches show that application of the logical condition replacement and optimal encoding of collections of microoperations results in significant hardware amount reducing, especially for complex FSM having more than 2000 transitions. This approach has one serious drawback, namely decrease in FSM performance in comparison with the primitive matrix realization of Mealy FSM. It is an effect of the cycle time increase due to increase for the number of levels in the resulting FSM realization (in comparison with the primitive matrix realization of Mealy FSM).

2.3 Optimization of Moore FSM Logic Circuit

Analysis of systems (1.8) and (1.10) shows that the combinational part of Moore FSM matrix realization (it is shown in Fig. 2.5) can be divided by two blocks. By analogy with Mealy FSM, let us denote as P Moore FSM the device whose structure is shown in Fig. 2.5 and let the denotation PY Moore FSM stand for the device from Fig. 2.15.

In Fig. 2.15, a matrix M_1 implements terms $F_h \in F$, corresponding to rows of Moore FSM structure table, while a matrix M_2 implements system (1.8). Matrices

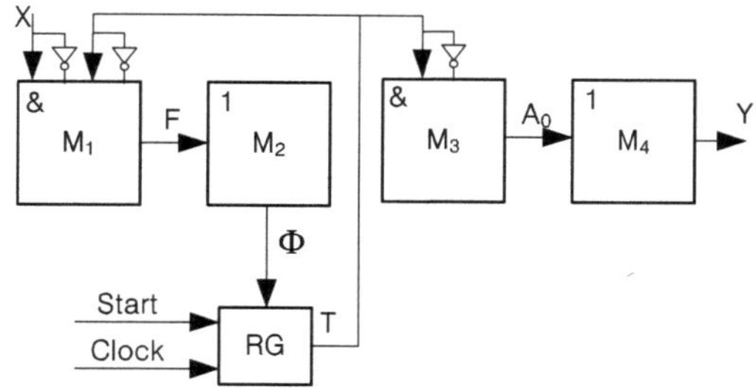

Fig. 2.15 Matrix realization of PY Moore FSM

M_3 andM_4 form a block BY. In this block, the matrix M_3 implements terms $A_m \in A_0$ from system (1.10), while the matrix M_4 implements microoperations $y_n \in Y$. Complexity of PY Moore FSM matrix realization is determined as a total area for matrices $M_1 - M_4$, namely:

$$\begin{aligned} S(M_1) &= 2(L+R)H; \mathrm{S(M_2) = HR}; \\ S(M_3) &= 2RMS(M_4) = MN; \\ S(M_{PY}) &= (2L+3R)H + (2R+N)M. \end{aligned} \tag{2.26}$$

For an FSM with average complexity ($L = 30$, $H = 2000$, $R = 8$, $L = 30$, $L = 30$), we can find that $S(M_{PY}) = 181200$. In case of P Moore FSM, interpretation of such kind GSA gives control units with the total area $S(M_T)_2 = 294800$. Therefore, the replacement of P Moore FSM by PY Moore FSM results in considerable area decreasing (in 1.63 times). It means that the efficiency of chip area usage increases too. If device shown in Fig. 2.6 is divided by blocks BP and BY, then it could be found that only 70 from 210 possible interconnections are used in the matrix M_1, 16 interconnections from 45 are used in the matrix M_2,and 15 interconnections from 30 are used in both matrices M_3 and M_4. It gives the value $E_{PY} = (70+16+15+15)/(210+45+30+30) = 0.37$, that is 12% more, than for the primitive matrix realization for Moore FSM S_6.

A chip area for matrices M_1 and M_2 can be decreased due to decrease of parameter H. It can be obtained if states of Moore FSM are encoded using the method from Section 1.2. Remind, this approach is named as *an optimal state encoding*. Let us discus application of this approach for the Moore FSM S_6.

As follows from Table 2.2, the Moore FSM S_6 includes $I = 3$ classes of the pseudoequivalent states, namely: $B_1 = \{a_1\}$, $B_2 = \{a_2, a_3\}$, $B_3 = \{a_4, a_5\}$. Let us encode the states $a_m \in A$ by the optimal codes (Fig. 2.16).

Fig. 2.16 Optimal state codes for Moore FSM S_6

T_1 \ T_2T_3	00	01	11	10
0	a_1	a_2	*	a_4
1	*	a_3	*	a_5

As follows from Fig. 2.16, the class B_1 corresponds to the generalized interval $\langle *, 0, 0\rangle$ of three-dimensional Boolean space, the class B_2 to $\langle *, *, 1\rangle$, and the class B_3 to $\langle *, 1, *\rangle$. Thus, the class B_1 is determined by the code $*00$, the class B_2 by the code $**1$, and the class B_3 by the code $*1*$. The transformed structure table of Moore FSM S_6 (Table 2.7) includes $H_0 = 9$ rows, where the symbol H_0 stands for the number of structure table rows for an equivalent Mealy FSM [1].

Table 2.7 Transformed structure table of Moore FSM S_6

B_i	$K(B_i)$	a_s	$K(a_s)$	X_h	Φ_h	h
B_1	$*00$	a_2	001	x_1	D_3	1
		a_3	101	$\bar{x}_1 x_2$	$D_1 D_3$	2
		a_4	010	$\bar{x}_1 \bar{x}_2$	D_2	3
B_2	$**1$	a_2	001	$x_1 x_2$	D_3	4
		a_1	000	$x_1 \bar{x}_3$	–	5
		a_5	110	$\bar{x}_1$	$D_1 D_2$	6
B_3	$*1*$	a_1	000	$x_3 x_4$	–	7
		a_4	010	$x_3 \bar{x}_4$	D_2	8
		a_4	010	$\bar{x}_3$	D_2	9

From Table 2.7, for example, the following Boolean function $D_1 = F_2 \vee F_6 = \bar{T}_2 \bar{T}_3 \bar{x}_1 x_2 \vee T_3 \bar{x}_1$ can be derived. The matrix circuit of Moore FSM S_6 is characterized by the following areas of matrices: $S(M_1) = (4+8) \cdot 9 = 108$, $S(M_2) = 9 \cdot 3 = 27$, $S(M_3) = 6 \cdot 5 = 30$, and $S(M_4) = 5 \cdot 6 = 30$. Therefore, the application of the optimal state encoding results in considerable decrease of the total area in comparison with PY Moore FSM. In considered example, the total area decreases from 315 to 195, that is near 1.6 times less.

The total area for matrices M_3 and M_4 (Fig. 2.15) can be decreased using the state encoding targeted in decrease for the number of terms in system Y. Let us name this approach as *a refined state encoding*. For the FSM S_6, one of the possible variants for the refined state encoding is shown in Fig. 2.17.

Fig. 2.17 Refined state codes for Moore FSM S_6

T_1 \ T_2T_3	00	01	11	10
0	a_1	a_2	a_5	*
1	a_4	a_3	*	*

As follows from Table 2.2, the system of microoperations for Moore FSM S_6 is the following one: $y_1 = y_4 = A_2 \vee A_3 \vee A_5$, $y_2 = A_3 \vee A_4$, $y_3 = A_2 \vee A_5$, $y_5 = A_2 \vee A_3$, $y_6 = A_2 \vee A_4 \vee A_5$. Taking into account the codes from Fig. 2.17, the system of microoperations can be represented as: $y_1 = y_4 = T_3$, $y_2 = T_1 \bar{T}_2$, $y_3 = \bar{T}_1 T_3$,

$y_5 = \bar{T}_2 T_3$, $y_6 = \bar{T}_1 T_3 \vee T_1 \bar{T}_3 = y_3 \vee T_1 \bar{T}_3$. This system is implemented using two matrices (Fig. 2.18).

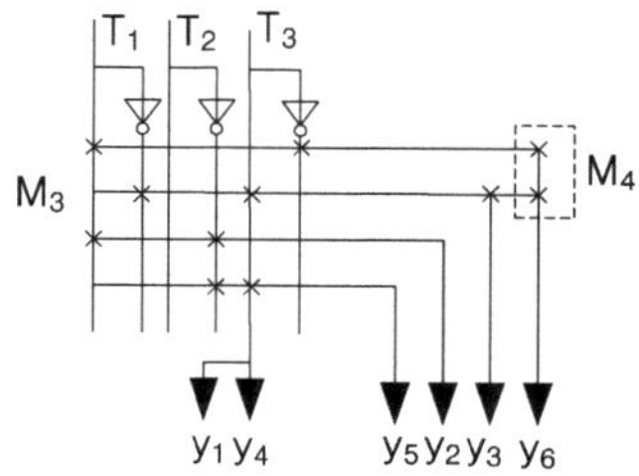

Fig. 2.18 Matrix realization for microoperations of FSM S_6

The total area of this matrix circuit can be calculated as $S(M_3) + S(M_4) = 5 \cdot 4 + 2 \cdot 1 = 22$. It means that the matrix M_3 has 5 inputs (the input T_2 is absent) and realizes 4 terms, but only two of them are used by the matrix M_4, implemented only the microoperation y_6. In case of the arbitrary state encoding, we have $S(M_3) + S(M_4) = 60$, that is 2.73 more than the area of these matrices for the refined state encoding.

Obviously, the aim of the optimal state encoding differs from the aim of the refined state encoding. Simultaneous decrease for areas of matrices M_1 - M_4 can be achieved by application of *a combined state encoding* [5, 6]. This method can be explained as the following. Let us construct the following systems of functions:

$$Y = Y(A); \tag{2.27}$$

$$B = B(A). \tag{2.28}$$

Let system (2.27) be determined by expression (1.10), while the elements of system (2.28) are represented as

$$B_i = \bigvee_{m=1}^{M} C_{im} A_m \quad (i = 1, \ldots, I), \tag{2.29}$$

where C_{im} is a Boolean variable equal to 1, iff $a_m \in B_i$. The combined state encoding is executed in such a manner, that the total number of terms is minimal for systems (2.27) and (2.28). This problem can be solved using, for example, the algorithm ESPRESSO [9].

For the FSM S_6, system (2.27) is more than once shown in our book, whereas system (2.28) can be derived from the partition Π_A as the following one: $B_1 = A_1$, $B_2 = A_2 \vee A_3$, $B_3 = A_4 \vee A_5$. One of the possible variants for combined state encoding is shown in Fig. 2.19.

From Fig. 2.19, it can be found that: $B_1 = \bar{T}_2 \bar{T}_3$, $B_2 = \bar{T}_2 T_3$, $B_3 = T_2$, $y_1 = y_4 = A_2 \vee A_3 \vee A_5 = T_3$, $y_2 = A_3 \vee A_4 = T_1$, $y_3 = A_2 \vee A_5 = \bar{T}_1 T_3$, $y_5 = A_2 \vee A_3 = \bar{T}_2 T_3$, $y_6 = A_2 \vee A_4 \vee A_6 = y_3 \vee T_3$. Obviously, the areas of matrices M_1 and M_2 are the same for both combined and optimal state encoding, thus $S(M_1) + S(M_2) = 135$. The matrix M_3 has 4 inputs (T_1, $\bar{T}_1$, $\bar{T}_2$, T_3), and only 3 outputs (y_3, y_5, T_3). Let us point out that microoperations y_1, y_2, y_4 are produced without additional transistors.

T_1 \ T_2T_3	00	01	11	10
0	a_1	a_2	a_5	*
1	*	a_3	*	a_4

Fig. 2.19 Combined state codes for Moore FSM S_6

The matrix M_4 has 2 inputs (y_3 and T_3), used to generate the microoperation y_6. The total area for these matrices can be found as $S(M_3)+S(M_4)=4\cdot3+2\cdot1=14$, and the total area for matrices M_1 - M_4 is equal to 149 in the case of FSM S_6. Therefore, total area of matrix realization is decreased from 315 (in the case of arbitrary state encoding) to 149, that is more than 2 times.

Let us point out that the combined state encoding could produce results, which are far from optimal for one or both parts of the matrix realization of PY Moore FSM. In this case the total area can be decreased using a transformer of state codes into codes of the classes of pseudoequivalent states [4, 8]. It results in P_CYMoore FSM shown in Fig. 2.20.

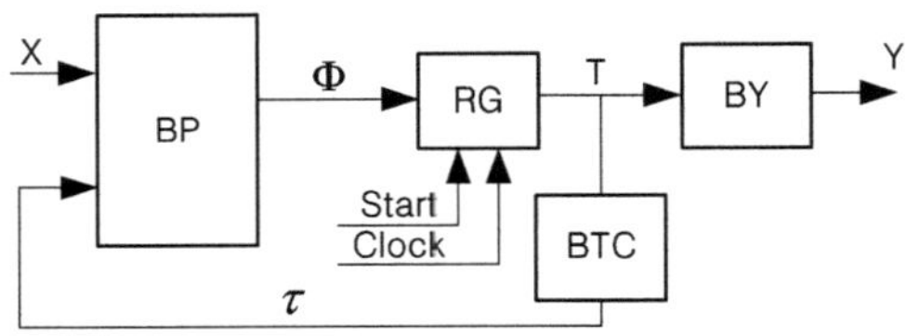

Fig. 2.20 Structure of P_CYMoore FSM

In P_CYMoore FSM, a block BP implements functions

$$\Phi=\Phi(\tau,X), \tag{2.30}$$

where τ is a set of variables used to code classes $B_i\in\Pi_A$. A code transformer BTC generates codes of classes $B_i\in\Pi_A$ on the base of codes for states $a_m\in B_i$, so the block BTC implements the Boolean system

$$\tau=\tau(T). \tag{2.31}$$

Consider an example of the P_CYMoore FSM S_7 design, where the FSM is set up by its transition table (Table 2.8).

The following values and sets can be derived from Table 2.8: $M=8$, $R=3$, $\Pi_A=\{B_1,B_2,B_3,B_4\}$, where $B_1=\{a_1\}$, $B_2=\{a_2,a_3,a_4\}$, $B_3=\{a_5,a_6,a_7\}$, $B_4=\{a_8\}$, $I=4$. Obviously, there is no such a state encoding variant which gives the transformed structure table with $H_0=9$ rows. Remind, this value corresponds to the number of rows in the structure table of equivalent Mealy FSM. The method of synthesis includes the following steps.

1. **Construction of systems** Y **and** B. For the Moore FSM S_7, we can construct the functions: $y_1=A_2\vee A_3$, $y_2=A_5\vee A_6$, $y_3=A_2\vee A_4\vee A_7$, $y_4=A_3\vee A_6$,

Table 2.8 Transition table of Moore FSM S_7

a_m	a_s	X_h	h
a_1 (–)	a_2	x_1	1
	a_3	$\bar{x}_1$	2
$a_2(y_1y_3y_5y_7)$	a_4	x_2	3
	a_5	$\bar{x}_2$	4
$a_3(y_1D_1y_7)$	a_4	x_2	5
	a_5	$\bar{x}_2$	6
$a_4(y_3)$	a_4	x_2	7
	a_5	$\bar{x}_2$	8
$a_5(y_2)$	a_7	x_3	9
	a_6	$\bar{x}_3x_4$	10
	a_8	$\bar{x}_3\bar{x}_4$	11
$a_6(y_2D_1y_7)$	a_7	x_3	12
	a_6	$\bar{x}_3x_4$	13
	a_8	$\bar{x}_3\bar{x}_4$	14
$a_7(y_3y_5y_7)$	a_6	x_3	15
	a_7	$\bar{x}_3x_4$	16
	a_8	$\bar{x}_3\bar{x}_4$	17
$a_8(y_6)$	a_3	x_5	18
	a_1	$\bar{x}_5$	19

$y_5 = A_2 \vee A_7$, $y_6 = A_7 \vee A_8$, $y_7 = A_2 \vee A_3 \vee A_6 \vee A_7$, $B_1 = A_1$, $B_2 = A_2 \vee A_3 \vee A_4$, $B_3 = A_5 \vee A_6 \vee A_7$, $B_4 = A_8$.

2. **State assignment.** For P$_C$YMoore FSM, the state encoding targets in hardware decrease for block of microoperations. Thus, the refined state encoding should be done. The outcome of this step is shown in Fig. 2.21.

Fig. 2.21 Refined state codes for Moore FSM S_7

T_1 \ T_2T_3	00	01	11	10
0	a_1	a_3	a_2	a_4
1	a_5	a_6	a_7	a_8

3. **Construction of functions describing the block BY.** The codes represented by Fig. 2.21 permit to get the following system:

$$
\begin{aligned}
y_1 &= \bar{T}_1T_3 = \Delta_1; & y_5 &= \Delta_4;\\
y_2 &= T_1\bar{T}_2 = \Delta_2; & y_6 &= T_1T_2 = \Delta_6;\\
y_3 &= \bar{T}_1T_2 \vee T_2T_3 = \Delta_3 \vee \Delta_4; & y_7 &= T_3 = \Delta_7.\\
y_4 &= \bar{T}_2T_3 = \Delta_5;
\end{aligned}
\qquad (2.32)
$$

The terms Δ_k from the system (2.32) form a set $F(Y)$.

4. **Construction of functions describing the block BTC.** Besides, the codes represented by Fig. 2.21 permit to get the following system:

$$\begin{aligned} B_1 &= \bar{T}_1\bar{T}_2\bar{T}_3 = \Delta_8; & B_3 &= T_1\bar{T}_2 \vee T_1T_3 = \Delta_2 \vee \Delta_{10}; \\ B_2 &= \bar{T}_1T_3 \vee \bar{T}_1T_2 = \Delta_1 \vee \Delta_3; & B_4 &= T_1T_2\bar{T}_3 = \Delta_9. \end{aligned} \quad (2.33)$$

Classes $B_i \in \Pi_A$ can be coded using

$$R_0 = \lceil \log_2 I \rceil \quad (2.34)$$

variables $\tau_r \in \tau$, where $|\tau| = R_0$. Encode the classes $B_i \in \Pi_A$ in a trivial way, namely: $K(B_1) = 00, \ldots, K(B_4) = 11$. Now we can find that $\tau_1 = B_3 \vee B_4$, $\tau_2 = B_2 \vee B_4$. It gives the following system to represent the system (2.31):

$$\begin{aligned} \tau_1 &= \Delta_2 \vee \Delta_9 \vee \Delta_{10}; \\ \tau_2 &= \Delta_1 \vee \Delta_3 \vee \Delta_9. \end{aligned} \quad (2.35)$$

The terms from the system (2.35) make up a set $F(TC)$.

5. **Construction of transformed structure table.** Let us construct the transformed structure table of the Moore FSM S_7. This table includes the columns B_i, $K(B_i)$, a_s, $K(a_s)$, X_h, Φ_h, h. For the FSM S_7, the codes $K(B_i)$ can be derived from Fig. 2.21.
The following form of system (2.30) is derived from Table 2.9:

$$\begin{aligned} D_1 &= F_4 \vee F_5 \vee F_6 \vee F_7; \\ D_2 &= F_1 \vee F_3 \vee F_5 \vee F_7; \\ D_3 &= F_1 \vee F_2 \vee F_5 \vee F_6 \vee F_8. \end{aligned} \quad (2.36)$$

The terms of system (2.30) are determined as the following conjunctions:

$$F_h = \bigwedge_{r=1}^{R_0} \tau_r^{l_{hr}} X_h \quad (h = 1, \ldots, H_0). \quad (2.37)$$

In (2.37), a variable $l_{hr} \in \{0,1\}$ is equal to the value of bit r for code $K(B_i)$, which is written in the row h of the table. From Table 2.9, for example, it can be found that: $F_= \bar{\tau}_1\bar{\tau}_2 x_1$, $F_2 = \tau_1\bar{\tau}_2\bar{x}_3 x_4$ and so on. Let us point out that this table includes $H_0 = 9$ rows, it is the absolute minimum for the Moore FSM S_7.

6. **Matrix realization of FSM circuit.** It is necessary 6 matrices to implement the logic circuit of P_CYMoore FSM (Fig. 2.22).
   In the matrix realization of P_CYMoore FSM, the block BP is implemented by matrices M_1 and M_2, whereas matrices M_3 and M_4 implement the block BY, in the same time the block BTC is implemented by matrices M_5 and M_6. Obviously, the blocks BTC and BY have the same inputs and, therefore, they can be combined in a single block consisting from two matrices.

For the Moore FSM S_7, the total area of matrices from Fig. 2.15 is equal to 361. For the P_CYMoore FSM S_7, we can found that: $S(M_1) = 10 \cdot 8 = 80$, $S(M_2) = 8 \cdot 3 = 24$, $S(M_3) = 6 \cdot 6 = 36$, $S(M_4) = 2 \cdot 1 = 2$, $S(M_5) = 6 \cdot 5 = 30$, $S(M_6) = 5 \cdot 2 = 10$, thus,

Table 2.9 Transformed structure table for P_CYMoore FSM S_7

B_i	$K(B_i)$	a_s	$K(a_s)$	X_h	Φ_h	h
B_1	00	a_2	011	x_1	D_2D_3	1
		a_3	001	$\bar{x}_1$	D_3	2
B_2	01	a_4	010	x_2	D_2	3
		a_5	100	$\bar{x}_2$	D_1	4
B_3	10	a_7	111	x_3	$D_1D_2D_3$	5
		a_6	101	$\bar{x}_3x_4$	D_1D_3	6
		a_8	110	$\bar{x}_3\bar{x}_4$	D_1D_2	7
B_4	11	a_3	001	x_5	D_3	8
		a_1	000	$\bar{x}_5$	–	9

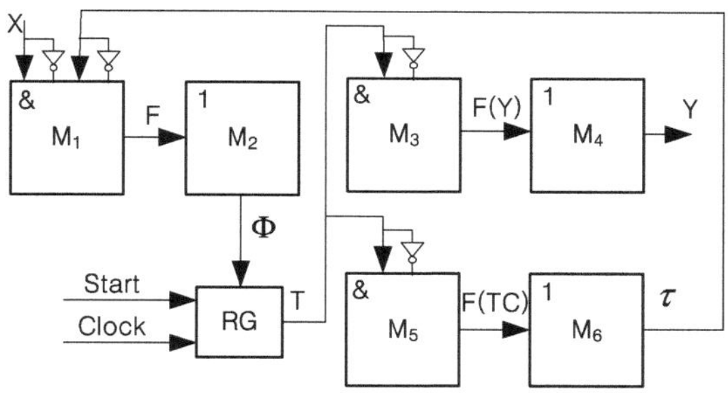

Fig. 2.22 Matrix realization of P_CYMoore FSM

the total area is equal to 182. These calculations show that the total area is decreased near two times due to replacement of the model PY by the model P_CY(in the case of FSM S_7).

The total area occupied by matrices M_1 and M_2 can be decreased using the approach of logical condition replacement. For the FSM S_7, we have $G = 2$, thus the logical conditions $x_1, \ldots, x_5$ can be replaced by the additional variables p_1, p_2 (Table 2.10).

Table 2.10 Replacement of logical conditions for P_CYMoore FSM S_7

a_m	p_1	p_2	a_m	p_1	p_2
a_1	x_1	–	a_5	x_3	x_4
a_2	–	x_2	a_6	x_3	x_4
a_3	–	x_2	a_7	x_3	x_4
a_4	–	x_2	a_8	x_5	–

Analysis of this table shows that content of columns p_1 and p_2 is the same for each state $a_m \in B_i$. Thus, application of the optimal state encoding permits decrease for the number of terms in system $P = P(X,T)$. The logical condition replacement turns P_CYMoore FSM into MP_CYMoore FSM (Fig. 2.23).

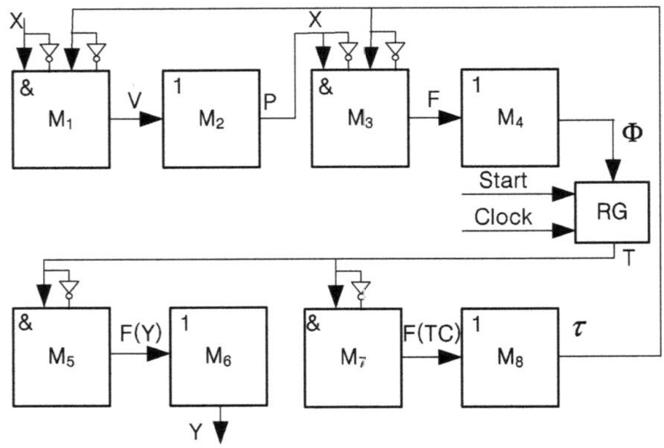

Fig. 2.23 Matrix realization of MP_CYMoore FSM

Functions of all matrices shown in Fig. 2.23 are clear from previous reading. But now the matrices M_1 and M_2 implement the system

$$P = P(\tau, X). \tag{2.38}$$

In contrast to system (2.15), forming for MPY Moore FSM, the functions of system (2.38) depend on variables $\tau_r \in \tau$. To form the system (2.38), it should be constructed the table of logical condition replacement, where states $a_m \in B_i$ are replaced by corresponding classes $B_i \in \Pi_A$ (Table 2.11).

Table 2.11 Transformed table of logical condition replacement for Moore FSM S_7

B_i	p_1	p_2	B_i	p_1	p_2
B_1	x_1	-	B_3	x_3	x_4
B_2	-	x_2	B_4	x_3	–

Using codes $K(B_i)$ from table 2.9, the following system (2.38) can be built for our example: $p_1 = \bar{\tau}_1\bar{\tau}_2x_1 \vee \tau_1\bar{\tau}_2x_3 \vee \tau_1\tau_2x_5$, $p_2 = \bar{\tau}_1\tau_2x_2 \vee \tau_1\bar{\tau}_2x_4$. The number of terms in system (2.38) can be decreased due to *an optimal class encoding* approach.

Let the following system be found for some Moore FSM S_8:

$$\begin{aligned} p_1 &= (B_1 \vee B_2)x_1 \vee (B_3 \vee B_4 \vee B_5)x_2; \\ p_2 &= (B_1 \vee B_2 \vee B_6)x_3 \vee (B_3 \vee B_4)x_4, \end{aligned} \tag{2.39}$$

Fig. 2.24 Optimal codes for classes of Moore FSM S_8

$\tau_1 \backslash \tau_2\tau_3$	00	01	11	10
0	B_1	B_2	B_6	*
1	B_3	B_4	B_5	*

having 10 terms. Let us encode the classes $B_i \in \Pi_A$ by analogy with the optimal state encoding of Moore FSM. This encoding outcome is shown in Fig. 2.24.

Using the codes from Fig. 2.24, we can transform the system (2.39) and get the following system of equations:

$$\begin{aligned} p_1 &= \bar{\tau}_1 \bar{\tau}_2 x_1 \vee \tau_1 x_2; \\ p_2 &= \bar{\tau}_1 x_3 \vee \tau_1 \bar{\tau}_2 x_4. \end{aligned} \tag{2.40}$$

For implementation of system (2.39), the matrix M_1 has $2R_0 + L = 10$ inputs and 10 outputs, whereas the matrix M_2 has 10 inputs and 2 outputs. The system (2.40) depends on variables τ_1, $\bar{\tau}_1$, $\bar{\tau}_2$, it means that now the matrix M_1 has 7 inputs and 4 outputs. The matrix M_2 has 4 inputs and 3 outputs. It means that the arbitrary class encoding for $\mathrm{MP_CY}$Moore FSM S_8 leads to the block BM with the area equal to $10 \cdot 10 + 10 \cdot 2 = 120$, whereas the optimal encoding for classes $B_i \in \Pi_A$ leads to the same block with the area $7 \cdot 4 + 4 \cdot 2 = 36$. Thus, applying of the optimal encoding permits to decrease the total area of the block BM by 3.3 times.

Thus, the total area of FSM matrix realization can be decreased due to increase for the number of levels and coding of some objects, such as states, classes of pseudoequivalent states, or collections of microoperations. Let the first approach be named as a *structural decomposition*, whereas the encoding belongs to *algorithmic methods*. As a rule, multilevel models of FSM include three different types of circuits (blocks):

1. The number of implemented terms is considerably less, than their total possible number. For example, structure tables for FSM with average complexity include $H \approx 2000$ rows, while there are $2^{L+R} \approx 2^{38}$ possible terms. Let us name these circuits and corresponding systems of functions as *irregular circuits* and *irregular functions* respectively.
2. The number of implemented terms is near 50% from their total possible number. This class of circuits includes, for example, blocks BY and BTC. Let us name these circuits and corresponding systems of functions as *regular circuits* and *regular functions* respectively.
3. Only direct (or only complement) values of logical conditions are used. This class of circuits includes, for example, the block BM. Let us name these circuits and corresponding systems of functions as *multiplexer circuits* and *multiplexer functions* respectively.

The methods of optimization discussed in this Chapter can be used for optimization of matrix FSM circuits, as well as FSM circuits implemented with standard VLSI chips. Obviously, peculiarities of specific types of VLSI circuits significantly affect the methods of FSM optimization. Let us discuss these features more thoroughly.

References

1. Adamski, M., Barkalov, A.: Architectural and Sequential Synthesis of Digital Devices. University of Zielona Góra Press, Zielona Góra (2006)
2. Baranov, S.: Logic and System Design of Digital Systems. TUT Press, Tallinn (2008)
3. Baranov, S.I.: Logic Synthesis of Control Automata. Kluwer Academic Publishers, Dordrecht (1994)
4. Barkalov, A., Titarenko, L., Chmielewski, S.: Decrease of hardware amount in logic circuit of moore FSM. Przegĺźd Telekomunikacyjny i Wiadomości Telokomunikacyjne (6), 750–752 (2008)
5. Barkalov, A., Titarenko, L., Chmielewski, S.: Optimization of moore control unit with refined state encoding. In: Proc. of the 15th Inter. Conf. MIXDES 2008, Poznań, Poland, pp. 417–420. Departament of Microeletronics and Computer Science, Technical University of Łódz (2008)
6. Barkalov, A., Titarenko, L., Chmielewski, S.: Optimization of moore fsm on system-on-chip using pal technology. In: Proc. of the International Conference TCSET 2008, pp. 314–317. Ministry of Education and Science of Ukraine, Lviv Polytechnic National University, Lviv, Publishing House of Lviv Polytechnic, Lviv-Slavsko (2008)
7. Barkalov, A., Węgrzyn, M.: Design of Control Units With Programmable Logic. University of Zielona Góra Press, Zielona Góra (2006)
8. Barkalov, A.A.: Principles of optimization of logic circuit of Moore FSM. Cybernetics and System Analysis (1), 65–72 (1998) (in Russian)
9. De Micheli, G.: Synthesis and Optimization of Digital Circuits. McGraw-Hill, New York (1994)
10. Navabi, Z.: Embedded Core Design with FPGAs. McGraw-Hill, New York (2007)
11. Shriver, B., Smith, B.: The anatomy of a High-performance Microprocessor: A Systems Perspective. IEEE Computer Society Press, Los Alamitos (1998)
12. Villa, T., Kam, T., Brayton, R., Sangiovanni-Vincentelli, A.: A Synthesis of Finie State Machines: Logic Optimization. Kluwer Academic Publishers, Boston (1998)
13. Villa, T., Saldachna, T., Brayton, R., Sangiovanni-Vincentelli, A.: Symbolic two-level minimization. IEEE Transactions on Computer-Aided Design 16(7), 692–708 (1997)

Chapter 3
Evolution of Programmable Logic

Abstract. The chapter discussed contemporary field-programmable logic devices and their evolution, starting from the simplest programmable logic devices such as PROM, PLA, PAL and GAL, and finishing with very sophisticated chips such as CPLD and FPGA. This analysis shows particular features of different logic elements and permits to optimize the FSM logic circuits, in which some particular elements are used. The analysis is accompanied by some examples for systems of Boolean functions implementation using PROM, PLA and PAL chips. The principle of functional decomposition oriented on FPGA chips is analysed in the last part of the chapter.

3.1 Simple Field-Programmable Logic Devices

This book deals mostly with synthesis methods oriented on logic devices, which are programmed by the end user. Such devices are named field-programmable logic devices (FPLD) [61,62]. The programmability of FPLD is intended at the hardware level contrary to microprocessors, which run programs but posses a fixed hardware. A FPLD is a general purpose chip whose hardware can be configured by the end user to implement a particular project. Such emphasis of our book is explained by domination of FPLD for design of modern digital devices. Some researches treat FPLD as representatives of Application Specific Integrated Circuits [81], but mostly FPLD are marked out as a separate class of digital devices [61].

The first representatives of FPLD are programmable read-only memory chips (PROM), which were produced by Harris Semiconductor in 1970 [61] They include a fixed array of AND gates (AND-array) followed by a programmable array of OR gates (OR-array) as it shown in Fig. 3.1.

In a PROM, the AND-array implements an address decoder DC, having S inputs and $q = 2^S$ outputs, where each output corresponds to an unique address of a memory cell. The content of OR-array is programmable and the sign "X" in Fig. 3.1 shows a programmable connection.

This architecture perfectly fits for implementation of a system of Boolean functions $Y = \{y_1, \ldots, y_N\}$ on Boolean variables $X = \{x_1, \ldots, x_L\}$, which is represented

A. Barkalov and L. Titarenko: Logic Synthesis for FSM-Based Control Units, LNEE 53, pp. 53–75.
springerlink.com

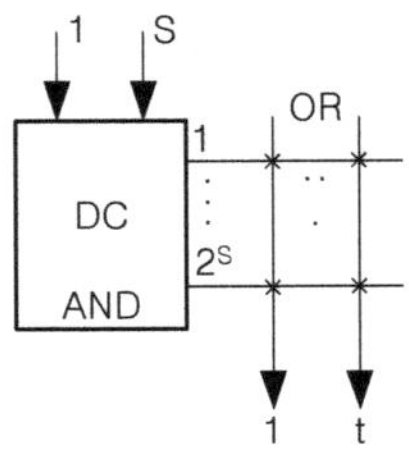

Fig. 3.1 Architecture of PROM

by a truth table [61]. In this case the system to be implemented can be viewed as a table with

$$H = 2^L \tag{3.1}$$

rows, where each row includes L input columns and N output columns. Let us denote this system of Boolean functions (SBF) as $Y(L,N)$ and discuss its implementation with PROM(S,t), where PROM(S,t) means that a PROM chip has S inputs and t outputs. Combinations of parameters S, t, L, N lead to the following implementations of SBF.

1. In case when $S \geq L, t \geq N$ a system $Y(L,N)$ can be implemented in a trivial way using only one chip of PROM(S,t). The implementation is shown in Fig. 3.2, where logical variables X are connected with address inputs of PROM, and functions Y are appeared on the outputs of PROM.

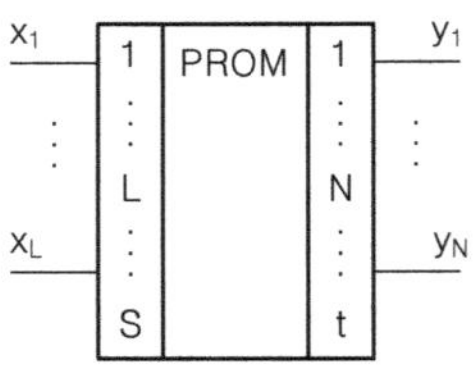

Fig. 3.2 Trivial implementation of SBF with PROM

2. In case when $S \geq L, t < N$ it is necessary

$$n_1 = \left\lceil \frac{N}{t} \right\rceil \tag{3.2}$$

chips of PROM(S,t) to implement a system $Y(L,N)$. Address inputs of all chips are connected with logical variables $x_l \in X$, and each chip generates up to t output functions $y_n \in Y$ (Fig. 3.3). Such an approach sometimes is named as "*an expansion of outputs of PROM*" [13,27].

The value of parameter i for Fig. 3.3 can be calculated as $i = t(n_1 - 1) + 1$.

3. If $S < L, t \geq N$, then the approach of expansion of inputs of PROM [13] is used and

$$n_2 = \lceil 2^L / 2^S \rceil = \lceil H/q \rceil \tag{3.3}$$

chips of PROM is necessary to implement a system $Y(L,N)$ (Fig. 3.4).

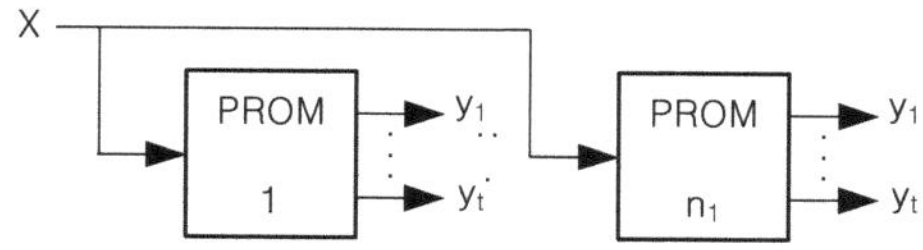

Fig. 3.3 Implementation of SBF with expansion of PROM outputs

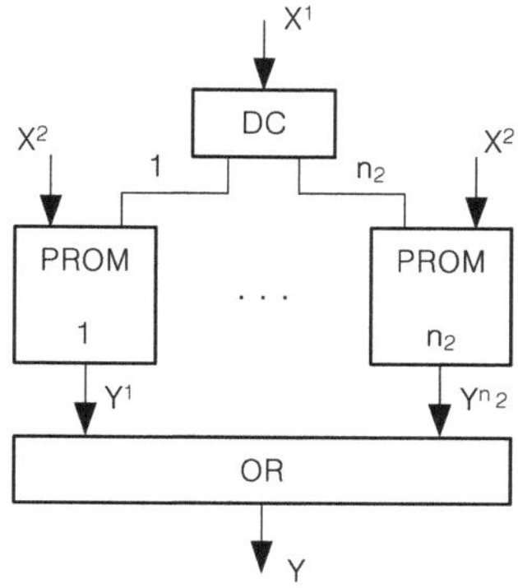

Fig. 3.4 Implementation of SBF with expansion of PROM inputs

In this circuit the $L-S$ leftmost bits of input assignment $\langle x_1,\ldots,x_L\rangle$ form a set of variables X^1, which are connected with inputs of a decoder DC having n_2 outputs. Outputs of the decoder are connected with enable inputs of corresponding PROMs, address inputs of all chips are connected with S rightmost bits of input assignment and these variables form a set X^2. The partial functions Y^i are generated as outputs of i-th microchip and these functions correspond to subtables of the truth table with rows from $q(i-1)$ till q_i. As it can be seen from Fig. 3.4, OR-gates are used to produce the final values of functions $y_n \in Y$. Such an approach is rather a theoretical one, because this level of the circuit could be implemented using three-stable outputs of PROM chips [61].

4. If $S < L, t < N$, then

$$n_3 = n_1 \cdot n_2 \tag{3.4}$$

chips of PROM is necessary for implementation of a SBF. In this case both methods of expansions of outputs and inputs are used together.

There are many ways for programming of FPLD [66], but the following ones are used mostly:

1. *Programming on the base of mask.* Programming is executed using a mask in a manufacturing process. This "program" cannot be changed. The memory devices used this type of programming are named read-only memories (ROM). Such an approach is used in case of ASIC, which targets on a mass production.
2. *One-time programming.* In this case programming is executed using a high voltage, it leads to PROM. This information cannot be altered or erased.

3. *Reprogramming with erasing of information*. In this case initial information can be completely erased and PROM can be reprogrammed. Such an approach is possible due to usage of floating-gate transistors. In case of PROM such devices are named as *Erasable* PROM (EPROM). Previous content is deleted by EPROM exposing to ultra-violet light (for several minutes). To do it, a device should be taking out from a printing board. Writing information into an EPROM is about a 1000 times slower than reading from a device. To get a real value for reprogramming time, the time of erasing should be taking into account.
4. *Reprogramming with electrical erasing*. In this case PROM can be electrically erased; because of it such chips are named EEPROM (*Electrically Erasable PROM*). An EEPROM can be erased and reprogrammed without removing from a printed board (as it was necessary for all previous cases). This feature is very useful for reconfiguring a design on-fly. An EEPROM can be reprogrammed from 10 to 20 000 times. Because both EPROMs and EEPROMs save their internal data while not powered, they belong to the class of non-volatile memories. Writing information into an EEPROM is about a 500 times slower than reading from a device.
5. *Partial reprogramming*. Such PLDs are divided into smaller fixed-size blocks that can be reprogrammed independently (erased and programmed). These devices are named *Flash Memory*. As a rule, they are used to keep either a system configuration (if they are internal devices) or to keep some temporary data (if they are external devices).

Thanks to regularity of their structure, chips of PROM find a wide application for implementation of tabular functions. The main drawback of PROM is doubling of their capacity if the number of inputs is incremented by 1. Besides, PROMs cannot be used for implementation of SBF satisfying to condition

$$H_1 \ll H, \tag{3.5}$$

where H_1 is the number of input assignments, such that at least one of the functions $y_n \in Y$ is equal to 1.

Programmable logic arrays (PLA) were introduced in the mid 1970s by Signetics [61] and they were oriented on implementation of SBF, when condition (3.5) takes place. The peculiarity of PLA is programmability of both AND- and OR-arrays (Fig. 3.5), that determines greater flexibility than in case of PROM.

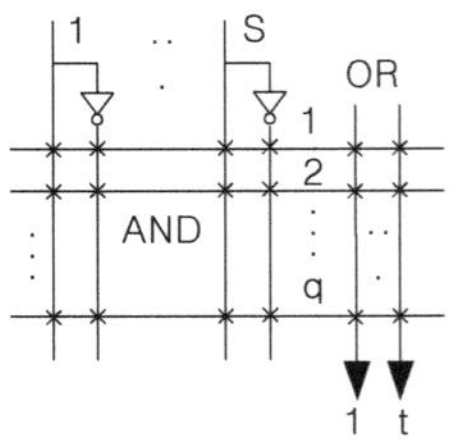

Fig. 3.5 Architecture of PLA

Thanks to programmability of both arrays, PLA can be applied to implement SBF represented as minimal sum-of-products [1,63]. But programmability of AND-array leads to increase of a chip area and decrease of the both resulting circuit speed and value of parameter q in comparison with PROM-implementations [61].

Let PLA with S inputs, t outputs and q terms be denoted as PLA(S,t,q) and let us discuss how they can be used for implementation of SBF $Y(L,N,H_1)$. There are the following combinations of SBF and PLA parameters, which are listed below.

1. If $S \geq L, t \geq N, q \geq H_1$, then SBF Y is implemented in a trivial way using one PLA chip. The structure of resulting circuit is similar to the structure shown in Fig. 2.2, where PLA should be used instead of PROM.
2. If $S \geq L, t < N, q \geq H_1$, then a logic circuit is implemented with n_1 PLA chips, where value n_1 is determined by (3.2), and the structure of this circuit is similar to the structure from Fig. 3.3.
3. If $S \geq L, t \geq N, q < H_1$, then the approach of "*expansion of PLA terms*" should be used [13, 27], and a circuit can be implemented using

$$n_4 = \left\lceil \frac{H_1}{q} \right\rceil \tag{3.6}$$

chips of PLA(S,t,q). Implementation of logic circuit in this case (Fig. 3.6) is similar to one, shown in Fig. 3.4, but decoder DC is absent, because inputs of all microchips are connected with the same logical conditions X.

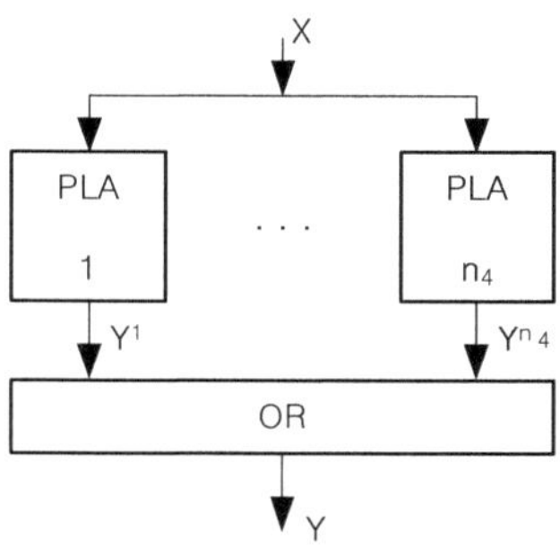

Fig. 3.6 Implementation of SBF with expansion of PLA terms

4. If $S \geq L, t < N, q < H_1$, then both abovementioned methods of expansion should be applied simultaneously. Minimization of hardware amount can be made with application of sophisticated design methods [27], based on the search of some partitions on the set of SBF terms.

More complex synthesis methods are used to implement a SBF Y, when the following condition holds:

$$S < L. \tag{3.7}$$

In this case a synthesis method depends on condition

$$F_{\max} \leq L, \tag{3.8}$$

where $F_{\max}$ is the maximal number of literals [1] in the terms of SBF Y. If this condition takes place, then an initial SBF can be implemented by a single-level circuit, which is shown in Fig. 3.7.

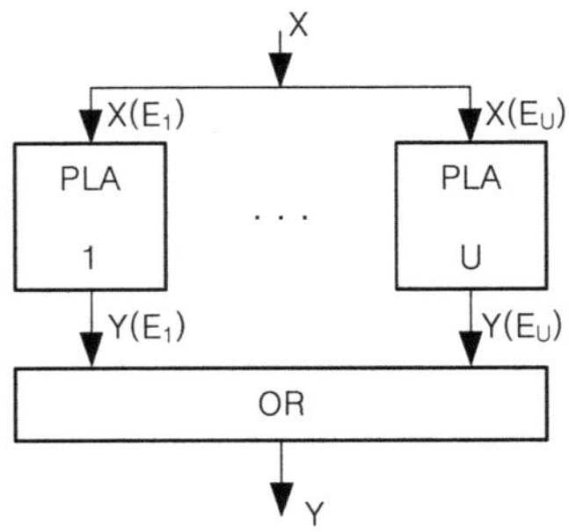

Fig. 3.7 Single-level implementation of SBF with PLA

To design the logic circuit, it should be found a partition Π_F of a set of terms F, where $|F| = H_1$, with the minimum number of blocks U [27]. Let $X(E_u)$ be a set of logical conditions, which form in the terms from a set $E_u \in \Pi_F = \{E_1, \ldots, E_U\}$, and $Y(E_u)$ be a set of functions depending on the terms $E_u \in \Pi_F$. The partition Π_F should satisfy to the following condition:

$$\begin{aligned} |X(E_u)| &\leq S, \\ |Y(E_u)| &\leq t, (u = 1, \ldots, U) \\ |E_u| &\leq q, \\ U &\rightarrow \min. \end{aligned} \tag{3.9}$$

Many different approaches are known, oriented on solution of the problem (3.9) with minimizing of value U [27]. If condition (3.8) is violated, then an SBF Y is implemented as a multilevel circuit [1], and it is connected with decrease of a resulted digital system performance.

It is clear, that PLA allows only the implementation of combinational circuits. If a sequential circuit should be implemented, then outputs of PLA should be connected with an external register. This disadvantage was eliminated with including of flip-flops at each output of PLA inside the chip. Such chips are named registered PLA or programmable logic sequencers (PLS) [61]. Design methods targeted on PLS use a decomposition of initial GSA by subgraphs in such a manner, that an FSM corresponding to each subgraph can be implemented using only one chip of PLS [27].

It is known, that practical digital devices are specified by SBF with limited number of terms, where condition (3.10) holds, where

$$|H(y_n)| \leq 16. \tag{3.10}$$

Here $H(y_n)$ is a set of terms, which are used as products of SOP of a Boolean function $y_n \in Y$. An analysis of this condition shows that PLA have redundancy of connections, because any term of PLA can be connected with any output of a chip.

Programmable array logic (PAL) chips, which were introduced by Monolithic Memories in 1978 [61], were oriented on implementation of SBF, satisfying (3.10). The peculiarity of PAL is existence of programmable AND-arrays and t fixed OR-arrays (Fig. 3.8). It results in increase for the number of inputs and outputs of PAL in comparison with a PLA chip of the same size.

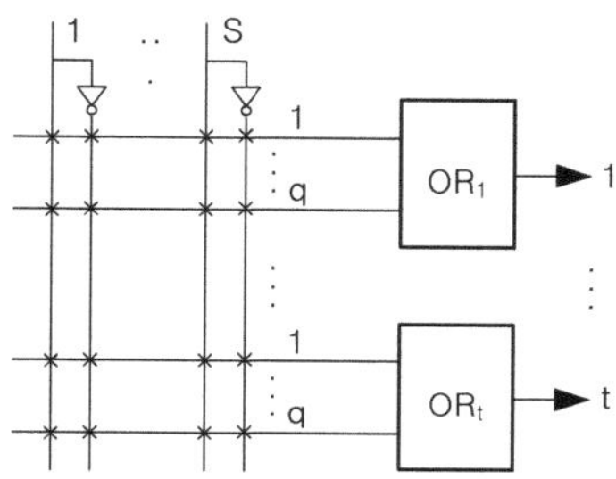

Fig. 3.8 Architecture of PAL

One of the new conceptions connected with PAL was the conception of a macrocell. The *macrocell* is a part of a chip connected with a single PAL output. For example, the chip shown in Fig. 3.8 includes t macrocells. To increase the area of such chips application, some additional elements were added to each output of PAL, such as flip-flops, logic gates and multiplexers. The macrocell has a feedback path from the output of the cell to the AND-array. The connections inside a macrocell were programmable too and it increases flexibility of PAL. Feedbacks in PAL chip permit to implement the functions with parenthesis [5, 6]. Macrocells have tristate outputs and there is a possibility to use the chip pins as bidirectional input-outputs. Besides, the tristable outputs permit usage of either direct or complement Boolean functions. Macrocells include a programmable embedded flip-flop, enabling FSM implementation without external memory registers.

Design methods for PAL are oriented on minimizing of the value $|H(y_n)|$ up to the some fixed value q, determining the number of AND-arrays connected with single OR-array [51–55, 55, 56, 58, 82, 83]. Appearance of PAL stimulated development of FSM design methods [1], which were rather different from designed methods with PLA chips [7, 9, 11, 43, 44, 74–77].

The growth of the number of PAL inputs results in drastic performance decrease and, hence, in limitations for their usage in practical designs [61]. Appearance of EECMOS (Electrically Erasable CMOS) technology permitted very simple reprogramming. Combining structure of PAL and EECMOS technology results in generic array logic (GAL) chips, which were introduced by Lattice in 1985 [61]. Let us point out, that chips of GAL are still manufactured in a standalone packages by Lattice, Atmel, TI, etc. A typical example of GAL device is the GAL16V8 chip, which has 16 inputs, 8 outputs and 20 pins. This device has 8 input pins and 8 bidirectional input/output pins, it means that these pins can be used either as inputs or as outputs.

Such chips as PLA, PLS, PAL, GAL belong to the class of Simple Programmable Logic Devices (SPLD), they have not more than 40 inputs/outputs and they are equivalent not more than 500 NAND-gates with two inputs [69], named as system gates [61].

3.2 Programmable Logic Devices Based on Macrocells

To implement complex logic controllers, it is necessary to have PAL chips with large number of terms per a macrocell, as well as with very large number of macrocells. Unfortunately, such chips are notable for very big propagation time and very small coefficient of chip area usage [66]. Development of semiconductor technology allowed quite different solution of this problem, when a single chip includes a collection of simple PAL macrocells connected using programmable connections. Such chips belong to the class of CPLD (Complex Programmable Logic Devices); the simplified structure of a typical CPLD is shown in Fig. 3.9.

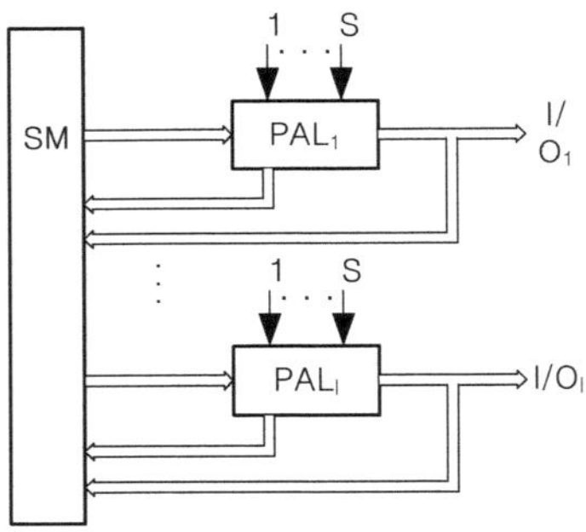

Fig. 3.9 Simplified architecture of CPLD

In this CPLD each macrocell PAL_i $(i = 1, \ldots, I)$ is connected with S fixed inputs of a chip and with programmable input/outputs IO_i. The block outputs can be used as input information for a switch matrix SM. The first CPLD were devices MegaPAL of MMI [61]. Now several companies such as Altera, Xilinx, Cypress, Atmel, Lattice manufacture CPLDs [61]. As example of a typical CPLD we can mention the Xilinx XC9500, where PLD resembling a 36V18 GAL device are used. Modern CPLDs contain some additional features, like JTAG support and interface to other logic standards. For different CPLD vendors, macrocells have different configurations [61]; it is interesting that different manufactures use different terminology to name the same things.

Let us discuss as a typical example a family MAX of Altera [2], where acronym MAX stands for Multiple Array Matrix. Let us choose a family MAX5000, based on EPROM technology. Macrocells of this family are combined in blocks, named LAB (Logic Array Block), which can interplay using programmable connections from PIA (Programmable Interconnect Array). For example, CPLD EPM5032 includes only single LAB. This chip includes a term expander ES to share terms among different macrocells, an input-output block I/O, a matrix of macrocells MCA, and a block of internal interconnections BII. The number of logic blocks LAB is increased with growth of chip complexity. This tendency is shown in Fig. 3.10, borrowed from [61].

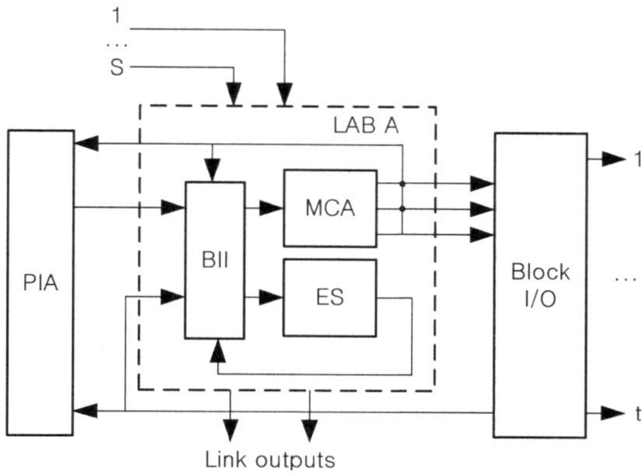

Fig. 3.10 Architecture of MAX5000 family members

Depending on a chip, the input number S varies from 8 to 20, the number of bidirectional input-outputs t from 4 to 16, and the number of blocks LAB from 1to 12. Internal expanders are used to increase the number of terms implemented by a macrocell. The macrocell has the following structure (Fig. 3.11).

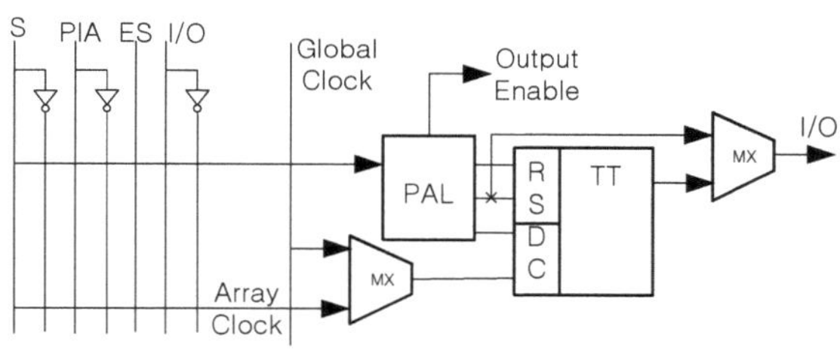

Fig. 3.11 Architecture of CPLD MAX5000 macrocell

In Fig. 3.11, the block PAL includes three programmable AND gates, connected with OR gate, as well as the AND gate to control both flip-flop TT and macrocell output (Output Enable). Additionally, each macrocell includes two multiplexers. The first of them is used to connect the macrocell output either with the combination output of PAL or registered output of TT flip-flop. The second multiplexer together with additional AND gate is used to control a synchronization mode of the flip-flop. The synchronization mode can be either common for all macrocells (Global Clock), or local for the given block (Array Clock). There are four types of macrocell inputs: fixed chip inputs (up to *S* inputs), outputs of matrix PIA, outputs of block ES, and outputs of input-output block I/O.

The expander block is used for increasing the numbers of terms implemented by the block PAL. For MAX5000 family, this block includes from 32 to 64 multi input AND gates, their inputs are connected with fixed chip inputs, and outputs of blocks PIA and BII.

With development of semiconductor technology, all parameters of CPLD (such as the number of pins, the number of macrocells and so on) are increased. For example, the chips of MAX7000S family are based on EEPROM technology and they can replace from 600 to 5000 system gates. The typical representative of this family is CPLD EMP7128S, having 8 LAB blocks, where each LAB includes 16 macrocells. Logic complexity of a chip is about 2500 system gates; the maximum frequency of its operation is 147,1MHz. The chip can operate with 5 V or 3.3 V. The CPLD of MAX9000 family are even more complex. For example, EMP9560 chip includes 560 macrocells, 772 flip-flops, 216 input-outputs. It is equal to 12000 system gates and operates with maximum frequency up to 118MHz.

One of the serious restrictions of CPLD based on PAL macrocells is the limited number of implemented product terms per macrocell. It restrains application of CPLD in such areas as mobile phones, computer games, and personal digital assistances. To overcome this drawback, the Cool Runner XPLA3 family was introduced by Xilinx. This family is based on macrocells of PLA type [89]. General overview of XPLA3 chip is shown in Fig. 3.12.

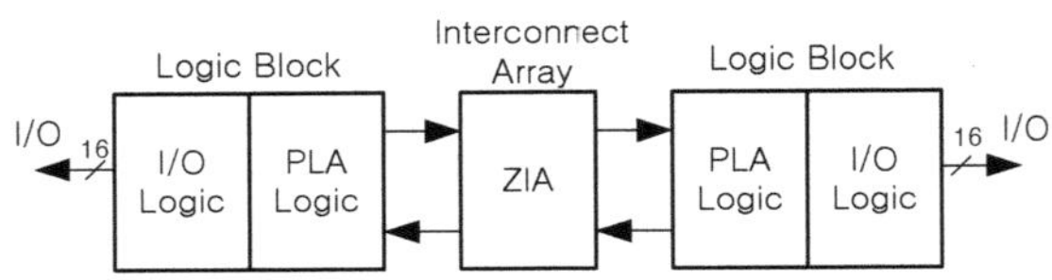

Fig. 3.12 Architecture of CPLD XPLA3 family

Different CPLD XPLA3 chips include up to 24 logic blocks, and each block contains 16 macrocells. Blocks interconnect through block ZIA (Zero-power Interconnect Array). Each logic block has 36 inputs from ZIA, whereas the number of inputs connected with a block of input-output logic (I/O Logic) can be different for different logic blocks from the same chip.

Each logic block can be viewed as a PLA block, having 36 inputs (outputs of ZIA) and 48 terms, which can be distributed among 16 OR gates. The outputs of OR gates are connected with multiplexers VFM (Variable Function Multiplexer). Thus, each logic block is equivalent to a PLA, having $S = 36$ inputs, $t = 16$ outputs, and $q = 48$ product terms. The terms of a logic block are generated by a matrix PTA (product term array); some outputs of PTA are used for special purposes. Architecture of the logic block is shown in Fig. 3.13.

The logic block generates 8 control terms PT[0–7] to asynchronous controlling flip-flops of macrocells $MC_i(i = 1,\dots,16)$, as well as states of their outputs. Inputs of macrocells MC_i are connected either with the term $PTi + 15$, or the disjunction of any terms from the set PT[0–47] using blocks OR_i and $VFM_i(i = 1,\dots,16)$. Terms PT[32–47] can be used for synchronization of macrocell flip-flops, whereas outputs PT[8–15] can be used to organize feedback as outputs of NAND gates.

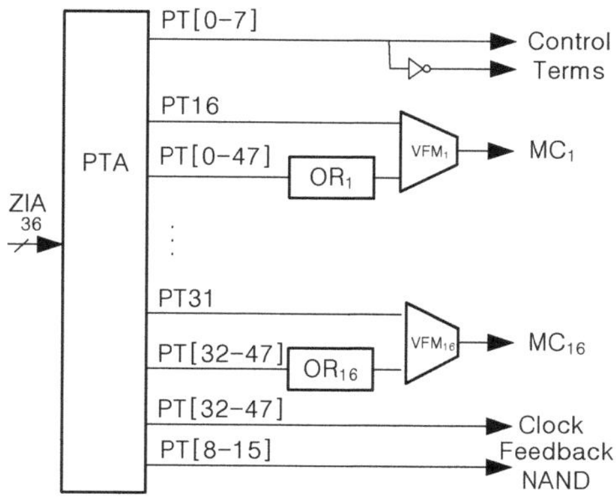

Fig. 3.13 Architecture of logic block for XPLA3 family

Macrocells of CoolRunner XPLA3 family include a memory element configured either as D FF or T FF, or as a latch. The flip-flop synchronization has 8 modes, including the global clock (system synchronization). The macrocell output can be either combinational, or registered, which is used as a feedback signal for the ZIA. The bidirectional macrocell input-output is connected with the block ZIA too. If this pin is used as an input, then macrocell output is set up in the third state using a special control term. In this case the combinational output cannot be used as feedback for the block ZIA. The input-output logic allows disconnection of pins, which are not used as inputs of ZIA.

Important features of the representatives of CoolRunner XPLA3 family are listed in Table 3.1; this table is based on [89].

Table 3.1 Characteristics of CoolRunner XPLA3 family

Chip	XCR3032XL	XCR3064XL	XCR3128XL	XCR3256XL	XCR3384XL	XCR3512XL
I	32	64	128	256	384	512
G	800	1600	3200	6400	9600	12800
t_p	5	6	6	7,5	7,5	7,5
f_{max}	200	145	145	140	127	127

In this table, the symbol I stands for the number of macrocells (the number of blocks is obtained automatically, dividing the number of macrocells by 16); G is the number of system (equivalent) gates; t_p is a propagation time, measured in nanoseconds; f_{max} is a maximal frequency of operation, measured in megahertz. The PLA-based macrocells are used in CPLD chips of Cool Runner II family by Xilinx [89].

Macrocells based on both PAL and PLA architectures are very efficient in implementing irregular and multiplexer functions, but they are not suitable enough for implementing regular functions. The truth table is the best way for presentation of regular functions (and their systems); it means that the best way for implementing regular functions is usage of memory blocks (PROM, RAM). Taking it into account, designers from Cypress use both PAL- and RAM-based macrocells in their Delta 39K family [39, 40].

Architecture of Delta 39K family includes a collection of clusters, where each cluster includes 8 logic blocks and 2 blocks of a cluster memory, named Cluster RAM Blocks (CRB). Each CRB has 8 Kbit of memory and can be configured as a memory block, operating in both synchronous or asynchronous modes and having the following characteristics: $8K \times 1$, $4K \times 2$, $2K \times 4$, $1K \times 8$. All 10 cluster blocks interplay through a matrix of programmable interconnections PIM (Programmable Interconnect Matrix). Each logic block includes 16 macrocells, based on PAL architecture. Besides, each cluster includes an additional block of channel memory with 4K bits; the number of its outputs can be different, namely, it can be equal to 1, 2, 4, and 8. There is a system of global interconnections, allowing interplay of different clusters.

This family is characterized by impressive values of parameters. For example, the Delta 39K200 chip includes 200000 system gates, 3072 macrocells, and 480Kb of RAM. Then, chips of Delta 39K(TM) family include 350000 gates; they have a propagation time around 7 nanoseconds, and operate with maximum frequency 233MHz. Obviously, these chips have effective tools for implementing all kind of Boolean functions, namely for implementation of regular, irregular, and multiplexer functions.

3.3 Programmable Devices Based on LUT Elements

Technology of field-programmable gate arrays (FPGA) is developed simultaneously with technology of CPLD [31, 59, 61, 62]. These chips can replace millions 2NAND gates; each logic block of FPGA is equivalent from 10 to 20 system gates [66]. They were introduced by designers of Xilinx in 1985 [89].

Consider some representatives of FPGA family produced by Xilinx, which are based on look-up table (LUT) elements. As a rule, LUT elements are based on RAM, having in average 4 inputs. Single LUT can implement an arbitrary Boolean function depended on $S \leq 4$ input variables and represented as a truth table. Together with reconfigurable flip-flops, LUT elements make a Configurable Logic Block (CLB). A simplified architecture of CLB is shown in Fig. 3.14.

In such a circuit, the LUT-element implements an arbitrary Boolean function $y = y(x_1, \ldots, x_S)$, the signal "Select" uses a multiplexer MX and chooses either combinational or register mode of PLB output, the pulse "Clock" is used for timing of the flip-flop TT.

Typical representatives of FPGA produced by Xilinx are the chips from the Spartan family, for example, Spartan-3. These chips are powered by 1,2V and they use the 90-nanometre technology. In 2003, the price for chip with 17000 CLBs, which is equivalent to 1000000 system gates, was only 20$.

On the one hand, LUT can be used as a 16-bit shift register. Separate registers can be combined together forming a long shift register chain. Contrariwise, each LUT element represents a RAM or PROM memory; these memory blocks can be

Fig. 3.14 Architecture of programmable logic block

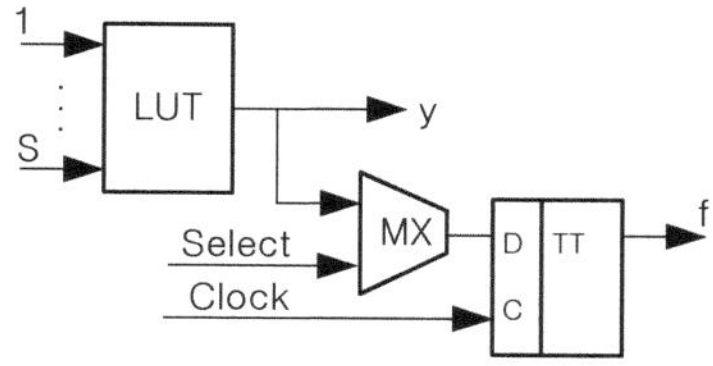

combined together to create a memory block with an arbitrary configuration. The chips of Spartan-3 family include up to 104 memory blocks, with 18Kb for each of them. Thus, these chips include up to 1,87Mb in their embedded blocks of RAM (BRAM). The frequency of operation for these chips can be variable (from 25 MHz till 325 MHz). They support 23 different input-output standards; this specific feature enables application of Spartan-3 chips in different fields of digital automatics, as well as for video- and multimedia systems. Some characteristics of Spartan-3 family are shown in Table 3.2.

Table 3.2 Characteristics of Spartan-3family

Device	XC3S50	XC3S200	XC3S400	XC3S1000	XC3S1500	XC3S2000	XC3S4000	XC3S5000
I	1728	4320	8064	17280	29952	46080	62208	74880
G	50K	20K	40K	1000K	1500K	2000K	4000K	5000K
BRAM	72K	216K	288K	432K	576K	720K	1728K	1872K
DRAM	12K	30K	56K	120K	208K	320K	432K	520K

In this table, the symbol I determines the number of macrocells per chip; the symbol G determines the number of system gates per chip; the symbol BRAM determines the number of embedded memory blocks; the symbol DRAM determines the number of bits when LUT elements are treated as a distributed memory.

There are many FPGA families produced by Xilinx [89]: XC9500, XC9500XL, XC9500XV, XCR3000XL (having up to 512 macrocells); Spartan, Spartan XL, and Spartan-3 (having up to 5000000 system gates); Virtex, Virtex E, Virtex II, and Virtex II Pro (having up to 4000000 system gates and up to 4 PowerPC microprocessors).

The second large-scale producer of FPGA chips is Altera [2]. Typical representatives of its FPGA are chips FLEX10K, where acronym FLEX stands for Flexible Logic Element MatriX. These devices can be reconfigured in 320 milliseconds. For example, the chip EPF10K70 of FLEX family is equal to 70000 system gates (taking into account embedded blocks BRAM). Logic of a project is implemented using blocks LAB; there are 468 blocks LAB arranged as a matrix having 52 columns and 9 rows. Each LAB block consists from 8 logic elements LE, that is the chip 3744 LEs. Moreover, the chip includes an additional column with 9 embedded memory blocks EAB (Embedded Array Block), and each EAB includes 2048 bits. Totally, the chip possesses up to 18432 bits of RAM.

Obviously, blocks EAB can be used for implementing systems of regular Boolean functions, where each EAB of FLEX10K family replaces up to 600 system gates. These blocks can be used for implementing parallel multipliers, sequential circuits (such as FSM), as well as different devices for signal processing, and so on. Each block can be used either separately, or together with others blocks. These blocks EAB can be viewed as an additional large LUT element. It permits to increase the system performance if calculation of complex combinatorial functions is in need. Blocks EAB can be used as synchronous blocks RAM, which can be adjusted to the global chip synchronization. As for Delta 39K, blocks EAB can be configured as 2048x1, 1024x2, 512x4, or 256x8 devices.

Eight logic elements LE and a local interconnect form single block LAB (obviously, the LAB of Altera corresponds to the CLB of Xilinx). Each LAB, as it is shown in Fig. 3.15, represents up to 100 system gates. Each LE includes one LUT element, having 4 inputs, a programmable flip-flop, and some additional logic used for organizing adders (carry logic or carry chain) and cascading of functions (cascade chain).

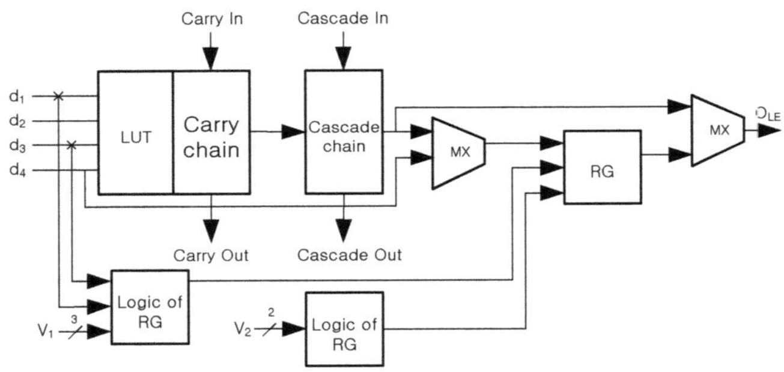

Fig. 3.15 Simplified architecture of logic element from FLEX10K family

Three variables are connected with an output O_{LE} of the logic element, namely: an input d_4 of the LE, a combinational output of the LUT, or a registered output of a programmable flip-flop RG. The output O_{LE} is connected with the matrix of local interconnect. The register RG can be programmed for D, T, JK, or RS operation. To control the RG (such as clear, preset, or synchronization), an internal logic represented by inputs $d_1$and d_3 is used, or it is controlled by elements of a set V_1, which includes the global clear signal. Elements of the set V_2 including global and local synchronization signals are used for synchronizing. A special block of Clock logic is used to generate synchronization pulses. For combinatorial functions, the RG is bypassed.

Technological progress leads to decrease of transistor sizes and, therefore, to increase of their number in a chip. For example, FPGA chips of Cyclone family produced by Altera include up to 20060 logic elements and 288K bits of RAM. In these devices, each LAB includes 10 logic elements. Different representatives of the Cyclone family have from 2910 to 20060 logic elements.

The embedded memory is represented by so called M4K RAM blocks; each of them includes 4K bits of memory. These blocks are reconfigurable and their outputs can include up to 36 bits (each word has 32 bytes and 4 bytes are used for parity control). Memory blocks operate with frequency up to 250 MHz. Some characteristics of Cyclone family are shown in Table 3.3.

Table 3.3 Characteristics of Cyclone family

Device	EP1C3	EP1C4	EP1C6	EP1C12	EP1C20
Number of LE	2910	4000	5980	12060	20060
Number of blocks RAM (128x36 bits)	13	17	20	52	64
Total capacity of RAM, bits	59904	78336	92160	239616	294912
Number of pins for a user	104	301	185	249	301

Examples of different representatives of FPGA family can be continued, but due to rapid technological advance such information is going out of date very quickly. In our opinion, a reader now has some preliminary knowledge about both CPLD and FPGA. To keep pace with technical progress, it is necessary to visit the web sites of such FPLD producers as Xilinx, Altera, Lattice, Atmel, Cypress [2, 4, 40, 59, 89].

3.4 Design of Control Units with FPLD

The main problem in design of control units is irregularity of their logic circuits. It does not permit usage of large library cells, in contrary to design of operational automata (data-path) having regular structures [5]. Thus, it is important to use models of control units having regular parts. In this case some part of a logic circuit can be implemented using such library cells as memory blocks, multiplexers, or decoders. The second problem is lack of universal methods for minimizing the hardware amount in logic circuits of control units. Optimization methods should take into account peculiarities of both logic elements in use and a control algorithm to be implemented. Let us discuss some of these features.

1. CPLD based on PLA macrocells. Design methods are reduced to modularization an initial logic circuit, that is to partitioning an initial logic circuit by subcircuits implemented using one PLA macrocell. It is necessary to perform the mutual minimization of the system of Boolean functions to be implemented. The subsystems with limited amount of product terms should be found. Design methods target in PLA can be found in [1,27,31,45]. To decrease the hardware amount, the structural decomposition was used; it allows usage of multiplexers and PROM chips jointly with PLA modules [27]. Reduction of combinational part could be reached due to usage of a counter instead of a register to interpret the linear parts of a control algorithm [8, 9, 12, 30, 67, 68]. Appearance of the chips CoolRunner family should renew an interest for development of these design methods.

2. CPLD based on PAL macrocells. Design methods are reduced to the separate minimization of functions representing a logic circuit of a control unit. It is desirable that the number of terms in the SOP of a Boolean function does not exceed the number of product terms per a macrocell. Otherwise, either the macrocell cascading should be executed, or logic expanders should be used. Both approaches lead to slowdown of a resulted design. The design methods have been developing starting from the first announcement about appearance of PAL chips [3, 31, 41, 42, 49, 51–56, 58, 82–84]. Some methods targeted on structural decomposition leading to combined use of PAL chips together with PROM chips [14–21, 28]. Now such methods can be used in designs connected with CPLD Delta 39K family by Cypress.
3. FPLD based on LUT elements. Methods of digital devices design with FPGA significantly distinguish from their counterparts targeted on CPLD. The principle of functional decomposition is the base for designing logic circuits with FPGA [33, 70, 72, 73, 78]. Design methods are developed permanently taking into account features of particular FPGA families (a practical approach), as well as the abstract conception of FPGA (a theoretical approach) [31, 37, 48, 60, 61, 72–74]. Besides, some methods are developed, which are targeted on the joint use of LUT elements and embedded memory blocks [33, 37, 72, 86]. Finally, a group of methods deal with LUT elements, EMBs and counters for interpretation of linear parts of control algorithms [10, 13, 22–26, 29, 31, 32, 86–88]. Consider the method of functional decomposition in more details. It is based on representation of a Boolean function $F = F(X)$ in the following form:

$$F(X) = H(X_0, G_1(X_1), \ldots, G_I(X_I)). \tag{3.11}$$

In (3.11), unification of the sets $X_i (i = 0, \ldots, I)$ gives an initial set X, as it is shown in Fig. 3.16.

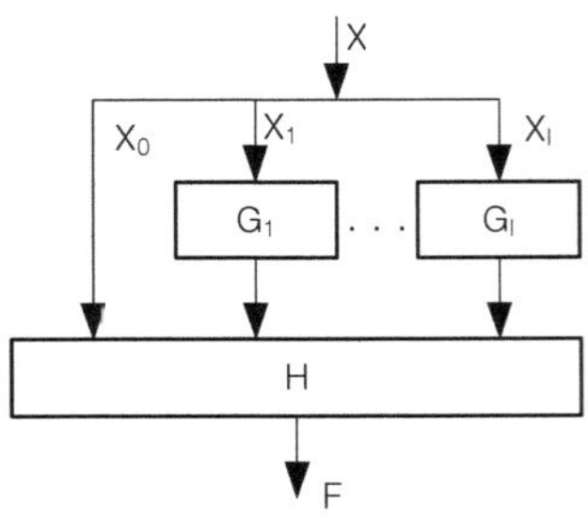

Fig. 3.16 Illustration of the principle of functional decomposition

In such a representation, there are coding functions G_i and a base function F. Let us point out that the decomposition is executed in such a manner that each from the functions $G_i (i = 0, \ldots, I)$, as well as the function H, could be implemented by a single LUT element.

Also, the functional decomposition can be used under the design of control units with CPLD [48]. It is shown in [56] that optimization methods targeted on FPGA can decrease hardware amount in logic circuits with CPLD. Results of the investigations described in [37, 86] show that decrease for amount of terms in SBF of control units yields in reduction of the number of LUT elements in logic circuits of these control units. Thus, it is reasonable to develop optimization methods targeted on some particular logic elements and then to check their usability for different types of FPLD chips.

The exceptional complexity of both CPLD and FPGA chips requires use of computer-aided design (CAD) tools for designing control units [49, 61]. It assumes development of formal methods for synthesis and verification of control units [5, 31, 46, 47, 50, 64, 66, 85, 91]. For example, a design process for FPLD from Xilinx includes the following steps:

1. **Specification of a project.** A design entry can be executed using the schematic editor (a design is represented by some circuit), or the state editor (a design is represented by a state diagram), or some program written on a hardware description language (HDL). Both VHDL and Verilog are the most popular HDLs [35, 36, 38, 71]. An initial specification should be verified using procedures of syntax and semantic analysis. After such a verification, the initial specification can be corrected and this step should be repeated.
2. **Logic synthesis.** During this step, the package FPGA Express executes synthesis and optimization of a control unit logic circuit. As an outcome, an FPGA Netlist file is generated with a list of chains for a control unit to be implemented. The file can be represented using either EDIF or XNF formats. The library cells from system and user libraries are used during this step.
3. **Simulation.** The functional correctness of a control unit is checked during this step. The step of simulation is executed without taking into account real propagation times in a chip. If an outcome of simulation is negative, then the previous steps should be repeated.
4. **Implementation of logic circuit.** Now the Netlist is translated into an internal format of CAD system and such physical objects as CLBs and chip pins are assigned for initial logic elements. This step is named a mapping. Next, such steps as placement and routing are executed. Now there are physical elements and connections among them. It allows finding out the real values of propagation times for an FPGA chip selected for implementation of a project with the control unit. The final outcome of this step is some data used to program a chip named as BitStream.
5. **Project verification.** The final simulation is performed, where the actual values of delays among the physical elements of the chip are used. If outcome of this step is negative (actual performance of control unit is less than it is needed), then previous steps of design process should be repeated to get some new results.
6. **Chip programming.** Obviously, that step is connected with writing a final bit stream into the chip.

Implementation of control units with CPLDs is as difficult, as in case of FPGA-based designs. Each producer of FPGA and CPLD chips has its own CAD to support the design process. The most known packages are, for example, MAX+PLUS,

MAX+PLUSII, WebPack, QuartusII. Some of them are free of charge and can be obtained through Internet. Besides, there are CAD tools of some producers specialized only on synthesis and verification, such as Symplicity Amplify Physical Synthesis Support, Leonardo Spectrum, Synopsis FPGA Compiler II and so on. Information about these tools can be found on the sites of corresponding producers.

Besides, there is a lot of different academic CAD tools, where can be found various methods for design and optimization of control units. As a rule, some intermediate software should be developed to connect academic and industrial tools. The intermediate software allows entering either Boolean systems describing some parts of control units or VHDL-models of these parts.

For example, the system SIS from Berkeley, USA [79, 80] has tools for single-level and multi-level design of digital devices. The well-known system MIS [34] is a base for development of the SIS. This system uses a special language KISS to specify a control unit. The language is an entry to the program ESPRESSO, executed minimization of logic functions by appropriate state assignment. A special algorithm STAMINA is used for minimizing the number of states. If control units are implemented with FPGA, then the state assignment is executed by an algorithm JEDI. Its counterpart for PLA case is an algorithm NOVA.

A design system DEMAIN [60, 73] is developed in Poland (Politechnika Warczawska) by Professor T. Łuba. This system deals with FPGA-based designs. The system generates some preliminary data used by the MAX+PLUSII of Altera. The base of the system is a set of algorithms target on decomposition of a system of Boolean functions describing a combinational part of a control unit.

Some other academic systems are known, such as ASYL [76, 77], targeting in designs with PLA and ROM, or ZUBR, targeting in designs with CPLD [82], as well as ATOMIC, targeting in designs of compositional microprogram control units with FPGA [86]. To compare design outcomes for different CAD, a standard set of tests (benchmarks) [90] is used. The set contains practical examples of sequential circuits, represented in the KISSII format.

In our book, either graph-schemes of algorithms [5, 6] or structure tables are used for specification of control units. The first form gives clearness of presentation, whereas the second form is very close to the KISSII format. As a rule, all presented methods are accompanied by examples. The examples are completed by some tables, which can be used to derive the systems of Boolean functions specified some parts of logic circuit of a control unit. We do not discuss the particular problems of logic circuit implementation using some specific chips. We escape from particular types of microchips; because of it we use only the symbol PLD (Programmable Logic Device) in logic circuits for discussed examples. This symbol can stand either for a group of LUT elements (if FPGAs are used to implement a circuit), or for a group of PAL- or PLA-based macrocells (if a circuit is implemented using CPLDs).

Among all known models of control units, the highest performance belongs to a single-level model (Fig. 3.17). Let us denote it as a P FSM. A block BP of P FSM can include more than one layer of logic elements; it depends on both complexity of a control algorithm to be interpreted and parameters of logic elements in use. But the P FSM is the single-level model because it includes only single block of

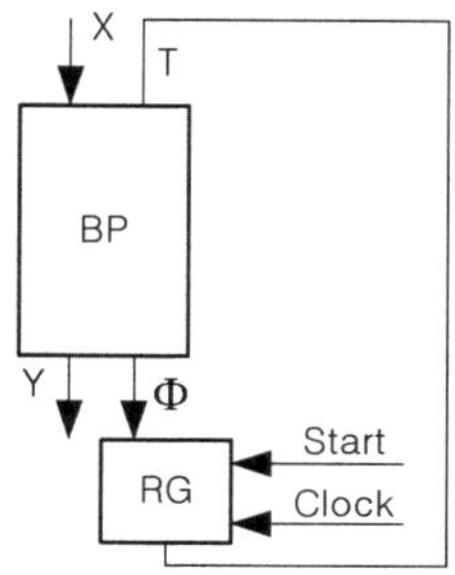

Fig. 3.17 Structural diagram of P FSM

combinational logic. Design methods of similar FSMs are discussed thoroughly in [27, 65, 66]; they are not the subject of our book.

In our book, the symbol S is used to denote any model of FSM. The symbol $S(\Gamma)$ emphasises the fact that FSM is synthesized to interpret some GSA Γ. Obviously, both an FSM and a GSA can have their serial numbers.

References

1. Adamski, M., Barkalov, A.: Architectural and Sequential Synthesis of Digital Devices. University of Zielona Góra Press, Zielona Góra (2006)
2. Altera Corporation Webpage, http://www.altera.com
3. Asahar, P., Devidas, S., Newton, A.: Sequential Logic Synthesis. Kluwer Academic Publishers, Boston (1992)
4. Atmel Corporation Webpage, http://www.atmel.com
5. Baranov, S.: Logic and System Design of Digital Systems. TUT Press, Tallinn (2008)
6. Baranov, S.I.: Logic Synthesis of Control Automata. Kluwer Academic Publishers, Dordrecht (1994)
7. Barkalov, A.: Multilevel pla schemes for microprogram automata. Cybernetics and system analysis 31(4), 489–495 (1995)
8. Barkalov, A., Beleckij, O., Nedal, A.: Applying of optimization methods of moore automaton for synthesis of compositional microprogram control unit. Automatic Control and Computer Sciences 33(1), 44–52 (1999)
9. Barkalov, A., Dzhaliashvili, Z., Salomatin, V., Starodubov, K.: Optimization of a microinstruction address scheme for microprogram control unit with pla and prom. Automatic Control and Computer Sciences 20(5), 83–87 (1986)
10. Barkalov, A., Kołopieńczyk, M., Titarenko, L.: Optimization of control memory size of control unit with codes sharing. In: Proc. of the IXth Inter. Conf. CADSM 2007, pp. 242–245. Lviv Polytechnic National University, Publishing House of Lviv Polytechnic National University, Lviv (2007)
11. Barkalov, A., Salomatin, V., Starodubov, K., Das, K.: Optimization of mealy automaton logic using programmable logic arrays. Cybernetics and system analysis 27(5), 789–793 (1991)
12. Barkalov, A., Titarenko, L.: Logic Synthesis for Compositional Microprogram Control Units. Springer, Berlin (2008)
13. Barkalov, A., Titarenko, L.: Synthesis of Operational and Control Automata. UNITECH, Donetsk (2009)

14. Barkalov, A., Titarenko, L., Barkalov Jr., A.: Moore fsm synthesis with coding of compatible microoperations fields. In: Proc. of IEEE East-West Design & Test Symposium - EWDTS 2007, Yerevan, Armenia, pp. 644–646. Kharkov National University of Radioelectronics, Kharkov (2007)
15. Barkalov, A., Titarenko, L., Chmielewski, S.: Optimization of logic circuit of moore fsm on CPLD. Pomiary Automatyka Kontrola 53(5), 18–20 (2007)
16. Barkalov, A., Titarenko, L., Chmielewski, S.: Optimization of moore fsm on CPLD. In: Proceedings of the Sixth Inter. conf. CAD DD 2007, Minsk, vol. 2, pp. 39–45 (2007)
17. Barkalov, A., Titarenko, L., Chmielewski, S.: Optimization of moore fsm on CPLD. International Journal of Applied Mathematics and Computer Science 17(4), 565–675 (2007)
18. Barkalov, A., Titarenko, L., Chmielewski, S.: Optimization of moore fsm on system-on chip. In: Proc. of IEEE East-West Design & Test Symposium – EWDTS 2007 (2007)
19. Barkalov, A., Titarenko, L., Chmielewski, S.: Decrease of hardware amount in logic circuit of moore fsm. Przeglźd Telekomunikacyjny i Wiadomości Telokomunikacyjne (6), 750–752 (2008)
20. Barkalov, A., Titarenko, L., Chmielewski, S.: Optimization of moore control unit with refined state encoding. In: Proc. of the 15th Inter. Conf. MIXDES 2008, Poznań, Poland, pp. 417–420. Departament of Microeletronics and Computer Science, Technical University of Łódz (2008)
21. Barkalov, A., Titarenko, L., Chmielewski, S.: Optimization of moore fsm on system-on-chip using pal technology. In: Proc. of the International Conference TCSET 2008, Lviv-Slavsko, Ukraina, pp. 314–317. Ministry of Education and Science of Ukraine, Lviv Polytechnic National University, Publishing House of Lviv Polytechnic, Lviv (2008)
22. Barkalov, A., Titarenko, L., Kołopieńczyk, M.: Optimization of circuit of control unit with code sharing. In: Proc. of IEEE East-West Design & Test Workshop - EWDTW 2006, Sochi, Rosja, pp. 171–174. Kharkov National University of Radioelectronics, Kharkov (2006)
23. Barkalov, A., Titarenko, L., Kołopieńczyk, M.: Optimization of control unit with code sharing. In: Proc. of the 3rd IFAC Workshop: DESDES 2006, Rydzyna, Polska, pp. 195–200. University of Zielona Góra Press, Zielona Góra (2006)
24. Barkalov, A., Wiśniewski, R.: Optimization of compositional microprogram control unit with elementary operational linear chains. Upravlauscie Sistemy i Masiny (5), 25–29 (2004)
25. Barkalov, A., Wiśniewski, R.: Optimization of compositional microprogram control units with sharing of codes. In: Proc. of the Fifth Inter. Conf. CADD'DD 2004, Minsk, Belorus, vol. 1, pp. 16–22. United Institute of the Problems of Informatics, Minsk (2004)
26. Barkalov, A., Wiśniewski, R.: Design of compositional microprogram control units with maximal encoding of inputs. Radioelektronika i Informatika (3), 79–81 (2004)
27. Barkalov, A., Węgrzyn, M.: Design of Control Units With Programmable Logic. University of Zielona Góra Press, Zielona Góra (2006)
28. Barkalov, A., Wêgrzyn, A., Barkalov Jr., A.: Synthesis of control units with transformation of the codes of objects. In: Proc. of the IXth Inter. Conf. CADSM 2007, Lviv - Polyana, Ukraine, pp. 260–261. Lviv Polytechnic National University, Publishing House of Lviv Polytechnic National University, Lviv (2007)
29. Barkalov, A.A., Wêgrzyn, M., Wiśniewski, R.: Partial reconfiguration of compositional microprogram control units implemented on fpgas. In: Proceedings of IFAC Workshop on Programmable Devices and Embedded Systems (Brno), pp. 116–119 (2006)

30. Barkalov, A.A.: Microprogram control unit as composition of automate with programmable and hardwired logic. Automatics and computer technique (4), 36–41 (1983) (in Russian)
31. Barkalov, A.A., Titarenko, L.A.: Design of control units with programmable logic devices. In: Korbicz, J. (ed.) Measurements, methods, systems and design, pp. 371–391. Wydawnictwo Komunikacji i Łączności, Warsaw (2007)
32. Barkalov, A.A., Wiśniewski, R.: Optimization of compositional microprogram control units implemented on system-on-chip. Theoretical and applied informatics (9), 7–22 (2005)
33. Borowik, G.: Synthesis of sequential devices into FPGA with embedded memory blocks. PhD thesis. WUT, Warszawa (2007)
34. Brayton, R., Rudell, R., Sangiovanni-Vincentelli, A., Wang, A.: MIS: a multi- level logic optimization system. IEEE Transactions on Computer-Aided Design 6, 1062–1081 (1987)
35. Brown, S., Vernesic, Z.: Fundamentals of Digital Logic with VHDL Design. McGraw-Hill, New York (2000)
36. Brown, S., Vernesic, Z.: Fundamentals of Digital Logic with Verilog Design. McGraw-Hill, New York (2003)
37. Bukowiec, A.: Synthesis of Finite State Machines for Programmable devices based on multi-level implementation. PhD thesis, University of Zielona Góra (2008)
38. Chu, P.: RTL Hardware Design Using VHDL: Coding for Efficiency, Portability and Scalability. Wiley-Interscience, Hoboken (2006)
39. Cypress Programmable Logic: Delta 39K. Data Sheet, http://cypress.com/pld/delta39k.html
40. Cypress Semiconductor Corporation, http://www.cypress.com
41. Czerwinski, R., Kania, D.: State assignment method for high speed FSM. In: Proc. of Programmable Devices and Systems, pp. 216–221 (2004)
42. Czerwinski, R., Kania, D.: State assignment for PAL-based CPLDs. In: Proc. of 8th Euromicro Sym. on Digital System Design, pp. 127–134 (2005)
43. Debnath, D., Sasao, T.: Multiple-valued minimization to optimize PLA with output EXOR gates. In: Proc. of IEEE Inter. Symp. on Multiple-Valued Logic, pp. 99–104 (1999)
44. Sasao, T., Debnath, D.: Doutput phase optimization for and-or-exor plas with decoders and its application to design of adders. IFICE Transactions on Information and Systems E88-D(7), 1492–1500 (2005)
45. Devadas, S., Ma, H.: Easily testable PLA-based finite state machines. IEEE Transactions on Computer-Aided Design of Integrated Circuits and Systems 9(6), 604–611 (1990)
46. Hassoun, S., Sasao, T.: Logic synthesis and verification. Kluwer Academic Publishers, Dordrecht (2002)
47. Hatchel, G., Somenzi, F.: Logic synthesis and verification algorithms. Kluwer Academic Publishers, Dordrecht (2000)
48. Hrynkiewicz, E., Kania, D.: Impact of decomposition direction on synthesis effectiveness. In: Proc. of Programmable Devices and Systems (PDS 2003), pp. 144–149 (2003)
49. Jenkins, J.: Design with FPGAs and CPLDs. Prentice Hall, New York (1995)
50. Kam, T., Villa, T., Brayton, R., Sangiovanni-Vincentelli, A.: A Synthesis of Finie State Machines: Functional Optimization. Kluwer Academic Publishers, Boston (1998)
51. Kania, D.: Two-level logic synthesis on PAL-based CPLD and FPGA using decomposition. In: Proc. of 25th Euromicro Conference, pp. 278–281 (1999)
52. Kania, D.: Two-level logic synthesis on PALs. Electronic Letters (17), 879–880 (1999)

53. Kania, D.: Coding capacity of PAL-based logic blocks included in CPLDs and FPGAs. In: Proc. of IFAC Workshop on Programmable Devices and Systems (PDS 2000), pp. 164–169. Elseveir Science, Amsterdam (2000)
54. Kania, D.: Decomposition-based synthesis and its application in PAL-oriented technology mapping. In: Proc. of 26th Euromicro Conference, pp. 138–145. IEEE Computer Society Press, Maastricht (2000)
55. Kania, D.: An efficient algorithm for output coding in PAL-based CPLDs. International Journal of Engineering [57], 325–328
56. Kania, D.: An efficient algorithm for output coding in PAL-based CPLDs. International Journal of Engineering [57], 325–328
57. Kania, D.: An efficient algorithm for output coding in PAL-based CPLDs. International Journal of Engineering 15(4), 325–328 (2002)
58. Kania, D.: An efficient approach to synthesis of multi-output boolean functions on PAL-based devices. In: IEEE Proc. – Computer and Digital Techniques, vol. 150, pp. 143–149 (2003)
59. Łuba, T., Jasiński, K., Zbierzchowski, B.: Spcialized digital circuits in PLD i FPGA structures. Wydawnictwo Komunikacji i Łączności (1997) (in Polish)
60. Łuba, T., Rawski, M., Jachna, Z.: Functional decomposition as a universal method for logic synthesis of digital circuits. In: Proc. of IX Inter. Conf. MIXDES 2002, pp. 285–290 (2002)
61. Maxfield, C.: The Design Warrior's Guide to FPGAs. Academic Press, Inc., Orlando (2004)
62. Maxfield, C.: FPGAs: Instant access. Newnes (2008)
63. McCluskey, E.: Logic Design Principles. Prentice Hall, Englewood Cliffs (1986)
64. De Micheli, G.: Synthesis and Optimization of Digital Circuits. McGraw-Hill, New York (1994)
65. Minns, P., Elliot, I.: FSM-based digital design using Verilog HDL. John Wiley and Sons, Chichester (2008)
66. Navabi, Z.: Embedded Core Design with FPGAs. McGraw-Hill, New York (2007)
67. Papachristou, C.: Hardware microcontrol schemes using PLAs. In: Proceeding of 14th Microprogramming Workshop, vol. 2, pp. 3–15 (1981)
68. Papachristou, C., Gambhir, S.: A microsequencer architecture with firmware support for modular microprogramming. ACM SIGMICRO Newsletters 13(4) (1982)
69. Parnel, K., Mechta, N.: Programmable Logic design Quick Start Hand Book. Xilinx (2003)
70. Patterson, D., Henessy, J.: Computer Organization and Design: The Hardware/Software Interface. Morgan Kaufmann, San Moteo (1998)
71. Pedroni, V.: Circuit Design with VHDL. MIT Press, Cambridge (2004)
72. Rawski, M., Luba, T., Jachna, Z., Tomaszewicz, P.: The influence of functional decomposition on modern digital design process. In: Design of Embedded Control Systems, pp. 193–203. Springer, Boston (2005)
73. Rawski, M., Selvaraj, H., Luba, T.: An application of functional decomposition in ROM-based FSM implementation in FPGA devices. Journal of System Architecture 51(6-7), 423–434 (2005)
74. Salcic, Z.: VHDL and FPLDs in Digital Systems Design, Prototyping and Customization. Kluwer Academic Publishers, Dordrecht (1998)
75. Sasao, T.: Switching Theory for Logic Synthesis. Kluwer Academic Publishers, Dordrecht (1999)
76. Saucier, G., Depaulet, M., Sicard, P.: ASYL: a rule-based system for controller synthesis. IEEE Transactions on Computer-Aided Design 6(11), 1088–1098 (1987)

77. Saucier, G., Sicard, P., Bouchet, L.: Multi-level synthesis on programmable devices in the ASYL system. In: Proceedings of Euro ASIC, pp. 136–141 (1990)
78. Scholl, C.: Functional Decomposition with Application to FPGA Synthesis. Kluwer Academic Publishers, Boston (2001)
79. Sentowich, E., Singh, K., Lavango, L., Moon, C., Murgai, R., Saldanha, A., Savoj, H., Stephan, P., Bryton, R., Sangiovanni-Vincentelli, A.: SIS: a system for sequential circuit synthesis. Technical report, University of California, Berkely (1992)
80. Sentowich, E., Singh, K., Lavango, L., Moon, C., Murgai, R., Saldanha, S., Savoj, H., Stephan, P., Bryton, R., Sangiovanni-Vincentelli, A.: SIS: a system for sequential circuit synthesis. In: Proc. of the Inter. Conf. of Computer Design (ICCD 1992), pp. 328–333 (1992)
81. Shriver, B., Smith, B.: The anatomy of a High-performance Microprocessor: A Systems Perspective. IEEE Computer Society Press, Los Alamitos (1998)
82. Solovjev, V., Czyzy, M.: Refined CPLD macrocells architecture for effective FSM implementation. In: Proc. of the 25th EUROMICRO Conference, Milan, Italy, vol. 1, pp. 102–109 (1999)
83. Solovjev, V., Czyzy, M.: The universal algorithm for fitting targeted unit to complex programmable logic devices. In: Proc. of the 25th EUROMICRO Conference, Milan, Italy, vol. 1, pp. 286–289 (1999)
84. Solovjev, V.V.: Design of Digital Systems Using the Programmable Logic Integrated Circuits. Hot line – Telecom, Moscow (2001) (in Russian)
85. Villa, T., Saldachna, T., Brayton, R., Sangiovanni-Vincentelli, A.: Symbolic two-level minimization. IEEE Transactions on Computer-Aided Design 16(7), 692–708 (1997)
86. Wiśniewski, R.: Synthesis of Compositional Microprogram Control Units for Programmable Devices. PhD thesis, University of Zielona Góra (2008)
87. Wiśniewski, R., Barkalov, A., Titarenko, L.: Optimization of address circuit of compositional microprogram unit. In: Proc. of IEEE East-West Design & Test Workshop - EWDTW 2006, Sochi, Rosja, pp. 167–170. Kharkov National University of Radioelectronics, Kharkov (2006)
88. Wiśniewski, R., Barkalov, A., Titarenko, L.: Synthesis of compositional microprogram control units with sharing codes and address decoder. In: Proc. of the Inter. Conf. MIXDES 2006, Gdynia, Polska, pp. 397–400. Departament of Microelectronics and Computer Science, Technical University of Łódz (2006)
89. Xilinx Corporation Webpage, http://www.xilinx.com
90. Yang, S.: Logic synthesis and optimization benchmarks user guide. Technical report, Microelectronic Center of North Carolina (1991)
91. Yanushkevich, S., Shmerko, V.: Introduction to Logic Design. CRC Press, Boca Raton (2008)

Chapter 4
Optimization for Logic Circuit of Mealy FSM

Abstract. The chapter is devoted to the hardware amount reduction in the logic circuit of Mealy FSM. The methods of logical condition replacement are analyzed, as well as different methods of encoding of collections of microoperations (maximal encoding and encoding of the classes of compatible microoperations). Next, the methods of structure table rows encoding are discussed. Each of these methods produces double-level circuit of Mealy FSM. The main part of the chapter is devoted to joint application of these methods, the main advantage of whose is possibility of standard library cells use for implementation of logic circuits for some blocks of an FSM model. For example, the logical condition replacement allows application of multiplexers, whereas the encoding of collections of microoperations permits to use embedded memory blocks. Standard decoders can be used in case of encoding of the classes of compatible microoperations. It increases FSM logic circuit regularity and leads to simplification of its design process.

4.1 Synthesis of FSM with Replacement of Logical Conditions

Usage of the logical condition replacement transforms P Mealy FSM shown in Fig. 3.17 into MP Mealy FSM (Fig. 4.1).

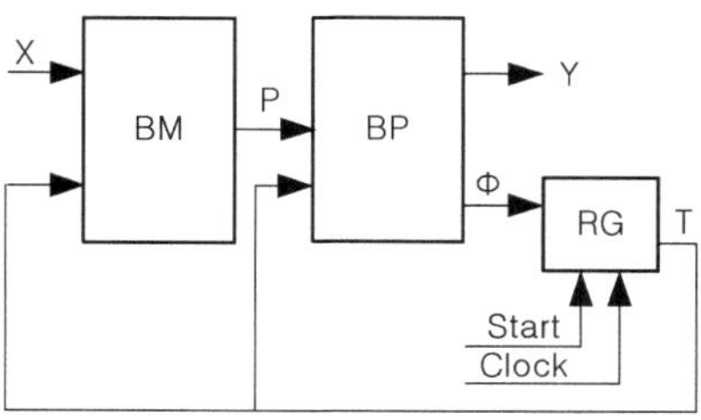

Fig. 4.1 Structure diagram of MP Mealy FSM

In MP Mealy FSM, a block BM replaces the matrices M_5 and M_6 shown in Fig. 2.8; it replaces logical conditions $x_l \in X$ by additional variables $p_g \in P$. A block BP replaces the matrices M_7 and M_8 (Fig. 2.8) and implements the following systems:

A. Barkalov and L. Titarenko: Logic Synthesis for FSM-Based Control Units, LNEE 53, pp. 77–102.
springerlink.com

$$Y = Y(P,T), \tag{4.1}$$
$$\Phi = \Phi(P,T). \tag{4.2}$$

The functions of these systems depend on variables

$$P = P(X,T), \tag{4.3}$$

generated by the block BM. Analysis of system (4.3), represented as (2.16), shows that system (4.3) uses only direct values of logical conditions. Therefore, functions $p_g \in P$ belong to the class of multiplexer functions and multiplexers can be used for their implementation. Let us point out that multiplexers are standard library cells implemented from basic cells of PLD in use and their usage accelerates the design process for the logic circuit of a control unit. Obviously, functions (4.1) and (4.2) are irregular and they are implemented using basic PLD cells. The synthesis method for MP Mealy FSM includes the following steps [2]:

1. Construction of table for replacement of logical conditions.
2. Construction of transformed structure table for MP Mealy FSM.
3. Implementation of systems (4.1) – (4.3) using PLD cells.

Let us discuss application of this method for optimization of the Mealy FSM S_9 represented by its structure table (Table 4.1).

As follows from Table 4.1, the Mealy FSM S_9 has $M = 10$ states, $L = 9$ logical conditions, and $N = 8$ microoperations. The transitions for states $a_m \in A$ depend on logical conditions forming the following subsets of the initial set of logical conditions X: subsets $X(a_1) = X(a_4) = X(a_9) = \emptyset$ (unconditional jumps), and $X(a_2) = \{x_1x_2x_3\}$, $X(a_5) = \{x_2x_5\}$, $X(a_7) = \{x_1x_7\}$, $X(a_6) = \{x_3x_5x_6\}$, $X(a_8) = \{x_5x_8\}$, $X(a_10) = \{x_3x_9\}$ (conditional jumps). Thus, it is enough $G = 3$ variables $p_g \in P$ to replace the logical conditions $x_l \in X$. The principle of logical condition replacement was discussed in Chapter 2. For our example, the table for logical condition replacement includes $G = 3$ columns and 10 rows (Table 4.2).

This table is the base for deriving of system (4.3), in the case of FSM S_9 this system is the following one:

$$\begin{aligned} p_1 &= (A_2 \vee A_7)x_1 \vee (A_5 \vee A_6 \vee A_8)x_5 \vee A_{10}x_9; \\ p_2 &= (A_2 \vee A_5)x_2 \vee A_3x_4 \vee A_6x_6; \\ p_3 &= (A_2 \vee A_6 \vee A_{10})x_3 \vee A_3x_5 \vee A_7x_7 \vee A_8x_8. \end{aligned} \tag{4.4}$$

As it was mentioned a bit earlier, the transformed structure table of MP Mealy FSM is constructed from its initial structure table. In this case the column X_h of initial structure table is replaced by the column P_h. For example, some subtable of transformed structure table for the state a_5 of the Mealy FSM S_9 is shown in Table 4.3.

The transformed structure table is used for deriving systems (4.1) – (4.2), depended on terms

$$F_h = A_mP_h. \tag{4.5}$$

Table 4.1 Structure table of Mealy FSM S_9

a_m	$K(a_m)$	a_s	$K(a_s)$	X_h	Y_h	Φ_h	h
a_1	0000	a_2	0001	1	y_1y_2	D_4	1
a_2	0001	a_3	0010	$x_1\bar{x}_2$	D_1	D_3	2
		a_4	0011	x_1x_2	y_2y_3	D_3D_4	3
		a_3	0010	$\bar{x}_1x_3$	y_5	D_3	4
		a_5	0100	$\bar{x}_1\bar{x}_3$	y_6	D_2	5
a_3	0010	a_2	0001	x_4	y_3	D_4	6
		a_3	0011	$\bar{x}_4x_3$	y_1y_9	D_2D_4	7
		a_5	0100	$\bar{x}_4\bar{x}_3$	y_5	D_2	8
a_4	0011	a_6	0101	1	D_1	D_2D_4	9
a_5	0100	a_3	0010	x_2	y_1y_3	D_3	10
		a_5	0100	$\bar{x}_2x_5$	y_5	D_2	11
		a_6	0101	$\bar{x}_2\bar{x}_5$	y_1y_6	D_2D_4	12
a_6	0101	a_2	0001	x_3x_6	D_1	D_4	13
		a_5	0100	$x_3\bar{x}_6$	y_2y_3	D_2	14
		a_7	0110	$\bar{x}_3x_5$	y_7	D_2D_3	15
		a_8	0111	$\bar{x}_3\bar{x}_5$	y_1y_8	$D_2D_3D_4$	16
a_7	0110	a_8	0111	x_1x_7	y_2D_1	$D_2D_3D_4$	17
		a_9	1000	$x_1\bar{x}_7$	y_3	D_1	18
		a_{10}	1001	$\bar{x}_1$	y_1y_2	D_1D_4	19
a_8	0111	a_3	0010	x_5	–	D_3	20
		a_9	1000	$\bar{x}_5x_8$	y_2D_1	D_1	21
		a_1	0000	$\bar{x}_5\bar{x}_8$	y_6	–	22
a_9	1000	a_{10}	1001	1	y_1y_8	D_1D_4	23
a_{10}	1001	a_1	0000	x_9	y_5	–	24
		a_2	0001	$\bar{x}_9x_3$	y_1y_3	D_4	25
		a_6	0101	$\bar{x}_9\bar{x}_3$	y_6	D_2D_4	26

Table 4.2 Table for logical condition replacement of FSM S_9

a_m	p_1	p_2	p_3	a_m	p_1	p_2	p_3
a_1	–	–	–	a_6	x_5	x_6	x_3
a_2	x_1	x_2	x_3	a_7	x_1	–	x_7
a_3	–	x_4	x_5	a_8	x_5	–	x_8
a_4	–	–	–	a_9	–	–	–
a_5	x_5	x_2		a_{10}	x_9	–	x_3

Table 4.3 Fragment of transformed structure table for MP Mealy FSM S_9

a_m	$K(a_m)$	a_s	$K(a_s)$	P_h	Y_h	Φ_h	h
a_5	0100	a_3	0010	p_2	y_1y_3	D_3	10
		a_5	0100	$\bar{p}_2p_1$	y_5	D_2	11
		a_6	0101	$\bar{p}_2\bar{p}_1$	y_1y_6	D_2D_4	12

In our example, there are the terms $F_{10} = A_5 p_2$, $F_{11} = A_5 \bar{p}_2 p_1$, and $F_{12} = A_5 \bar{p}_2 \bar{p}_1$, where $A_5 = \bar{T}_1 T_2 \bar{T}_3 \bar{T}_4$. Using these terms, the following parts of SOP can be derived from Table 4.3: $y_1 = F_{10} \vee F_{11}, y_3 = D_3 = F_{10}, y_5 = F_{11}, y_6 = D_4 = F_{12}, D_2 = F_{11} \vee F_{12}$.

The following approach can be used to implement the block BM:

1. Each function $p_g \in P$ corresponds to one multiplexer MX_g, having R control inputs and 2^R data inputs.
2. For all multiplexers, control inputs are connected with state variables $T_r \in T$ of Mealy FSM.
3. If a variable $p_g \in P$ replaces a logical condition $x_l \in X$ for a state $a_m \in A$, then this logical condition is connected with data input of multiplexer MX_g, activated by the state code $K(a_m)$.

To design the logic circuit of the block BM, it is enough to replace states $a_m \in A$ by their codes in the table of logical condition replacement. After such a changing, the table of replacement corresponds to G tables, where each table determines one of the multiplexers MX_g. For the MP Mealy FSM S_9, the block BM includes three multiplexers (Fig. 4.2).

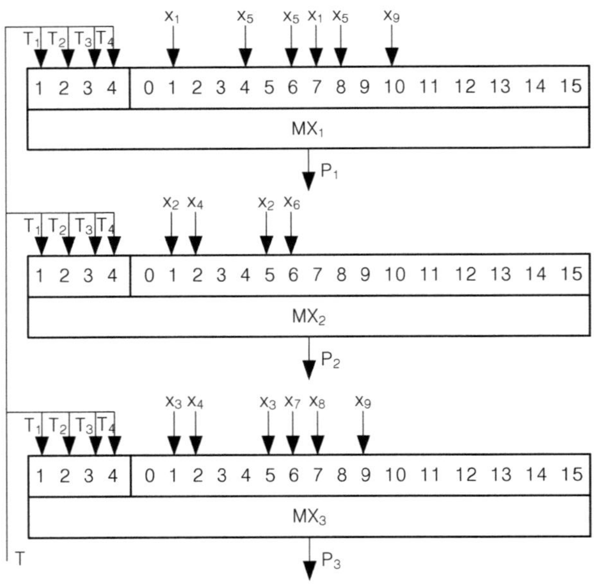

Fig. 4.2 Circuit of block BM for MP Mealy FSM S_9

There is obvious correspondence between the block BM (Fig. 4.2) and Table 4.2. As follows from Fig. 4.2, only 6 from available 16 data inputs of the multiplexer MX_1 are used (they are connected with logical conditions), whereas only 4 for MX_2, and only 6 for MX_3. Thus, only 37% from potentials of both multiplexers MX_1 and

MX_3 are used, whereas only 25% is used for MX_2. Thus, only 33% of available data inputs are used, that is a very poor outcome. To increase the rate of data inputs' usage, it is necessary to decrease the number of control inputs per a multiplexer. The following methods can be used to solve this problem [2, 4]:

1. Multiplexer state encoding.
2. State code transformation into multiplexer state codes.
3. State code transformation into codes of logical conditions.

Let us discuss the main idea of the multiplexer state coding. Let us split the set of states A on classes A_C and A_U in such a manner, that sets $X(a_m) \neq \emptyset$ for states $a_m \in A_C$, whereas $X(a_m) = \emptyset$ for states $a_m \in A_U$. Besides, the initial state a_1 always belongs to the set A_C. Binary codes of states $a_m \in A_C$ should correspond to decimal equivalents from zero (for the state a_1) till $M_C - 1$, where $M_C = |A_C|$. Remained codes are used for states $a_m \in A_U$, and they can be coded in an arbitrary order.

For the FSM S_9, the sets $A_C = \{a_1, a_2, a_3, a_5, \ldots, a_8, a_{10}\}$ and $A_U = \{a_4, a_9\}$ can be found, that gives $M_C = 8$. The value of parameter M_C gives the number of data inputs for multiplexers of the block BM. Encode states $a_m \in A$ as it shown in the Karnaugh map (Fig. 4.3).

Fig. 4.3 Multiplexer state codes for MP Mealy FSM S_9

T_1T_2 \ T_3T_4	00	01	11	10
00	a_1	a_2	a_3	a_5
01	a_6	a_7	a_8	a_{10}
11	*	*	*	*
10	a_4	a_9	*	*

Now the states $a_m \in A_C$ correspond to the part of Karnaugh map with $T_4 = 0$, therefore these states are determined by variables $T_1 - T_3$. It means that all multiplexers of the block BM for FSM S_9 have three control and eight data inputs per a multiplexer. As it follows from Fig. 4.4, now it is used 6 from 8 data inputs of both MX_1 and MX_3 (75% of data inputs), whereas only 50% out of data inputs is used for MX_2(that is, 4 out of 8 inputs). In average, 67% out of all data inputs is used.

If a method of state encoding in use targets on the hardware decrease for the block BP, it is reasonable to use the methods from second group, which belong to the methods of *object code transformation* indextransformation of!object codes. In this case, a special state code transformer CCS (Fig. 4.5) should be used. It generates some additional variables $z_r \in Z$ used as control inputs of multiplexers of the block BM.

Let the model of Mealy FSM with transformation of state codes into multiplexer state codes be denoted as MPC Mealy FSM, whereas the symbol MPL denotes FSM with transformation of state codes into codes of logical conditions. In both models of FSM, states $a_m \in A$ are encoded to solve some other problems distinguished from

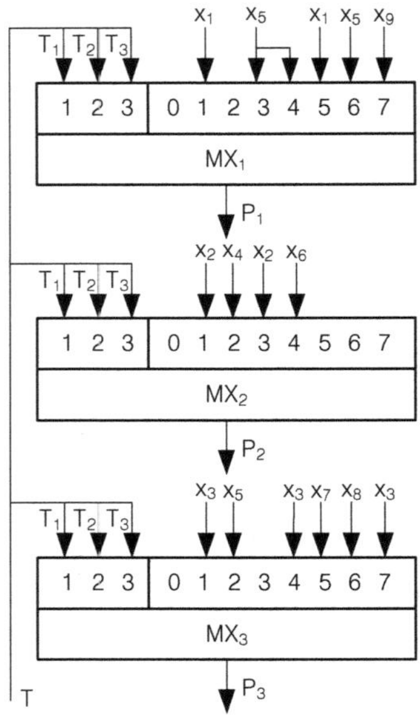

Fig. 4.4 Block of logical condition replacement for MP Mealy FSM S_9 with multiplexer state coding

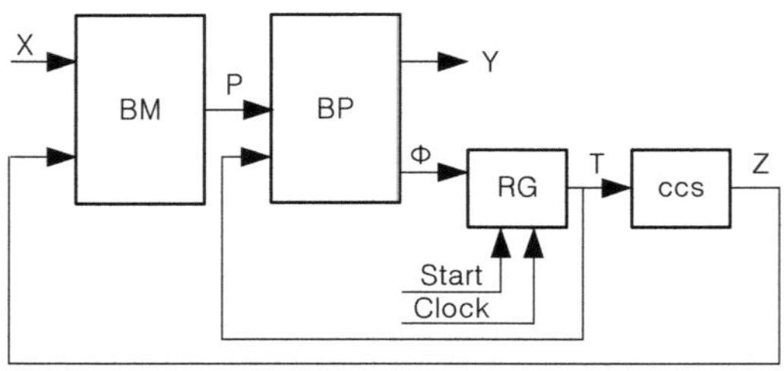

Fig. 4.5 Structural diagram of Mealy FSM with state code transformer

the hardware optimization for the block BM. In both cases, the initial state $a_1 \in A$ is included into the set A_C if and only if (iff) $X(a_1) \neq \emptyset$.

The synthesis method for MPC Mealy FSM includes the following additional steps:

1. Coding of states $a_m \in A_C$ by multiplexer binary codes $C(a_m)$ with R_C bits, where

$$R_C = \lceil \log_2 M_C \rceil . \tag{4.6}$$

2. Construction of a table for code transformer CCS.
3. Implementation of the block CCS using given logic elements.

In the discussed case, there are $M_C = 7$, $R_C = 3$. It means that states are encoded using the variables from the set $Z = \{z_1, z_2, z_3\}$. Coding can be executed in an arbitrary order, though its outcome can decrease the number of control inputs for some multiplexers. Let $A(MX_g)$ be a set of states, such that the multiplexer $MX_g (g = 1, \dots, G)$

transforms logical conditions determining transitions from these states. In the discussed case, the following set $A(MX_2) = \{a_2, a_3, a_5, a_6\}$ can be found, such that it is enough two variables for its components encoding. Let us encode the states of FSM S_9 using the following multiplexer codes:$C(a_2) = 000$, $C(a_3) = 001$, $C(a_5) = 010$, $C(a_6) = 011$, $C(a_7) = 100$, $C(a_8) = 101$, and $C(a_{10}) = 110$. Now the states $a_m \in A(MX_2)$ are determined by the variables z_2, z_3; it yields in the following circuit for the logical condition replacement (Fig. 4.6).

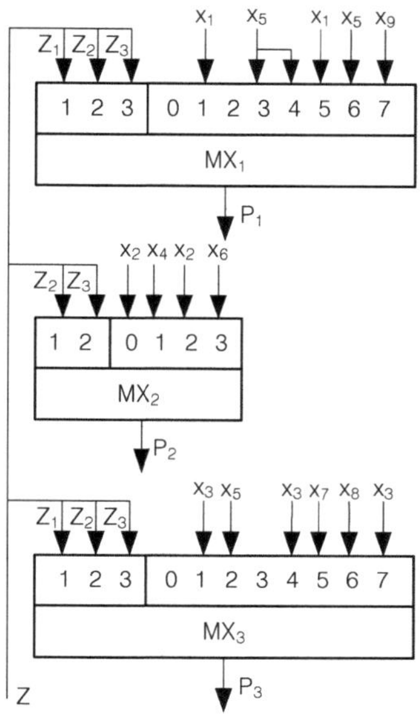

Fig. 4.6 Block BM for MPC Mealy FSM S_9

In this circuit, the multiplexer MX_2 uses 100% of its data inputs, while both MX_1 and MX_3 only 75%. Therefore, average use of data inputs is increased up to 83%. Of course, it is connected with introduction of the code transformer CCS consuming some area of a chip.

The table of code transformer CCS includes columns a_m, $K(a_m)$, $C(a_m)$, Z_m, m. In this table, the code $K(a_m)$ is used as an input of the block CCS, whereas the code $C(a_m)$ is its output. For the FSM S_9, this table (Table 4.4) includes $M = 10$ rows. The column Z_m includes variables $z_r \in Z$, equal to 1 in the code $C(a_m)$.

Obviously, the best way for implementation of this table is use of a PROM chip having R inputs and R_C outputs. If a PROM-based implementation is not possible due to absence of free resources of a chip in use, then this table is used to derive the following SOP represented the system $Z = Z(T)$:

$$z_r = \bigvee_{m=1}^{M} C_{mr} A_m \quad (r = 1, \ldots, R_C). \tag{4.7}$$

Table 4.4 Table of code transformer CCS for FSM S_9

a_m	$K(a_m)$	$C(a_m)$	Z_m	m
a_1	0000	000	–	1
a_2	0001	001	z_3	2
a_3	0010	010	z_2	3
a_4	0011	–	–	4
a_5	0100	011	z_2z_3	5
a_6	0101	100	z_1	6
a_7	0110	101	z_1z_3	7
a_8	0111	110	z_1z_2	8
a_9	1000	–	–	9
a_{10}	1001	111	$z_1z_2z_3$	10

In (4.7) , the variable $C_{mr} \in \{01\}$, let us point out that $C_{mr} = 1$, iff the bit r of the code $C(a_m)$ is equal to 1 ($r = 1,\ldots,R_C$). Obviously, system (4.7) can be minimized. For example, the following form can be derived from the Karnaugh map shown in Fig. 4.7: $z_1 = T_2T_4 \vee T_2T_3 \vee T_1T_4$. Depending on logic elements in use, either joint or separate minimization should be carried out for the system. The first approach is applied if PLA -based macrocells are used, whereas the second one targets on PAL-implementation. Obviously, such an approach is applied for any system of Boolean functions; let us just remember about it.

Fig. 4.7 Karnaugh map for function z_1

T_1T_2 \ T_3T_4	00	01	11	10
00	0	0	0	0
01	0	1	1	1
11	*	*	*	*
10	0	1	*	*

The logic circuit of MPC Mealy FSM S_9 is shown in Fig. 4.8. In this circuit, the symbol MX shows that multiplexer functions are implemented, the symbol PLD corresponds to implementation of irregular functions, whereas the symbol PROM informs about implementation of regular functions. Obviously, in reality only LUT elements (or PAL macrocells) are used to implement logic circuits for multiplexer and irregular functions. We do not discuss the problems of logic circuits' physical realization. Remind, all examples discussed here are ended by construction of some tables describing blocks of FSM and corresponding systems of Boolean functions. It is quite enough to start using of commercial CAD.

Let $X(P_g)$ be a set of logical conditions from the column P_g ($g = 1,\ldots,G$), where $L_g = |X(P_g)|$. Obviously, it is enough

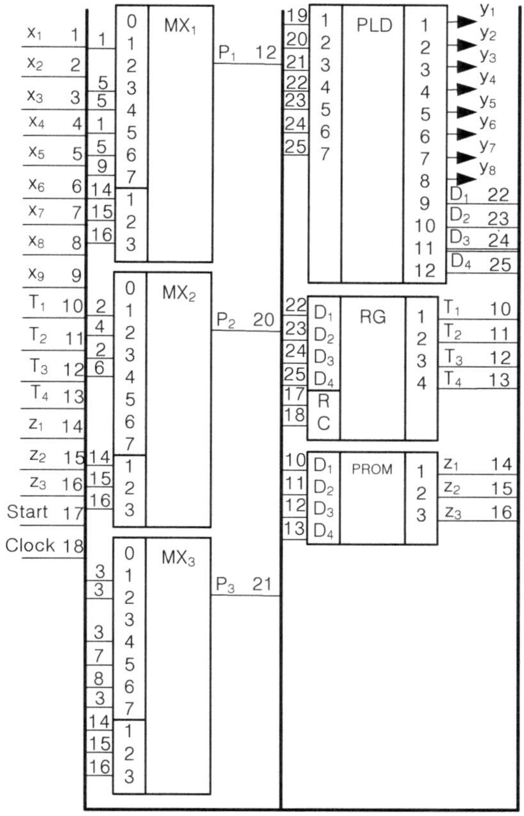

Fig. 4.8 Logic circuit of MPC Mealy FSM S_9

$$R_g = \lceil \log_2 L_g \rceil \tag{4.8}$$

variables to encode the logical conditions $x_l \in X(P_g)$. To encode the logical conditions $x_l \in X$ it is enough

$$R_L = R_1 + \ldots + R_G \tag{4.9}$$

variables, forming a set Z.

The method of state code transformation into the codes of logical conditions is based on replacement of state variables $T_r \in T$ by variables $z_r \in Z$, where $|Z| = R_L$. Let us denote such an FSM as MPL Mealy FSM. Structural diagrams are the same for both MPL and MPC models of Mealy FSM. Additionally, the design method for MPC Mealy FSM includes the step of encoding of logical conditions by some binary codes $K_g(x_l)$.

For the FSM S_9, we can get the following sets: $X(p_1) = \{x_1 x_5 x_9\}$, $L_1 = 3$, $X(p_2) = \{x_2 x_4 x_6\}$, $L_2 = 3$, $X(p_3) = \{x_3, x_5, x_7, x_8\}$, $L_3 = 4$. It gives $R_1 = R_2 = R_3 = 2$ and $R_L = 6$, $Z = \{z_1, \ldots, z_6\}$. Let us encode the logical conditions for FSM S_9 as shown in Table 4.5.

Table 4.5 Codes of logical conditions for MPL Mealy FSM S_9

$X(p_1)$	z_1	z_2	$X(p_2)$	z_3	z_4	$X(p_3)$	z_5	z_6
x_1	0	0	x_2	0	0	x_3	0	0
x_5	0	1	x_4	0	1	x_5	0	1
x_9	1	0	x_6	1	0	x_7	1	0
–	1	1	–	1	1	x_8	1	1

The logic circuit of block BM for the MPL Mealy FSM S_9 is implemented using three multiplexers (Fig. 4.9).

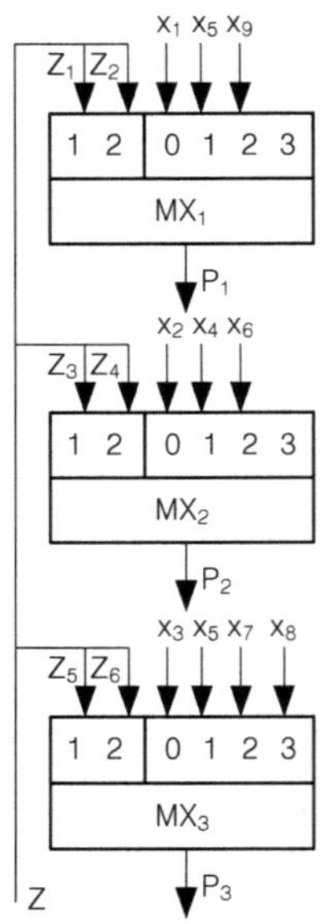

Fig. 4.9 Logic circuit of block BM for MPL Mealy FSM S_9

To implement the logic circuit for the block CCS, it is necessary to construct a corresponding table with columns a_m, $K(a_m)$, $K_1(x_l), \dots, K_G(x_l)$, Z_m, m. For the MPL Mealy FSM S_9, this block is specified by Table 4.6. Logic circuit of MPL Mealy FSM can be implemented in the same way as it is done for MPC Mealy FSM.

4.2 Synthesis of FSM with Encoding of Collections of Microoperations

Two different approaches are possible under encoding of collections of microoperations, namely:

Table 4.6 Specification of block CCS for MPL Mealy FSM S_9

a_m	$K(a_m)$	$K_1(x_l)$	$K_2(x_l)$	$K_3(x_l)$	Z_m	m
a_1	0000	–	–	–	–	1
a_2	0001	00	00	00	–	2
a_3	0010	–	01	01	z_4z_6	3
a_4	0011	–	–	–	–	4
a_5	0100	01	00	–	z_2	5
a_6	0101	01	10	00	z_2z_3	6
a_7	0110	00	–	10	z_5	7
a_8	0111	01	–	11	$z_2z_5z_6$	8
a_9	1000	–	–	–	–	9
a_{10}	1001	10	–	00	z_1	10

1. Collections $Y_t \subseteq Y$ are encoded by binary codes $K(Y_t)$ having minimal bit capacity R_3, determined by (2.21). This approach turns P Mealy FSM shown in Fig. 3.17 into PY Mealy FSM (Fig. 4.10).

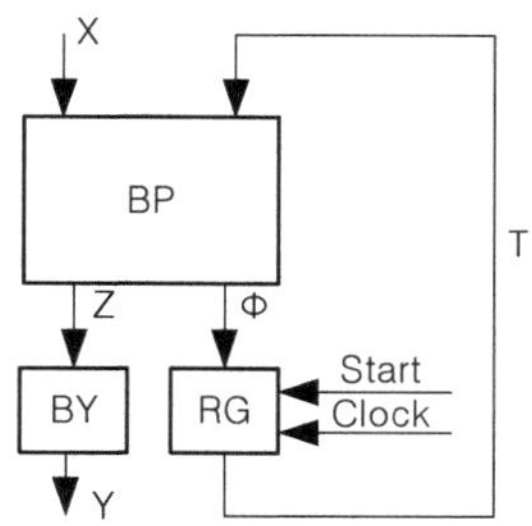

Fig. 4.10 Structural diagram of PY Mealy FSM

For PY Mealy FSM, a block BP corresponds to matrices M_{11} and M_{12} (Fig. 2.11); it generates functions Φ and Z determined by expressions (1.3) and (2.22) respectively. A block BY replaces matrices M_9 and M_{10} (Fig. 2.11); it generates data-path microoperations represented as (2.23). Due to regularity of system (2.23), the logic circuit of block BY can be implemented using either PROM or RAM chips.

2. The set of microoperations Y is divided by the classes of compatible microoperations [2] and represented as

$$Y = Y^1 \cup \ldots \cup Y^K. \tag{4.10}$$

Remind, microoperations $y_i, y_j \in Y$ are compatible, iff they do not written in the same operator vertex of an interpreted GSA [3]. Let $N_k = |Y^k|$, then microoperations $y_n \in Y^k$ are encoded by binary codes $K(y_n)$ having R_k bits, where

$$R_k = \lceil \log_2(N_k + 1) \rceil. \tag{4.11}$$

If interpreted GSA includes some collections of microoperations without representatives of the class k, then 1 is added to N_k in (4.11).

To encode the microoperations, it is enough

$$R_D = R_1 + \ldots + R_K \quad (4.12)$$

variables forming a set $Z = Z^1 \cup \ldots \cup Z^K$. The variables $z_r \in Z^k$ are used for encoding of microoperations $y_n \in Y^k$; let us point out that

$$Z^i \cap Z^j = \emptyset \quad (i \neq j; i, j \in \{1, \ldots, K). \quad (4.13)$$

After encoding, the system Y can be represented as the following collection of systems:

$$\begin{aligned} Y^1 &= Y(Z^1); \\ &\vdots \\ Y^K &= Y(Z^K). \end{aligned} \quad (4.14)$$

Microoperations $y_n \in Y^k$ are generated by a decoder DC_k, having R_k inputs and N_k outputs $(k = 1, \ldots, K)$.

The totality of these decoders forms a block BD. It turns P Mealy FSM into PD Mealy FSM [2] with the structural diagram shown in Fig. 4.11.

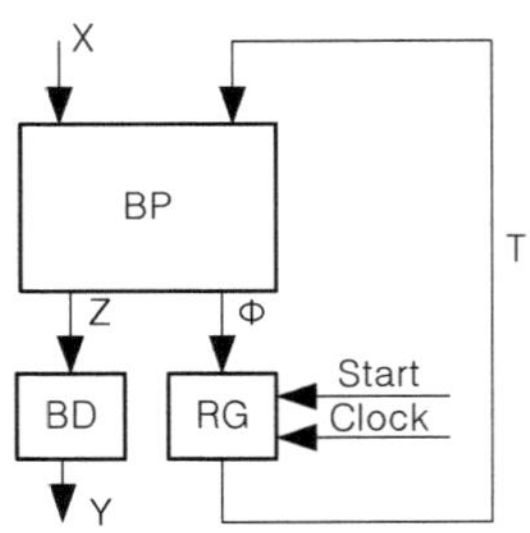

Fig. 4.11 Structural diagram of PD Mealy FSM

Let us discuss an example of synthesis for the PY Mealy FSM S_{10}, represented by its structure table (Table 4.7).

1. **Encoding of collections of microoperations.** As it can be found from Table 4.7, there are $T_0 = 8$ collections of microoperations, where $Y_1 = \{y_1, y_2\}$, $Y_2 = \{y_3\}$, $Y_3 = \{y_4, y_5\}$, $Y_4 = \{y_6, y_7\}$, $Y_5 = \{y_2, y_8\}$, $Y_6 = \{y_7\}$, $Y_7 = \{y_7, y_9\}$, $Y_8 = \emptyset$. As follows from (2.21), it is enough $R_3 = 3$ variables $z_r \in Z$ for encoding of these collections. Let us encode collections $Y_t \subseteq Y$ in a trivial way: $K(Y_1) = 000, \ldots, K(Y_8) = 111$.
2. **Construction of transformed structure table.** As it was mentioned in Chapter 2, the transformed structure table includes the column Z_h, replacing the column Y_h from the initial structure table. The column Z_h contains variables $z_r \in Z$, equal to 1 for the code $K(Y_t)$ from the row h of the initial structure table $(h = 1, \ldots, H)$.

Table 4.7 Structure table for Mealy FSM S_{10}

a_m	$K(a_m)$	a_s	$K(a_s)$	X_h	Y_h	Φ_h	h
a_1	000	a_2	001	x_1	y_1y_2	D_3	1
		a_3	010	$\bar{x}_1x_2$	y_3	D_2	2
		a_4	011	$\bar{x}_1\bar{x}_2$	D_1y_5	D_2D_3	3
a_2	001	a_3	010	x_3	y_1y_2	D_2	4
		a_5	100	$\bar{x}_3$	y_3	D_1	5
a_3	010	a_6	101	1	y_6y_7	D_1D_3	6
a_4	011	a_2	001	x_2	y_2y_8	D_3	7
		a_1	000	$\bar{x}_2x_3$	y_1y_2	–	8
		a_6	101	$\bar{x}_2\bar{x}_3$	y_7	D_1D_3	9
a_5	100	a_2	001	x_1x_2	D_1y_5	D_3	10
		a_3	010	$x_1\bar{x}_2$	y_7y_9	D_2	11
		a_1	000	$\bar{x}_1x_3$	–	–	12
		a_5	100	$\bar{x}_1\bar{x}_3$	y_3	D_1	13
a_6	101	a_1	000	1	y_1y_2	–	14

For the FSM S_{10}, the transformed structure table is represented by Table 4.8. Obviously, the number of rows for both tables is the same, only contents of some columns are different.

Table 4.8 Transformed structure table of PY Mealy FSM S_{10}

a_m	$K(a_m)$	a_s	$K(a_s)$	X_h	Z_h	Φ_h	h
a_1	000	a_2	001	x_1	–	D_3	1
		a_3	010	$\bar{x}_1x_2$	z_3	D_2	2
		a_4	011	$\bar{x}_1\bar{x}_2$	z_2	D_2D_3	3
a_2	001	a_3	010	x_3	–	D_2	4
		a_5	100	$\bar{x}_3$	z_3	D_1	5
a_3	010	a_6	101	1	z_2z_3	D_1D_3	6
a_4	011	a_2	001	x_2	z_1	D_3	7
		a_1	000	$\bar{x}_2x_3$	–	–	8
		a_6	101	$\bar{x}_2\bar{x}_3$	z_1z_3	D_1D_3	9
a_5	100	a_2	001	x_1x_2	z_2	D_3	10
		a_3	010	$x_1\bar{x}_2$	z_1z_2	D_2	11
		a_1	000	$\bar{x}_1x_3$	$z_1z_2z_3$	–	12
		a_5	100	$\bar{x}_1\bar{x}_3$	z_3	D_1	13
a_6	101	a_1	000	1	–	–	14

Using the transformed structure table, system (2.22) is constructed, which can be represented as the following one:

$$z_r = \bigvee_{h=1}^{H} C_{rh} A_m X_h \quad (r = 1, \ldots, R_3). \tag{4.15}$$

In expression (4.15), a Boolean variable $C_{rh} = 1$, iff the variable z_r presents in the row h of transformed ST($h = 1, \ldots, H$). For example, the following SOP $z_1 = F_7 \vee F_8 \vee F_{11} \vee F_{12}$ can be derived from Table 4.8. System (1.3) is used for designing the block BP too. For example, the following equation can be derived from Table 4.8: $D_1 = F_5 \vee F_6 \vee F_9 \vee F_{13}$.

3. **Specification of block BY.** This block is represented by a table reflected the dependence of microoperations from variables $z_r \in Z$. This table is constructed in a trivial way (Table 4.9) and can be used for programming of PROM.

Table 4.9 Specification of block BY PY Mealy FSM S_{10}

z_1	z_2	z_3	y_1	y_2	y_3	y_4	y_5	y_6	y_7	y_8	y_9
0	0	0	1	1	0	0	0	0	0	0	0
0	0	1	0	0	1	0	0	0	0	0	0
0	1	0	0	0	1	1	1	0	0	0	0
0	1	1	0	0	0	0	0	1	1	0	0
1	0	0	0	1	0	0	0	0	0	1	0
1	0	1	0	0	0	0	0	0	1	0	0
1	1	0	0	0	0	0	0	0	1	0	1
1	1	1	0	0	0	0	0	0	0	0	0

If the logic circuit of block BY is implemented using some macrocells, then system $Y(Z)$ is represented as the following SOP:

$$y_n = \bigvee_{t=o}^{T_0} C_{nt} Z_t \quad (r = 1, \ldots, R_3). \tag{4.16}$$

In (4.16), the Boolean variable$C_{nt} = 1$, iff $y_n \in Y_t$. For example, the following SOP $y_2 = Z_1 \vee Z_5 = \bar{z}_2 \bar{z}_3$ can be derived from Table 4.9 (after minimization). Logic circuit of PY Mealy FSM is implemented on the base of these tables (and derived systems of minimized Boolean functions). The logic circuit of PY Mealy FSM S_{10} is shown in Fig. 4.12.

Now, let us discuss the example of logic synthesis for PD Mealy FSM S_{11}, represented by its structure table (Table 4.10).

1. **Partitioning of the set of microoperations by classes of compatible microoperations**. This step is executed using rather complex combinatorial algorithms [1], which are programmed for some CAD systems. For FSM S_{11}, it is easy to get three following classes: $Y^1 = \{y_1, y_4, y_7\}$, $Y^2 = \{y_2, y_6, y_8\}$, $Y^3 = \{y_3, y_5\}$. So, there are the following values and sets: $R_1 = 2$, $Z^1 = \{z_1, z_2\}$, $R_2 = 2$, $Z^2 = \{z_3, z_4\}$, $R_3 = 2$, $Z^3 = \{z_5, z_6\}$, $R_D = 6$, and $Z = \{z_1, \ldots, z_6\}$.

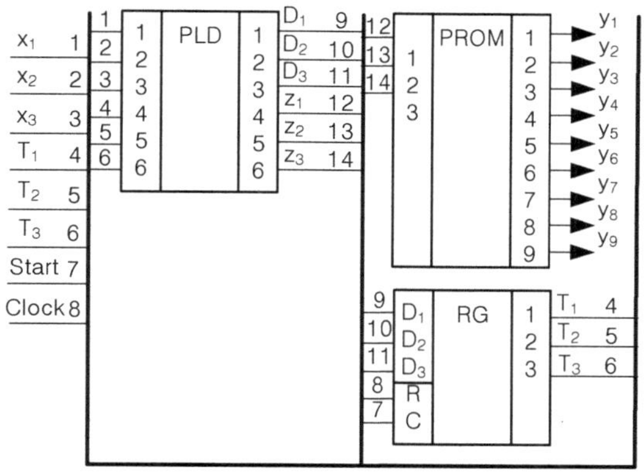

Fig. 4.12 Logic circuit of PY Mealy FSM S_{10}

Table 4.10 Structure table of Mealy FSM S_{11}

a_m	$K(a_m)$	a_s	$K(a_s)$	X_h	Y_h	Φ_h	h
a_1	000	a_2	001	x_1	$y_1y_2y_3$	D_3	1
		a_3	010	$\bar{x}_1$	y_1y_6	D_2	2
a_2	001	a_3	010	x_2	$y_3D_1y_6$	D_2	3
		a_4	011	$\bar{x}_2\bar{x}_3$	D_1y_8	D_2D_3	4
		a_2	001	$\bar{x}_2\bar{x}_3$	y_5y_7	D_3	5
a_3	010	a_5	100	1	y_1y_5	D_1	6
a_4	011	a_3	010	x_2	y_8	D_2	7
		a_5	100	$\bar{x}_2x_4$	y_3D_1	D_1	8
		a_6	101	$\bar{x}_2\bar{x}_4$	y_7y_8	D_1D_2	9
a_5	100	a_2	001	x_5	y_1y_6	D_3	10
		a_1	000	$\bar{x}_5$	–	–	11
a_6	101	a_1	000	1	y_5y_7	–	12

2. **Encoding of compatible microoperations.** If standard decoders are used for implementing the logic circuit of block BD, then compatible microoperations can be encoded in a trivial way. It is true, because codes of microoperations have no influence on the hardware amount in the logic circuit. For the FSM S_{11}, the outcome of trivial encoding is shown in Table 4.11. Obviously, the logic circuit for block BD has the same hardware amount for any codes of compatible microoperations.

   In this table, the column $K(Y^1)$ contains codes $K(y_n)$ of microoperations $y_n \in Y^1$ and so on. The symbol "$\emptyset$" corresponds to lack of microoperations of the given class in some collection of microoperations $Y_t(t = 1, \dots, T_0)$.
3. **Transformation of initial structure table.** This step is executed in the same manner, as it was done for PY Mealy FSM. For the discussed example, Table 4.12 can be formed.

Table 4.11 Codes of compatible microoperations for PD Mealy FSM S_{11}

Y^1	$K(Y^1)$ z_1z_2	Y^2	$K(Y^2)$ z_3z_4	Y^3	$K(Y^3)$ z_5z_6
$\emptyset$	0 0	$\emptyset$	0 0	$\emptyset$	0 0
y_1	0 1	y_2	0 1	y_3	0 1
y_4	1 0	y_6	1 0	y_5	1 0
y_7	1 1	y_8	1 1		

Table 4.12 Transformed structure table of PD Mealy FSM S_{11}

a_m	$K(a_m)$	a_s	$K(a_s)$	X_h	Z_h	Φ_h	h
a_1	000	a_2	001	x_1	$z_2z_4z_6$	D_3	1
		a_3	010	$\bar{x}_1$	z_2z_3	D_2	2
a_2	001	a_3	010	x_2	$z_1z_3z_6$	D_2	3
		a_4	011	$\bar{x}_2\bar{x}_3$	$z_1z_3z_4$	D_2D_3	4
		a_2	001	$\bar{x}_2\bar{x}_3$	$z_1z_2z_5$	D_3	5
a_3	010	a_5	100	1	z_2z_5	D_1	6
a_4	011	a_3	010	x_2	z_3z_4	D_2	7
		a_5	100	$\bar{x}_2x_4$	z_1z_6	D_1	8
		a_6	101	$\bar{x}_2\bar{x}_4$	$z_1z_2z_3z_4$	D_1D_2	9
a_5	100	a_2	001	x_5	z_2z_3	D_3	10
		a_1	000	$\bar{x}_5$	–	–	11
a_6	101	a_1	000	1	$z_1z_2z_5$	–	12

System (4.15) is derived from this table, having R_D functions. For example, the following SOP $z_1 = F_3 \vee F_4 \vee F_5 \vee F_8 \vee F_9$ can be derived from Table 4.12.

The logic circuit of PD Mealy FSM (in our case this circuit is shown in Fig. 4.13 is implemented using either obtained tables or systems of Boolean functions, which can be derived from them.

As follows from Fig. 4.13, the decoder DC_1 is used for implementation of microoperations $y_n \in Y^1$, the decoder DC_2 for $y_n \in Y^2$, whereas the decoder DC_3 is absent, because each from the microoperations of the third class depends only on one variable ($y_3 = z_6, y_5 = z_5$). Because decoders are the standard library cells, their application yields in simplification and acceleration of a design process.

4.3 Synthesis of FSM with Encoding of Rows of Structure Table

The main outcome of encoding of collections of microoperations is decrease for the number of the block BP outputs from $R+N$ (P Mealy FSM) till $R+R_3$ (PY Mealy FSM) or $R+R_D$ (PD Mealy FSM). It leads to decrease for the number

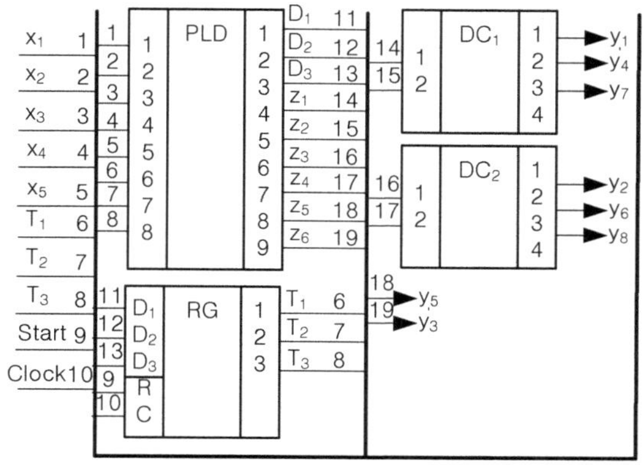

Fig. 4.13 Logic circuit of PD Mealy FSM S_{11}

of macrocells, used to implement irregular functions. The *method of encoding of structure table rows* [3] targets on solution of this problem too.

Let us encode the row h of ST by a binary code $K(F_h)$ having R_F bits, where

$$R_F = \lceil \log_2 H \rceil . \tag{4.17}$$

Let us use variables $z_r \in Z$, where $|Z| = R_F$, for such an encoding. It results in the model of BF Mealy FSM, shown in Fig. 4.14.

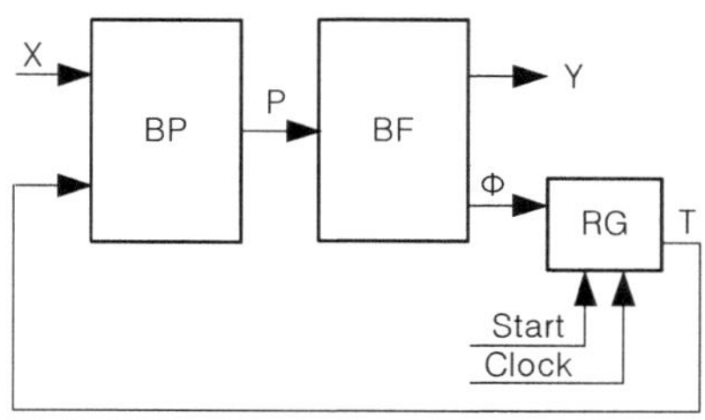

Fig. 4.14 Structure diagram of PF Mealy FSM

In PF Mealy FSM, the block BP implements system (4.15), which includes R_F functions. A block PF implements systems Y and Φ, represented as:

$$y_n = \bigvee_{h=1}^{H} C_{nh} Z_h \quad (h = 1, \ldots, N), \tag{4.18}$$

$$\phi_r = \bigvee_{h=1}^{H} C_{rh} Z_h \quad (r = 1, \ldots, R). \tag{4.19}$$

In systems (4.18)–(4.19) the symbol Z_h stands for a conjunction of variables $z_r \in Z$, corresponded to the code $K(F_h)$:

$$Z_h = \bigwedge_{r=1}^{R_F} z_r^{l_{rh}} . \tag{4.20}$$

In (4.20), the symbol $l_{rh} \in \{0,1\}$ stands for value of the bit r of the code $K(F_h)$ corresponded to the row h of ST, and $z_r^0 = \bar{z}_r$, $z_r^1 = z_r$ $(r = 1,\dots,R_F)$.

Let us discuss an example of PF Mealy FSM synthesis for the FSM S_{10}, represented by Table 4.7.

1. **Encoding of structure table rows.** As follows from Table 4.7, the ST includes $H = 14$ rows and, therefore, $R_F = 4$, and $Z = \{z_1,\dots,z_4\}$. Let us encode the rows in a trivial way: $K(F_1) = 0000$, ..., $K(F_{14}) = 1101$.
2. **Construction of transformed structure table.** The transformation is reduced to moving away the columns $a_s - \Phi_h$ of initial ST and replacement them by the columns $K(F_h)$ and Z_h. The transformed ST of PF Mealy FSM S_{10} is represented by Table 4.13.

Table 4.13 Transformed structure table of PF Mealy FSM S_{10}

a_m	$K(a_m)$	$K(F_h)$	X_h	Z_h	h
a_1	000	0000	x_1	–	1
		0001	$\bar{x}_1x_2$	z_4	2
		0010	$\bar{x}_1\bar{x}_2$	z_3	3
a_2	001	0011	x_3	z_3z_4	4
		0100	$\bar{x}_3$	z_2	5
a_3	010	0101	1	z_2z_4	6
a_4	011	0110	x_2	z_2z_3	7
		0111	$\bar{x}_1x_4$	$z_2z_3z_4$	8
		1000	$\bar{x}_2\bar{x}_3$	z_1	9
a_5	100	1001	x_1x_2	z_1z_4	10
		1010	$x_1\bar{x}_2$	z_2z_3	11
		1011	$\bar{x}_1x_3$	$z_1z_3z_4$	12
		1100	$\bar{x}_1\bar{x}_3$	z_1z_2	13
a_6	101	1101	1	$z_1z_2z_4$	14

This table is the base for deriving the system Z. For example, the following SOP $z_1 = \bar{T}_1T_2T_3\bar{x}_2\bar{x}_3 \vee T_1\bar{T}_2$ can be derived from Table 4.13 (after minimization).

3. **Specification of block BF.** This block can be specified by a table with the columns $K(F_h)$, $y_1,\dots,y_N$, $D_1,\dots,D_R$, h (Table 4.14 for FSM S_{10}). This table in constructed in a trivial way.

   Obviously, the simplest way for implementation of the logic circuit of the block BF is usage either of PROM or RAM chips with inputs $z_r \in Z$.
4. **Synthesis of FSM logic circuit** is executed using the obtained tables and systems of Boolean functions. For the PF Mealy FSM S_{10}, the logic circuit is shown in Fig. 4.15.

Table 4.14 Table of block BF for PF Mealy FSM S_{10}

$K(F_h)$	y_1	y_2	y_3	y_4	y_5	y_6	y_7	y_8	y_9	D_1	D_2	D_3	h
0000	1	1	1	0	0	0	0	0	1	0	0	1	1
0001	0	0	1	0	0	0	0	1	0	0	1	0	2
0010	0	0	1	1	0	0	0	1	1	0	1	1	3
0011	1	1	0	0	0	0	0	1	0	0	1	0	4
0100	0	0	1	0	0	0	1	0	0	1	0	0	5
0101	0	0	0	0	0	1	1	0	0	1	0	1	6
0110	0	1	0	0	0	0	0	1	0	0	0	1	7
0111	1	1	0	0	0	0	0	0	0	0	0	0	8
1000	0	0	0	0	0	0	1	0	0	1	0	1	9
1001	0	0	0	1	1	0	0	0	0	0	0	1	10
1010	0	0	0	0	0	0	1	0	1	0	1	0	11
1011	0	0	0	0	0	0	0	0	0	0	0	0	12
1100	0	0	1	0	0	0	0	0	0	1	0	0	13
1101	1	1	0	0	0	0	0	0	0	0	0	0	14

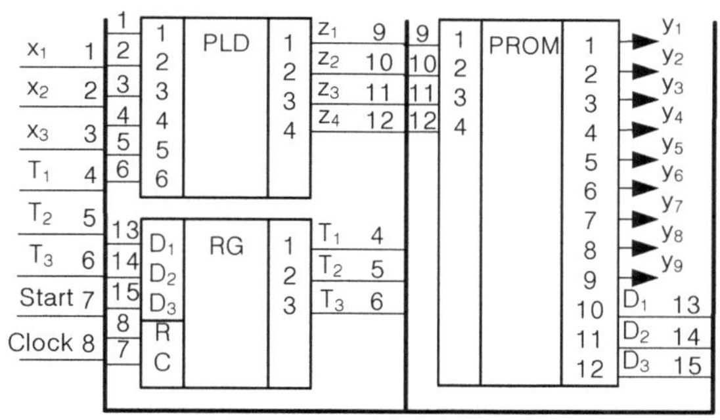

Fig. 4.15 Logic circuit of PF Mealy FSM S_{10}

4.4 Synthesis of FSM Multilevel Logic Circuits

Combined application of methods discussed in this Chapter allows obtaining three- and four-levels models of Mealy FSM [2]. All possible multilevel models are represented by Table 4.15.

Table 4.15 Multilevel models of Mealy FSM

LA	LB	LC	LD
M MC ML	P	F D Y	D Y

The process of generation of three-level Mealy FSM logic circuit structures can be interpreted as a word-formation process, where the level LA is a prefix of the word, the level LB as its base, the level LC either as its suffix (for the block BF) or ending (for blocks BF, BD, and BY), and the level LD as its ending (for some particular cases of PF Mealy FSM). For example, the word LA*LB*LC can stand for either MPY- or MLPD Mealy FSM. Four-level models are based on encoding of ST rows; they always include all four levels. For example, the word LA*LB*LC*LD determines MPFD Mealy FSM. It means that synthesis method includes the methods of logical condition replacement, encoding of structure table rows, and encoding of the classes of compatible microoperations. Obviously, synthesis methods for multi-level models can be viewed as combining of corresponding methods for two-level model synthesis. Let us discuss some examples.

Let the Mealy FSM S_{12} be specified by its structure table (Table 4.16). Let us discuss an example of synthesis for the MPD Mealy FSM S_{12}.

Table 4.16 Structure table of Mealy FSM S_{12}

a_m	$K(a_m)$	a_s	$K(a_s)$	X_h	Y_h	Φ_h	h
a_1	000	a_2	001	x_1	$y_1y_2y_3$	D_3	1
		a_3	010	$\bar{x}_1$	y_1y_6	D_2	2
a_2	001	a_2	001	x_2x_3	$y_3D_1y_6$	D_3	3
		a_3	010	$x_2\bar{x}_3$	D_1y_8	D_2	4
		a_4	011	$\bar{x}_2$	y_5y_7	D_2D_3	5
a_3	010	a_5	100	1	y_1y_5	D_1	6
a_4	011	a_5	100	$x_3\bar{x}_4$	y_8	D_1	7
		a_1	000	x_3x_4	$y_3D_1y_6$	–	8
		a_7	110	$\bar{x}_3$	D_1y_8	D_1D_2	9
a_5	100	a_6	101	x_5	y_1y_6	D_1D_3	10
		a_7	110	$\bar{x}_5$	–	D_1D_2	11
a_6	101	a_1	000	1	y_5y_7	–	12
a_7	110	a_2	001	x_5x_6	y_1y_6	D_3	13
		a_5	100	$x_5\bar{x}_6$	y_8	D_1	14
		a_3	010	$\bar{x}_5$	y_1y_5	D_2	15

Obviously, the model of MPD Mealy FSM should include the blocks BM, BP, and BD (Fig. 4.16).

The block BM implements logical condition replacement and generates functions (4.3). The block BP generates functions (4.2), as well as functions

$$Z = Z(P,T). \tag{4.21}$$

Functions (4.21) control the block BD, which implements functions (4.14). The following procedure can be used to synthesize the logic circuit of MPD Mealy FSM.

1. **Logical condition replacement.** For the FSM S_{12}, the following sets are derived from Table 4.16: $X(a_1) = \{x_1\}$, $X(a_2) = \{x_2,x_3\}$, $X(a_3) = X(a_6) = \emptyset$,

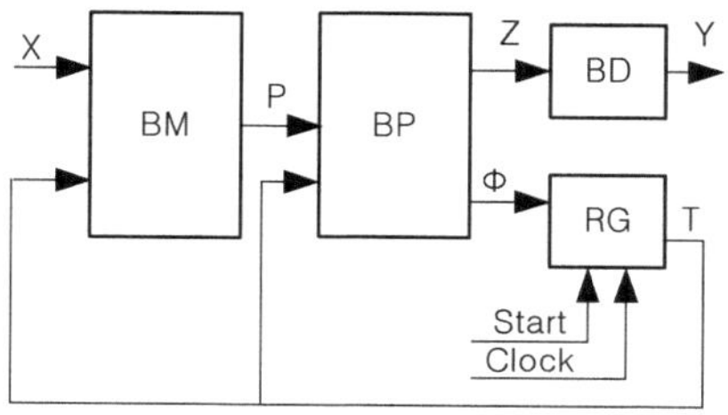

Fig. 4.16 Structural diagram of MPD Mealy FSM

$X(a_4) = \{x_3, x_4\}$, $X(a_5) = \{x_5\}$, and $X(a_7) = \{x_5, x_6\}$. It means that $G = 2$ and determines the set $P = \{p_1, p_2\}$. The table of logical condition replacement for FSM S_{12} (Table 4.17) is constructed using all rules discussed in previous Sections.

Table 4.17 Table of logical condition replacement for MPD Mealy FSM S_{12}

a_m	a_1	a_2	a_3	a_4	a_5	a_6	a_7
p_1	x_1	x_2	–	x_4	x_5	–	x_5
p_2	–	x_3	–	x_3	–	–	x_6

As it follows from Table 4.17, the multiplexer MX_1 has three control inputs, whereas only two control inputs are enough for the multiplexer MX_2. Thus, the codes of states $a_m \in A$ should be recoded using the method of multiplexer encoding. The outcome of such a recoding is shown in Fig. 4.17.

Fig. 4.17 Multiplexer codes for Mealy FSM S_{12}

T_1 \ T_2T_3	00	01	11	10
0	a_1	a_2	a_7	a_4
1	a_3	a_5	*	a_6

2. **Encoding of the classes of compatible microoperations.** For the FSM S_{12}, the set Y can be divided by three classes of compatible microoperations, namely: $Y^1 = \{y_1, y_4, y_7\}$, $Y^2 = \{y_2, y_6, y_8\}$, $Y^3 = \{y_3, y_5\}$. It is enough $R_D = 6$ variables $z_r \in Z$ for encoding of microoperations; let us point out that microoperations $y_n \in Y^3$ are encoded using one-hot codes. The final codes are represented by Table 4.18.
3. **Transformation of FSM structure table.** To transform an initial structure table, the column X_h is replaced by the column P_h, and the column Y_h by the column Z_h. The first replacement is executed in the manner used for MP Mealy FSM, while the second for PD Mealy FSM (Table 4.19).

Table 4.18 Codes of microoperations for MPD Mealy FSM S_{12}

Y^1	$K(Y^1)$	Y^2	$K(Y^2)$	Y^3	$K(Y^3)$
	z_1z_2		z_3z_4		z_5z_6
$\emptyset$	0 0	$\emptyset$	0 0	$\emptyset$	0 0
y_1	0 1	y_2	0 1	y_3	0 1
y_4	1 0	y_6	1 0	y_5	1 0
y_7	1 1	y_8	1 1		

Table 4.19 Transformed structure table of MP Mealy FSM S_{12}

a_m	$K(a_m)$	a_s	$K(a_s)$	P_h	Z_h	Φ_h	h
a_1	000	a_2	001	p_1	$z_2z_4z_6$	D_3	1
		a_3	010	$\bar{p}_1$	z_2z_3	D_1	2
a_2	001	a_2	001	p_1p_2	$z_1z_3z_6$	D_3	3
		a_3	100	$p_1\bar{p}_2$	$z_1z_3z_4$	D_1	4
		a_4	010	$\bar{p}_1$	$z_1z_2z_5$	D_2	5
a_3	100	a_5	101	1	z_2z_5	D_1D_3	6
a_4	010	a_5	101	$p_2\bar{p}_1$	z_3z_4	D_1D_3	7
		a_1	000	p_2p_1	z_1z_6	–	8
		a_7	011	$\bar{p}_2$	$z_1z_2z_3z_4$	D_2D_3	9
a_5	101	a_6	110	p_1	z_2z_3	D_1D_2	10
		a_7	011	$\bar{p}_1$	–	D_2D_3	11
a_6	110	a_1	000	1	$z_1z_2z_5$	–	12
a_7	011	a_2	001	p_1p_2	z_2z_3	D_3	13
		a_5	101	$p_1\bar{p}_2$	z_3z_4	D_1D_3	14
		a_3	100	$\bar{p}_1$	z_2z_5	D_1	15

The table is used to derive functions $z_r \in Z$, $D_r \in \Phi$. For example, the following Boolean equations for functions $z_1 = F_3 \vee F_4 \vee F_5 \vee F_8 \vee F_9 \vee F_{12} = A_2 \vee A_3 p_2 p_1 \vee A_3 \bar{p}_2 \vee A_6$; $D_1 = F_2 \vee F_4 \vee F_6 \vee F_7 \vee F_{10} \vee F_{14} \vee F_{15}$ can be derived from Table 4.19.

The logic circuit of MPD Mealy FSM S_{12} is shown in Fig. 4.18. In this circuit, the block BP includes two multiplexers and is implemented using information from Table 4.17, as well as codes shown in Fig. 4.16. The block BP is designed on the base of transformed ST (Table 4.19); the block BD includes two decoders and is implemented using Table 4.18.

Let us discuss an example of design for the MPFY Mealy FSM S_{12}, represented by its structure table (Table 4.16). The structural diagram of MPFY Mealy FSM is shown in Fig. 4.19. In this model, the block BM replaces logical conditions and generates functions $P(X,T)$; the block BP generates variables $z_r \in Z_P$ used for encoding of rows of transformed structure table; the block BF generates variables $z_r \in Z_F$ used for encoding of collections of microoperations, as well as input memory functions $\phi_r \in \Phi$. The block BY implements the system Y depended on variables $z_r \in Z_F$. The

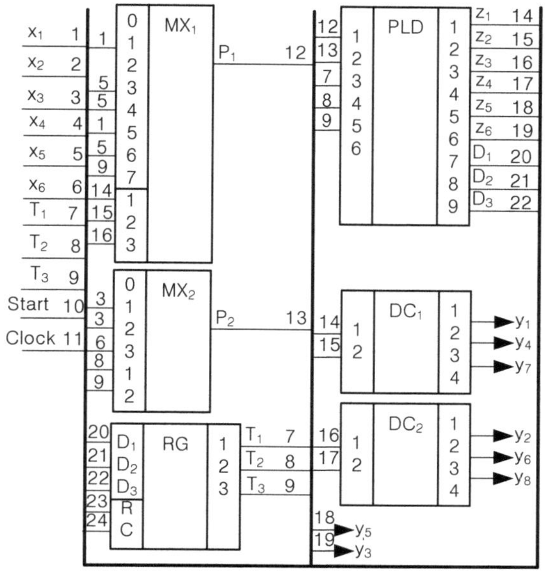

Fig. 4.18 Logic circuit of MPD Mealy FSM S_{12}

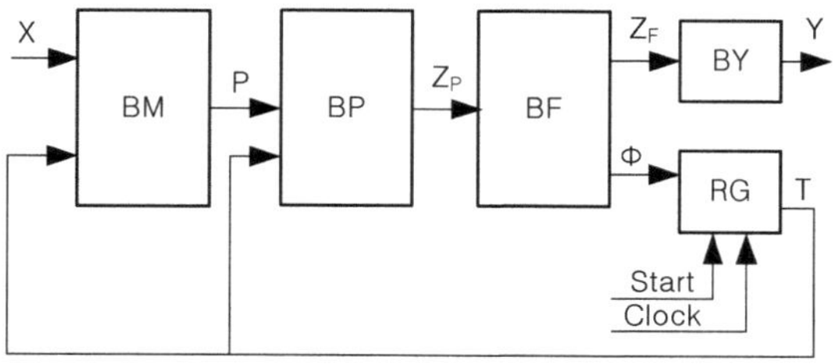

Fig. 4.19 Structural diagram of MPFY Mealy FSM

following systems of Boolean functions should be found to design the logic circuit of MPFY Mealy FSM:

$$P = P(X, T), \tag{4.22}$$

$$Z_P = Z_P(P, T), \tag{4.23}$$

$$Z_F = Z_F(Z_P), \tag{4.24}$$

$$\Phi = \Phi(Z_P), \tag{4.25}$$

$$Y = Y(Z_F). \tag{4.26}$$

1. **Logical condition replacement.** This step has been discussed for the MPD Mealy FSM S_{12}. Obviously, the outcome for this step does not depend on the model in use; instead, it is determined by the initial structure table. As in previous case, states codes are shown in Fig. 4.17, whereas Table 4.17 shows the outcome of logical condition replacement. This table determines system (4.22).

2. **Encoding of structure table rows.** For the Mealy FSM S_{12}, the structure table includes $H = 15$ rows, therefore, it is enough $R_F = 4$ variables for encoding of its rows. Thus, we have the set $Z_P = \{z_1, \ldots, z_4\}$. Let us encode the structure table rows in a trivial way: $K(F_1) = 0000, \ldots, K(F_{15}) = 1110$.
3. **Encoding of collections of microoperations.** The FSM S_{12} includes $T_0 = 8$ collections of microoperations: $Y_1 = \emptyset$, $Y_2 = \{y_1, y_2, y_3\}$, $Y_3 = \{y_1, y_6\}$, $Y_4 = \{y_3, y_4, y_6\}$, $Y_5 = \{y_4, y_8\}$, $Y_6 = \{y_5, y_7\}$, $Y_7 = \{y_1, y_5\}$, $Y_8 = \{y_8\}$. It is enough $R_3 = 3$ variables for such an encoding, therefore, we have the set $Z_F = \{z_5, z_6, z_7\}$. In the discussed example, the block BY is represented by Table 4.20.

Table 4.20 Table of block BY for MPFY Mealy FSM S_{12}

z_5	z_6	z_7	$Y(Z_F)$	t	z_5	z_6	z_7	$Y(Z_F)$	t
0	0	0	–	1	1	0	0	y_4y_8	5
0	0	1	$x_1y_2y_3$	2	1	0	1	y_5y_7	6
0	1	0	x_1y_6	3	1	1	0	x_1y_5	7
0	1	1	$y_3y_4y_6$	4	1	1	1	y_8	8

The simplest way for implementation of the block BY is use of PROM (or RAM) chips with address inputs z_5 – z_7. This table can be used for deriving SOP functions $y_n \in Y$. Let the symbol Z_t stand for a conjunction of variables $z_r \in Z_F$, corresponding to the code $K(Y_t)$. In this case, the following SOP $y_1 = Z_2 \vee Z_3 \vee Z_7 = \bar{z}_5\bar{z}_6z_7 \vee z_6z_7$, for example, can be derived from Table 4.20.
4. **Transformation of initial structure table.** To construct such a table, the column X_h should be replaced by the column P_h (as it is for MP Mealy FSM), and columns a_s - Φ_h are replaced by the column Z_h (as it is for PF Mealy FSM). For the FSM S_{12}, the outcome of transformation is shown in Table 4.21.

   This table is used for deriving system (4.23). For example, the following Boolean function for the block BP $z_1 = F_9 \vee \ldots \vee F_{15} = \bar{T}_1T_2\bar{T}_3\bar{p}_2 \vee A_5 \vee A_6 \vee$ $\vee A_7 = \bar{T}_1T_2\bar{T}_3\bar{p}_2 \vee T_1T_3 \vee T_1T_2 \vee T_2T_3$ can be derived from Table 4.21.
5. **Specification of block BF.** The block BF is represented by the table with columns $K(F_h)$ (inputs of memory blocks),Φ, Z_F (outputs of memory blocks). In the discussed case, this table includes $H = 15$ rows (Table 4.22). This table is used for deriving systems (4.24) and (4.25). Obviously, these systems can be represented as SOPs, as it is shown for the system (4.26).
6. **Logic circuit implementation.** Logic circuits for blocks BM, BP, BF, and BY are implemented during this step. Next, these circuits are combined together to form a final logic circuit. The logic circuit for block BM is implemented using both Table 4.17 and state codes shown in Fig. 4.16. The logic circuit for block BP is implemented using the transformed structure table (Table 4.21). The logic circuit for block BF is implemented using Table 4.22. The logic circuit for block BY is implemented using Table 4.20. Joining these circuits into the final logic circuit is easy enough; we do not discuss this step.

Table 4.21 Transformed structure table of MPFY Mealy FSM S_{12}

a_m	$K(a_m)$	P_h	Z_h	h
a_1	000	p_1	–	1
		$\bar{p}_1$	z_4	2
a_2	001	$p_1 p_2$	z_3	3
		$p_1 \bar{p}_2$	$z_3 z_4$	4
		$\bar{p}_1$	z_2	5
a_3	100	1	$z_2 z_4$	6
a_4	010	$p_2 p_1$	$z_2 z_3$	7
		$p_2 \bar{p}_1$	$z_2 z_3 z_4$	8
		$\bar{p}_2$	z_1	9
a_5	101	p_1	$z_1 z_4$	10
		$\bar{p}_1$	$z_1 z_3$	11
a_6	110	1	$z_1 z_3 z_4$	12
a_7	011	$p_1 p_2$	$z_1 z_2$	13
		$p_1 \bar{p}_2$	$z_1 z_2 z_4$	14
		$\bar{p}_1$	$z_1 z_2 z_3$	15

Table 4.22 Specification of block BF for MPFY Mealy FSM S_{12}

$K(F_h)$	Φ	Z_F	h
0000	D_3	z_7	1
0001	D_1	z_6	2
0010	D_3	$z_6 z_7$	3
0011	D_1	z_5	4
0100	D_2	$z_5 z_7$	5
0101	$D_1 D_3$	$z_5 z_6$	6
0110	$D_1 D_3$	$z_5 z_6 z_7$	7
0111	–	$z_6 z_7$	8
1000	$D_2 D_3$	z_5	9
1001	$D_1 D_3$	z_6	10
1010	$D_1 D_2$	–	11
1011	–	$z_5 z_7$	12
1100	D_3	z_6	13
1101	$D_1 D_3$	$z_5 z_6 z_7$	14
1110	D_1	$z_5 z_6$	15

Logic circuits for any of FSM represented by Table 4.15 can be designed using the same approach. Some examples of designs can be found in [2].

References

1. Adamski, M., Barkalov, A.: Architectural and Sequential Synthesis of Digital Devices. University of Zielona Góra Press, Zielona Góra (2006)
2. Barkalov, A., Titarenko, L.: Synthesis of Operational and Control Automata. UNITECH, Donetsk (2009)
3. Barkalov, A., Węgrzyn, M.: Design of Control Units With Programmable Logic. University of Zielona Góra Press, Zielona Góra (2006)
4. Barkalov, A., Zelenjova, I.: Optimization of replacement of logical conditions for an automaton with bidirectional transitions. Automatic Control and Computer Sciences 34(5), 48–53

Chapter 5
Optimization for Logic Circuit of Moore FSM

Abstract. The chapter is devoted to original synthesis and optimization methods oriented on Moore FSM logic circuit implemented with CPLD. These methods are based on results of joint investigations conducted by the authors and their PhD students Cololo S. (Ukraine) and Chmielewski S. (Poland). These methods deal with both homogenous and heterogeneous CPLD chips. In the first case, only PAL- or PLA- based macrocells are used for logic circuit implementation. In the second case, the logic circuit is implemented using both PAL-based macrocells and embedded memory blocks. The hardware amount reduction is based on use of several sources (up to three) to represent the codes of classes of pseudoequivalent states. The methods assume joint minimization of Boolean expressions for input memory functions and microoperations of Moore FSM. The last part of the chapter is devoted to joint application of proposed methods and logical condition replacement.

5.1 Optimization for Two-Level FSM Model

If logic circuit of Moore FSM is implemented using standard FPLD chips, then optimization methods targeted on optimization of matrix FSM model should be adapted to take into account some particular features of these microchips. Let us discuss a two-level Moore FSM model shown in Fig. 5.1.

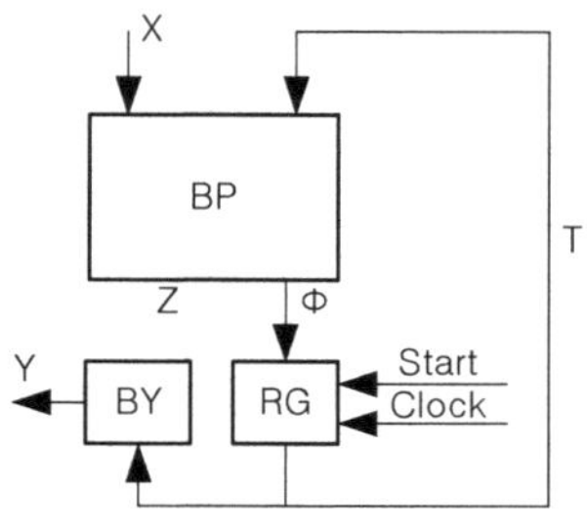

Fig. 5.1 Structural diagram of PY Moore FSM

A. Barkalov and L. Titarenko: Logic Synthesis for FSM-Based Control Units, LNEE 53, pp. 103–127.
springerlink.com

In PY Moore FSM, a block BP generates input memory functions

$$\Phi = \Phi(T, X), \tag{5.1}$$

whereas a block BY generates microoperations

$$Y = Y(T). \tag{5.2}$$

The block BP corresponds to matrices M_1 and M_2 (Fig. 2.15); functions (5.1) are represented as the system (1.8). The block BY corresponds to matrices M_3 and M_4 (Fig. 2.15); functions (5.2) are represented as the system (1.10).

As it is shown in Chapter 2, the methods of optimal, refined, and combined state encoding can be used to optimize a logic circuit of Moore FSM. This very problem can be solved using the method of transformation of state codes into the codes of classes of pseudoequivalent states. Let us discuss application of these methods for the Moore FSM S_{13}, represented by its structure table (Table 5.1).

Table 5.1 Structure table of Moore FSM S_{13}

a_m	$K(a_m)$	a_s	$K(a_s)$	X_h	Φ_h	h
a_1 (–)	000	a_2	001	1	D_3	1
a_2 (y_1y_2)	001	a_3	010	x_1	D_2	2
		a_4	011	$\bar{x}_1x_2$	D_2D_3	3
		a_5	100	$\bar{x}_1\bar{x}_2$	D_1	4
a_3 (y_3y_6)	010	a_6	101	$\bar{x}_3$	D_1D_3	5
		a_7	110	x_3	D_1D_2	6
a_4 (y_2y_4)	011	a_6	101	$\bar{x}_3$	D_1D_3	7
		a_7	110	x_3	D_1D_2	8
a_5 (y_5y_6)	100	a_6	101	$\bar{x}_3$	D_1D_3	9
		a_7	110	x_3	D_1D_2	10
a_6 (y_1y_2)	101	a_1	000	1	–	11
a_7 (y_3y_6)	110	a_1	000	1	–	12

As follows from Table 5.1, the Moore FSM S_{13} is described by the following sets: $X = \{x_1, x_2, x_3\}$, $Y = \{y_1, \ldots, y_6\}$, $A = \{a_1, \ldots, a_7\}$, $\Phi = \{D_1, D_2, D_3\}$, $T = \{T_1, T_2, T_3\}$. It gives the following values of its main parameters: $L = 3$, $N = 6$, $M = 7$, $R = 3$. As in case of Mealy FSM, the structure table is used for deriving systems of Boolean functions $y_n \in Y$ and $\phi_r \in \Phi$. For example, the following systems $D_1 = F_1 \vee \ldots F_{10} = A_3 \vee A_4 \vee A_5 = \bar{T}_1T_2 \vee T_1\bar{T}_2\bar{T}_3$ (after minimizing), and $y_1 = A_2 \vee A_6 = \bar{T}_2T_3$ can be derived from Table 5.1. If logic circuit of the block BY is implemented using some memory blocks such as PROM or RAM, then this block is specified by the table with columns: a_m, $K(a_m)$, $Y(a_m)$, and m, where m is the number of the table row (it is Table 5.2 for the Moore FSM S_{13}).

Table 5.2 Specification of block BY of Moore FSM S_{13}

a_m	$K(a_m)$	$Y(a_m)$	m
a_1	000	–	1
a_2	001	y_1y_2	2
a_3	010	y_3y_6	3
a_4	011	y_2D_1	4
a_5	100	y_1y_6	5
a_6	101	y_1y_2	6
a_7	110	y_3y_6	7

For PY Moore FSM, the logic circuit is implemented in a trivial way using the structure table, as well as the table for block BY. The logic circuit of PY Moore FSM S_{13} is shown in Fig. 5.2.

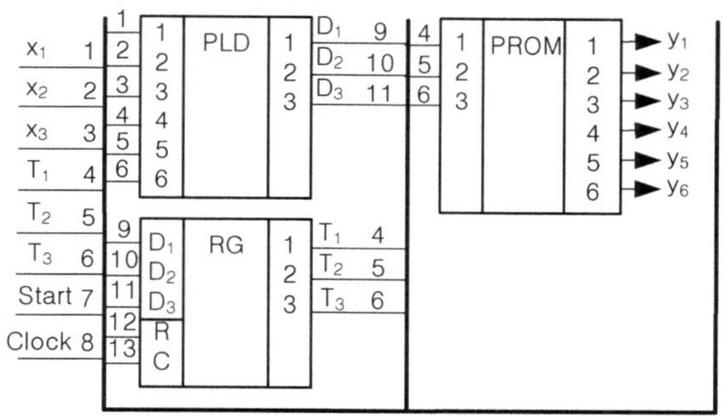

Fig. 5.2 Logic circuit of PY Moore FSM S_{13}

As it is discussed in Chapter 2, the optimization methods for Moore FSM [4, 11] are based on existence of the classes of pseudoequivalent states $B_i \in \Pi_A$, where $\Pi_A = \{B_1, \ldots, B_I\}$ is a partition of the state set A by the classes of pseudoequivalent states. For the Moore FSM S_{13}, there is the partition $\Pi_A = \{B_1, B_2, B_3, B_4\}$, where $B_1 = \{a_1\}$, $B_2 = \{a_2\}$, $B_3 = \{a_3, a_4, a_5\}$, $B_4 = \{a_6, a_7\}$.

Let the symbol P_0Y stand for Moore FSM with optimal state encoding. Structure diagrams are the same for both P_0Y and PY Moore FSMs (Fig. 5.1). Let us construct system (2.29), using the following equations for the Moore FSM S_{13}:

$$\begin{aligned} B_1 &= A_1; & B_3 &= A_3 \vee A_4 \vee A_5; \\ B_2 &= A_2; & B_4 &= A_6 \vee A_7. \end{aligned} \tag{5.3}$$

In case of the optimal state encoding, the state assignment for states $a_m \in A$ is executed in such a manner, that the number of terms in system (2.29) is minimal. Obviously, the absolute minimum is equal to the number of classes in the partition.

Let us encode states $a_m \in A$ by optimal codes shown in the Karnaugh map (Fig. 5.3); the well-known algorithm ESPRESSO [12] can be used for the optimal state encoding.

Fig. 5.3 Optimal state codes for Moore FSM S_{13}

T_1 \ T_2T_3	00	01	11	10
0	a_1	a_2	a_6	a_7
1	a_3	a_4	a_5	*

As follows from Fig. 5.3, the system (5.3) is represented as the following one:

$$\begin{aligned} B_1 &= \bar{T}_1\bar{T}_2\bar{T}_3; & B_3 &= T_1; \\ B_2 &= \bar{T}_1\bar{T}_2T_3; & B_4 &= \bar{T}_1T_2. \end{aligned} \tag{5.4}$$

As follows from (5.4), the absolute minimal value of terms is reached. Now, the class B_1 corresponds to the code $K(B_1) = 000$, the class B_2 to $K(B_2) = 001$, the class B_3 to $K(B_3) = 1{*}{*}$, and the class B_4 to $K(B_4) = 01{*}$.

The transformed structure table of P_0Y Moore FSM includes columns B_i, $K(B_i)$, a_s, $K(a_s)$, X_h, Φ_h, h. For the P_0Y Moore FSM S_{13}, this table includes $H_0 = 7$ rows (Table 5.3).

Table 5.3 Transformed structure table of P_0Y Moore FSM S_{13}

B_i	$K(B_i)$	a_s	$K(a_s)$	X_h	Φ_h	h
B_1	000	a_2	001	x_1	D_3	1
B_2	001	a_3	100	x_1	D_1	2
		a_4	101	$\bar{x}_1x_2$	D_1D_3	3
		a_5	111	$\bar{x}_1\bar{x}_2$	$D_1D_2D_3$	4
B_3	$1{*}{*}$	a_6	011	x_3	D_2D_3	5
		a_7	010	$\bar{x}_3$	D_2	6
B_4	$01{*}$	a_1	000	1	–	7

Transformed ST is used to derive functions (5.1), represented in the following manner:

$$D_r = \bigvee_{h=1}^{H} C_{rh}\Big(\bigwedge_{r=1}^{R} T_r^{l_{rh}}\Big)X_h. \tag{5.5}$$

In this expression, C_{rh} is a Boolean variable equal to 1, iff the row h of the table includes input memory function $D_r = 1$; $l_{rh} \in \{0, 1, *\}$ is the value of the bit r for the code $K(B_i)$ from the row h of the table, $T_r^0 = \bar{T}_r$, $T_r^1 = T_r$, $T_r^* = 1$ $(r = 1, \ldots, R; h = 1, \ldots, H)$. In the discussed example, the following expression $D_3 = F_1 \vee F_3 \vee F_4 \vee F_5 = \bar{T}_1\bar{T}_2\bar{T}_3 \vee \bar{T}_1\bar{T}_2T_3\bar{x}_1 \vee T_1x_3$ can be derived from Table 5.3.The system (5.5) is used for implementation of logic circuit of the block BP. A table similar to Table 5.2

is used for implementation of logic circuit of the block BY. Obviously, logic circuits for both models of FSM S_{13} (PY and P_0Y) are identical, but the block BP includes fewer terms in the second model.

This method has some drawbacks, namely [1]:

1. The number of the block BP inputs, as a rule, exceeds their minimal possible number R_0, determined by (2.24). Remind that $R_0 = \lceil \log_2 I \rceil$ and this value is equal to the number of bits for state codes of equivalent Mealy FSM.
2. The optimal state encoding does not guarantee an achievement of the minimal number of terms I in system (2.29). Sometimes, it is not possible at all. For example, there is no such optimal encoding variant for an FSM with the partition $B_1 = \{a_1\}$, $B_2 = \{a_2, a_3, a_4\}$, when the system (2.29) includes only 2 terms. If the absolute minimum is not reached, then the number of rows for transformed ST is greater than for equivalent Mealy FSM.

The method of refined state encoding targets on optimization of logic circuit of the block BY implemented without use of PROM or RAM chips. For the Moore FSM S_{13}, the system (5.2) is represented as:

$$\begin{aligned} y_1 &= A_2 \vee A_6; & y_4 &= A_4;\\ y_2 &= A_2 \vee A_4 \vee A_6; & y_5 &= A_5;\\ y_3 &= A_3 \vee A_{75}; & y_6 &= A_3 \vee A_5 \vee A_7. \end{aligned} \tag{5.6}$$

Use of the refined state encoding leads to the model of P_RYMoore FSM, let us point out that its structural diagram is the same as for PY Moore FSM. As an outcome of refined state encoding, there is the system (5.2) with the following features:

1. Each Boolean function $y_n \in Y$ is implemented using only one LUT element (if FPGA chips are used to implement the block BY).
2. The SOP for each Boolean function $y_n \in Y$ includes not more than q terms, where q is the number of terms per one PAL macrocell (if CPLD chips with PAL-based macrocells are used to implement the block BY).
3. The total number of terms in the system (5.2) is decreased up to the point, when there is a partition of the set Y with the minimal number $\lceil N/t \rceil$ of subsystems such that each of them is implemented using one PLA macrocell, where t is the number of PLA outputs (if CPLD chips with PLA-based macrocells are used to implement the block BY).

To get the above mentioned properties, well-known methods can be used presented in [1, 12]. One of the possible variants of the refined state encoding is shown in Fig. 5.4.

The following Boolean equations $y_1 = \bar{T}_1 T_3$, $y_2 = \bar{T}_1 T_3 \vee \bar{T}_1 T_2$, $y_3 = T_1 \bar{T}_2$, $y_4 = T_2 \bar{T}_3$, $y_5 = T_1 T_2$, $y_6 = T_1$ can be derived from Fig. 5.4. Obviously, the minimal number of terms for system (5.2) coincides with the number of microoperations, N. In the discussed example, the system (5.6) includes $N = 6$ terms; it means that the absolute minimum is reached. Of course, there is no guarantee that this minimum will be reached. It depends strongly on characteristics of the initial GSA.

Fig. 5.4 Refined state codes for Moore FSM S_{13}

T_1 \ T_2T_3	00	01	11	10
0	a_1	a_2	a_6	a_4
1	a_3	a_7	a_5	*

Because this method targets on minimization of the block BY, it does not guarantee decreasing the number of ST rows up to the point, which is possible for the optimal state encoding. For example, the refined state codes from Karnaugh map (Fig. 5.4) results in the following transformation of initial system (5.3):

$$\begin{aligned} B_1 &= \bar{T}_1\bar{T}_2\bar{T}_3; & B_3 &= T_1\bar{T}_3 \vee T_2\bar{T}_3 \vee T_1T_2; \\ B_2 &= \bar{T}_1\bar{T}_2T_3; & B_4 &= \bar{T}_1T_2T_3 \vee T_1\bar{T}_2T_3. \end{aligned} \tag{5.7}$$

As it can be found from the system (5.7), the transformed ST includes three subtables with transitions from the states from class B_3, and two subtables for the class B_4. It means that the transformed ST includes 12 rows.

The method of combined state encoding targets on joint minimization for systems $Y = Y(A)$ and $B = B(A)$. There is no any effective method for this task solving. We can suppose that it should be an iterative algorithm. For example, the states are encoded by optimal codes, as the first step. Next, these codes are rearranged in corresponding subtables of the common Karnaugh map to optimize the system $Y(A)$. For the Moore FSM S_{13}, one of the possible variants of combined state encoding is shown in Fig. 5.5.

Fig. 5.5 Combined state codes for Moore FSM S_{13}

T_1 \ T_2T_3	00	01	11	10
0	a_1	*	a_4	a_2
1	a_7	a_3	a_5	a_6

Using these codes, the systems (5.3) and (5.6) can be transformed and represented as the following:

$$\begin{aligned} & & y_1 &= T_1\bar{T}_3; \\ B_1 &= \bar{T}_1\bar{T}_2\bar{T}_3; & y_2 &= \bar{T}_1T_3 \vee T_2\bar{T}_3; \\ B_2 &= \bar{T}_1T_2\bar{T}_3; & y_3 &= T_1T_3; \\ B_3 &= T_3; & y_4 &= \bar{T}_1T_3; \\ B_4 &= T_1\bar{T}_3; & y_5 &= T_1T_2T_3; \\ & & y_6 &= T_1\bar{T}_2 \vee T_1T_3. \end{aligned} \tag{5.8}$$

As it follows from system (5.8), the transformed ST includes $H_0 = 7$ rows (because each function of the system (5.3) is represented by a SOP with only one term). In the same time, the number of terms in the system (5.6) is equal to $N + 1 = 7$, that is only one term more, than it is in the case of refined state encoding shown in Fig. 5.4.

If the approach of combined state encoding is used, then PY Moore FSM turns into P_KY Moore FSM with the structure identical to the structure shown in Fig. 5.1. Synthesis methods for each model (PY, P_0Y, P_RY, and P_KY) includes the same steps, namely:

1. Finding of partition Π_A for the set of states A by the classes of pseudoequivalent states.
2. Construction of Boolean systems $B(A)$ and $Y(A)$.
3. Appropriate state encoding (arbitrary, optimal, refined, or combined).
4. Specification of the block BY.
5. Construction of the transformed structure table.
6. Implementation of FSM logic circuit using the given logic elements.

Let us point out that the system $Y(A)$ is not constructed, if logic circuit of the block BY is implemented using embedded memory blocks. If the approach of arbitrary state encoding is used, then steps 1, 2, and 5 of the synthesis method are eliminated. All these methods have the same drawback, namely, they do not guarantee reduction for the number of transformed ST rows up to H_0 simultaneously with decrease for the number of terms in system of microoperations up to N. Such an outcome is guaranteed if a designer uses the approach of *transformation of state codes into codes of the classes of pseudoequivalent states*.

In this case PY Moore FSM turns into P_CY Moore FSM, with a structural diagram shown in Fig. 2.20. Let us consider application of the design method for Moore FSM S_{14} represented by its structure table (Table 5.4). Let us use PAL macrocells with two terms ($q = 2$) to implement logic circuits of blocks BP, BY, and BTC.

1. **Construction of the partition Π_A.** As follows from Table 5.4, this set includes $I = 4$ classes, that is $\Pi_A = \{B_1, \ldots, B_4\}$. The pseudoequivalent states $a_m \in A$ are distributed among the classes $B_i \in \Pi_A$ in the following manner: $B_1 = \{a_1\}$, $B_2 = \{a_2, a_3, a_4\}$, $B_3 = \{a_5, a_6, a_7\}$, $B_4 = \{a_8, a_9, a_{10}\}$.

   Obviously, it is enough two variables ($R_0 = 2$) for encoding of the classes $B_i \in \Pi_A$ and it determines the set $\tau = \{\tau_1, \tau_2\}$. Let us encode the classes in a trivial way: $K(B_1) = 00, \ldots, K(B_4) = 11$.
2. **Construction of systems $B(A)$ and $Y(A)$.** As follows from Table 5.4 and partition Π_A, systems (2.29) and (5.2) are represented in this particular case as the following:

$$\begin{aligned} B_1 &= A_1; \\ B_2 &= A_2 \vee A_3 \vee A_4; \\ B_3 &= A_5 \vee A_6 \vee A_7; \\ B_4 &= A_8 \vee A_9 \vee A_{10}; \end{aligned} \qquad \begin{aligned} y_1 &= A_2 \vee A_5 \vee A_{10}; \\ y_2 &= A_2 \vee A_4 \vee A_8; \\ y_3 &= A_3 \vee A_5 \vee A_9; \\ y_4 &= A_4 \vee A_8; \\ y_5 &= A_6; \\ y_6 &= A_9; \\ y_7 &= A_{10}. \end{aligned} \tag{5.9}$$

Table 5.4 Structure table of Moore FSM S_{14}

a_m	$K(a_m)$	a_s	$K(a_s)$	X_h	Φ_h	h
a_1 (–)	0000	a_2	0001	x_1	D_4	1
		a_3	0010	$\bar{x}_1x_2$	D_3	2
		a_4	0011	$\bar{x}_1\bar{x}_2$	D_3D_4	3
a_2 (y_1y_2)	0001	a_5	0100	x_2	D_2	4
		a_6	0101	$\bar{x}_2x_4$	D_2D_4	5
		a_7	0110	$\bar{x}_2\bar{x}_4$	D_2D_3	6
a_3 (y_3)	0010	a_5	0100	x_2	D_2	7
		a_6	0101	$\bar{x}_2x_4$	D_2D_4	8
		a_7	0110	$\bar{x}_2\bar{x}_4$	D_2D_3	9
a_4 (y_2y_4)	0011	a_5	0100	x_2	D_2	10
		a_6	0101	$\bar{x}_2x_4$	D_2D_4	11
		a_7	0110	$\bar{x}_2\bar{x}_4$	D_2D_3	12
a_5 (y_1y_3)	0100	a_8	0111	x_1	$D_2D_3D_4$	13
		a_9	1000	$\bar{x}_1x_5$	D_1	14
		a_{10}	1001	$\bar{x}_1\bar{x}_5$	D_1D_4	15
a_6 (y_5)	0101	a_8	0111	x_1	$D_2D_3D_4$	16
		a_9	1000	$\bar{x}_1x_5$	D_1	17
		a_{10}	1001	$\bar{x}_1\bar{x}_5$	D_1D_4	18
a_7 (y_3)	0110	a_8	0111	x_1	$D_2D_3D_4$	19
		a_9	1000	$\bar{x}_1x_5$	D_1	20
		a_{10}	1001	$\bar{x}_1\bar{x}_5$	D_1D_4	21
a_8 (y_2y_4)	0111	a_1	0000	1	1	22
a_9 (y_3y_6)	1000	a_1	0000	1	1	23
a_{10} (y_1y_7)	1001	a_1	0000	1	1	24

3. **Combined state encoding.** This state assignment targets on maximal possible simplification for functions $Y(A)$. Next, these state codes are rearranged to optimize equations from $B(A)$. One of the possible encoding variants is represented by the Karnaugh map shown in Fig. 5.6.

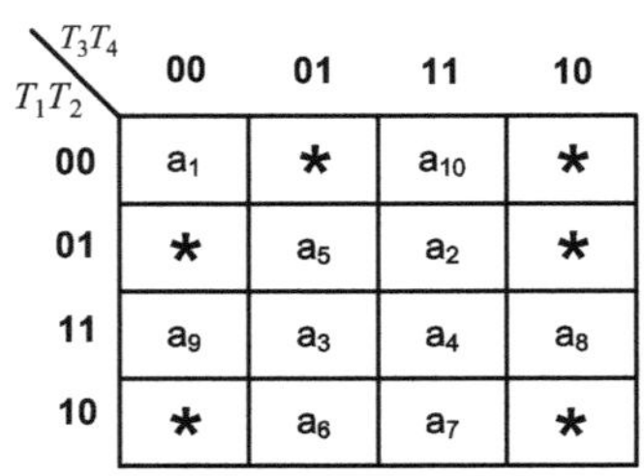

Fig. 5.6 Combined state codes for Moore FSM S_{14}

Taking these codes into account, we can minimize equations from initial system (5.9). After minimization, we can get the following system corresponding to initial system (5.9):

$$
\begin{array}{ll}
 & y_1 = T_1\bar{T}_4; \\
B_1 = \bar{T}_2\bar{T}_4; & y_2 = T_2T_3; \\
B_2 = T_2T_3T_4 \vee T_1T_2T_4; & y_3 = T_2\bar{T}_3; \\
B_3 = \bar{T}_1T_2\bar{T}_3 \vee T_1\bar{T}_2; & y_4 = T_1T_2T_3; \\
B_4 = T_1\bar{T}_4 \vee \bar{T}_1\bar{T}_2T_4; & y_5 = T_1\bar{T}_2\bar{T}_3; \\
 & y_6 = T_1\bar{T}_3\bar{T}_4; \\
 & y_7 = \bar{T}_1\bar{T}_2T_3.
\end{array}
\qquad (5.10)
$$

Each equation of the system (5.10) includes up to two terms and, therefore, can be implemented using only one PAL-based macrocell with $q = 2$.

4. **Specification of block BY.** There is no need in such a table, because the system of microoperations to be implemented in the discussed example has already represented as a SOP.
5. **Construction of transformed structure table.** This step is executed using the approach we have discussed in previous sections. For the P_CY Moore FSM S_{14}, the transformed ST includes $H_0 = 10$ rows (Table 5.5).

Table 5.5 Transformed structure table of P_CY Moore FSM S_{14}

B_i	$K(B_i)$	a_s	$K(a_s)$	X_h	Φ_h	h
B_1	00	a_2	0011	x_1	$D_2D_3D_4$	1
		a_3	1101	$\bar{x}_1x_2$	$D_1D_2D_4$	2
		a_4	1111	$\bar{x}_1\bar{x}_2$	$D_1D_2D_3D_4$	3
B_2	01	a_5	0101	x_2	D_2D_4	4
		a_6	1001	$\bar{x}_2x_4$	D_1D_4	5
		a_7	1011	$\bar{x}_2\bar{x}_4$	$D_1D_3D_4$	6
B_3	10	a_8	1110	x_1	$D_1D_2D_3$	7
		a_9	1100	$\bar{x}_1x_5$	D_1D_2	8
		a_{10}	0011	$\bar{x}_1\bar{x}_5$	D_3D_4	9
B_4	11	a_1	0000	1	–	10

This table is used to derive Boolean equations from the system $D_r \in \Phi$. For example, the following equation $D_4 = F_1 \vee \ldots \vee F_6 \vee F_9$ can be derived from Table 5.5. It can be minimized, and the final expression $D_4 = \bar{\tau}_1 \vee \tau_1\bar{\tau}_2\bar{x}_1\bar{x}_5$ is used to design a corresponding part of FSM logic circuit.

6. **Construction of Boolean system for block BCT.** System (5.10) includes functions $B(T)$, whereas the block BCT generates functions $\tau(T)$. The following approach can be used for construction of the system $\tau(T)$. If a variable τ_r is equal to 1 for the class code $K(B_i)$, then the SOP of function τ_r includes all terms belonging to function $B_i(T)$, where $r = 1,\ldots,R_0$; $i = 1,\ldots,I$.

For the Moore FSM S_{14}, terms of the class B_1 are not included into SOPs of functions $\tau_r \in \tau$ because $K(B_1) = 00$. The class codes $K(B_2) = 01$, $K(B_3) = 10$, $K(B_4) = 11$ determine the following SOPs: $\tau_1 = B_3 \vee B_4$, $\tau_2 = B_2 \vee B_4$. Our analysis of the Karnaugh map from Fig. 5.6 shows that the function τ_2 can be minimized and represented as $\tau_2 = T_1T_2 \vee \bar{T}_1T_3$. In the same time, the function τ_1 cannot be minimized.

Obviously, codes for classes $B_i \in \Pi_A$ can be assigned in such a way, that system $\tau(T)$ includes the minimal possible number of terms. The well-known algorithm ESPRESSO [12] can be used to solve this problem. For example, if the classes are encoded as $K(B_1) = 11$, $K(B_2) = 00$, $K(B_3) = 01$, $K(B_4) = 10$, then the following minimized SOPs can be represented as $\tau_1 = B_1 \vee B_4$, $\tau_2 = B_1 \vee B_3$. Basing on Fig. 5.6, the following final SOPs can be obtained: $\tau_1 = \bar{T}_1\bar{T}_2 \vee T_1\bar{T}_4$, $\tau_2 = \bar{T}_1\bar{T}_3 \vee T_1T_2T_4$. Now a logic circuit for each function $\tau_r \in \tau$ can be implemented using only one PAL-based macrocell with the number of terms $q = 2$. Let us point out that in the initial outcome of state encoding the logic circuit for function τ_1 is implemented using one macrocell with $q = 2$, whereas function τ_2 includes 4 terms and its logic circuit consumes 3 macrocells with $q = 2$.

7. **Implementation of FSM logic circuit.** This step is executed using systems of functions obtained for blocks BP, BY, and BCT. For the Moore FSM S_{14}, this circuit is shown in Fig. 5.7.

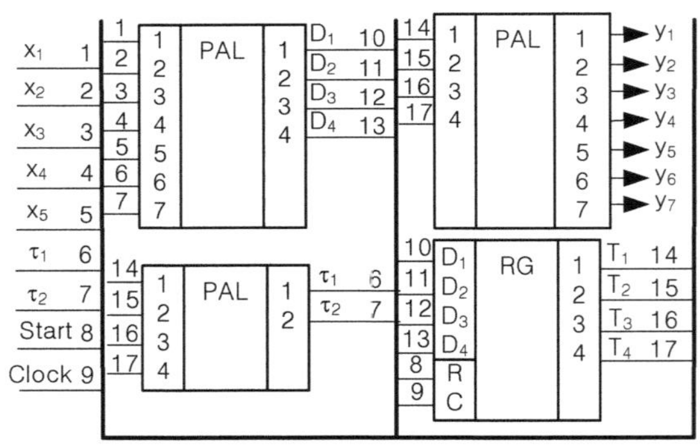

Fig. 5.7 Logic circuit of P_CY Moore FSM S_{14}

The following section is devoted to application of this approach for CPLD chips with PAL-based macrocells and embedded memory blocks BRAM.

5.2 FSM Synthesis for CPLD with Embedded Memory Blocks

The CPLDs produced by Cypress have two specific features:

1. Their PAL-based macrocells have many inputs, it means they are macrocells with "wide fan-in" [1].

2. Their embedded memory blocks can be configured, changing the numbers of outputs t and words q. The product of these numbers

$$Q = q \cdot t \tag{5.11}$$

is a constant.

These specifics are used in [2–10] to minimize the hardware amount in logic circuit of $P_C Y$ Moore FSM. The main idea of this approach is using more than one source for codes of the classes of pseudoequivalent states. There are three possible code sources, such as the register RG, the block BY, and the code transformer BCT. Let us consider a binary vector $\langle \mathrm{RG}, \mathrm{BY}, \mathrm{BTC} \rangle$, where 1 in some position means use of the corresponding block as a source of the code $K(B_i)$. There are the following vectors $\langle \mathrm{RG}, \mathrm{BY}, \mathrm{BTC} \rangle$:

1. The vector $\langle 1,0,1 \rangle$. In this case class codes are generated by blocks RG and BTC; it leads to $P_{C1}Y$ Moore FSM (Fig. 5.8).
2. The vector $\langle 1,1,0 \rangle$. In this case class codes are generated by blocks RG and BY; it leads to $P_{C2}Y$ Moore FSM (Fig. 5.9).

Fig. 5.8 Structural diagram of $P_{C1}Y$ Moore FSM

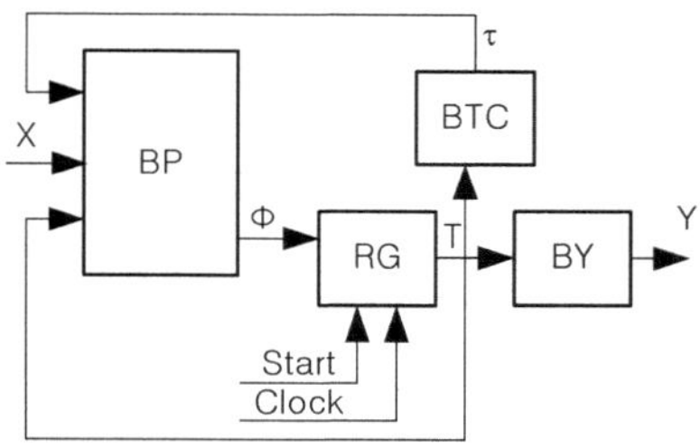

Fig. 5.9 Structural diagram of $P_{C1}Y$ Moore FSM

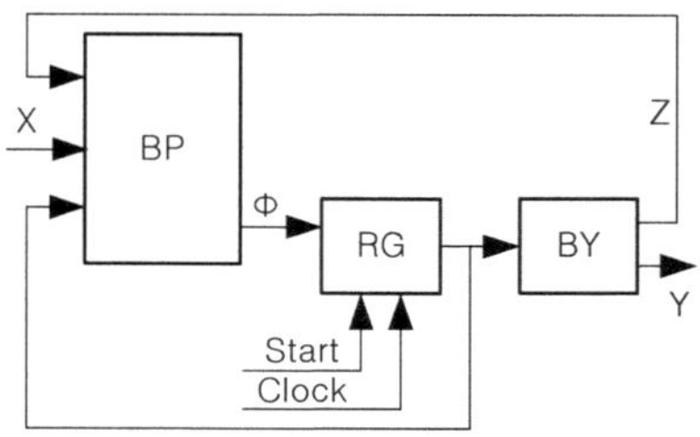

For $P_{C2}Y$ Moore FSM, variables $z_r \in Z$ represent class codes for classes $B_i \in \Pi_A$ such that the block BY is their source.

3. The vector $\langle 0,1,1 \rangle$. In this case class codes are generated by blocks BY and BTC; it leads to $P_{C3}Y$ Moore FSM (Fig. 5.10).
4. The vector $\langle 1,1,1 \rangle$. In this case class codes are generated by blocks RG, BY, and BTC; it leads to $P_{C4}Y$ Moore FSM (Fig. 5.11).

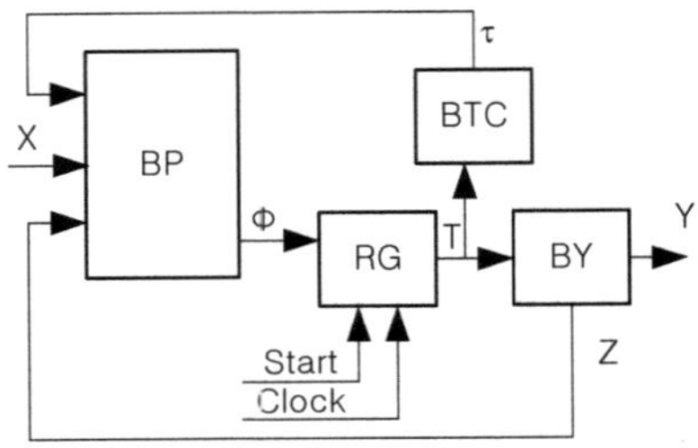

Fig. 5.10 Structural diagram of $P_{C3}Y$ Moore FSM

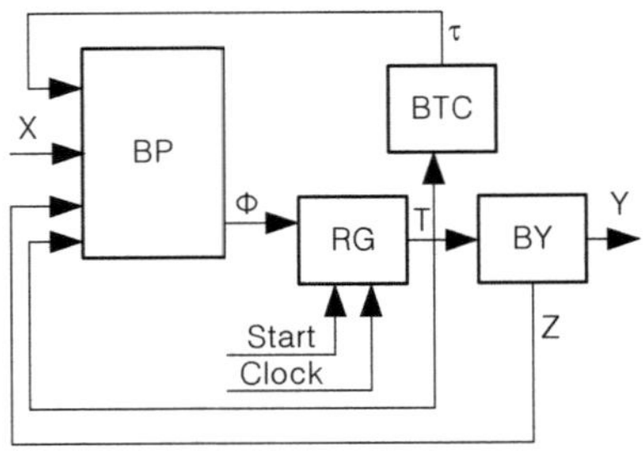

Fig. 5.11 Structural diagram of $P_{C4}Y$ Moore FSM

Let $\Pi_{\mathrm{RG}} \subseteq \Pi_A$ be a set of classes $B_i \in \Pi_A$, represented by a single generalized interval of the R - dimensional Boolean space. In this case the register RG is a source for codes of classes $B_i \in \Pi_{\mathrm{RG}}$. Let $\Pi_{TC} = \Pi_A \backslash \Pi_{\mathrm{RG}}$ be a set of classes, such that there is necessity of their state codes transformation. It is enough R_{TC} variables to encode classes $B_i \in \Pi_{\mathrm{TC}}$, where

$$R_{TC} = \lceil \log_2 I_{TC} \rceil . \tag{5.12}$$

In (5.12), we use the number of classes $B_i \in \Pi_{TC}$, namely $I_{TC} = |\Pi_{TC}|$.

As it is mentioned above, each block BRAM can be configured and possible fixed numbers of outputs create some set $T(\mathrm{BRAM})$. Up-to-day technology are characterized by the set $T(\mathrm{BRAM}) = \{1,2,4,8,16,32\}$. The block BY has R inputs, therefore a standard block BRAM should be configured in such a way, that is includes 2^R words. In theory, the word width is determined as

$$t_0 = \lceil Q/2^R \rceil . \tag{5.13}$$

But in reality, the nearest number from the set $T(\mathrm{BRAM})$ should be selected as a real word width. This number should be less or equal to t_0.

The block BY generates N microoperations. If $Q \geq 2^R$, then n_{BY} blocks BRAM is enough to implement the logic circuit of this block, where

$$n_{BY} = \lceil N/t_F \rceil . \tag{5.14}$$

In this case all BRAMs from the block BY have totally $n_{BY} \cdot t_F$ outputs, though R_{BY} outputs are not used for generation of microoperations, where

$$R_{BY} = n_{BY} \cdot t_F - N. \quad (5.15)$$

If condition

$$R_{BY} \geq R_{TC} \quad (5.16)$$

takes place, then the block BTC is absent, because class codes for classes $B_i \in \Pi_{TC}$ are generated by the block BY. If condition (5.16) is violated, then R_{BY} bits from total number of code bits for classes $B_i \in \Pi_{TC}$ are generated by the block BY, whereas the remainder $(R_{BY} - R_{TC})$ bits are generated by the block BTC.

In general, the synthesis method for P_{Cj}Y Moore FSM, where $j = 1,\ldots,4$, includes the following steps:

1. Construction of partition $\Pi_A = \{B_1,\ldots,B_I\}$ of the set of states A by the classes of pseudoequivalent states.
2. Construction of the system $B(A)$.
3. Optimal state encoding targets on minimizing the number of terms in system $B(A)$. Construction of the set Π_{RG}.
4. If $\Pi_{RG} = \Pi_A$, then PY Moore FSM is designed. Otherwise, the set Π_{TC} is formed and values of parameters R_{TC} and R_{BY} are calculated.
5. If condition (5.16) takes place, then P_{C2}Y Moore FSM is designed. In this case, if the set $\Pi_{RG} = \emptyset$, then there is no connections between the register RG and block BY.
6. If condition (5.16) is violated and in the same time $R_{BY} \neq 0$, then P_{C4}Y Moore FSM is designed. If $R_{BY} = 0$, then P_{C1}Y Moore FSM is designed.
7. If condition (5.16) is violated and $\Pi_A = \emptyset$, then P_{C3}Y Moore FSM is designed.

To design any from these models, a designer should construct the following items: the transformed structure table (to specify the block BP), the table with content of BRAM (to specify the block BY), and system $\tau = \tau(T)$ to specify the block BCT. Let us discuss some synthesis examples for the Moore FSM S_{15}, represented by its transformed table of transitions (Table 5.6). Let us point out that this table differ from the classical table of transitions, because in the column "a_m" states $a_m \in A$ are replaced by the classes $B_i \in \Pi_A$, where $a_m \in B_i$.

Let the following partition $\Pi_A = \{B_1,\ldots,B_7\}$ be constructed for the Moore FSM S_{15}, where $B_1 = \{a_1\}$, $B_2 = \{a_5,a_{12}\}$, $B_3 = \{a_{11},a_{13},a_{14}\}$, $B_4 = \{a_3,a_6\}$, $B_5 = \{a_2,a_4\}$, $B_6 = \{a_7,a_8\}$, $B_7 = \{a_9,a_{10}\}$. This partition can be represented by the following system:

$$\begin{aligned} B_1 &= A_1; & B_5 &= A_2 \vee A_4; \\ B_2 &= A_5 \vee A_{12}; & B_6 &= A_7 \vee A_8; \\ B_3 &= A_{11} \vee A_{13} \vee A_{14}; & B_7 &= A_9 \vee A_{10}. \\ B_4 &= A_3 \vee A_6; \end{aligned} \quad (5.17)$$

As follows from Table 5.6, there is the set of states $A = \{a_1,\ldots,a_{14}\}$, it determines the following values and sets: $R = 4$, $T = \{T_1,\ldots,T_4\}$, $\Phi = \{D_1,\ldots,D_4\}$. Let us use the method of optimal state encoding for the Moore FSM S_{15} (Fig. 5.12).

Table 5.6 Transformed table of transitions of Moore FSM S_{15}

B_i	a_s	X_h	h	B_i	a_s	X_h	h
B_1	a_2	x_1	1	B_5	a_3	x_3	11
	a_3	$\bar{x}_1x_2$	2		a_9	$\bar{x}_3$	12
	a_5	$\bar{x}_1\bar{x}_2$	3	B_6	a_2	x_4	13
B_2	a_1	x_3	4		a_{11}	$\bar{x}_4x_5$	14
	a_{10}	$\bar{x}_3$	5		a_{12}	$\bar{x}_4\bar{x}_5x_6$	15
B_3	a_4	x_2	6		a_{14}	$\bar{x}_4\bar{x}_5\bar{x}_6$	16
	a_7	$\bar{x}_2x_3$	7	B_7	a_1	x_3x_6	17
	a_6	$\bar{x}_2\bar{x}_3x_4$	8		a_3	$x_3\bar{x}_6$	18
	a_{13}	$\bar{x}_2\bar{x}_3\bar{x}_4$	9		a_{10}	$\bar{x}_3x_2$	19
B_4	a_8	1	10		a_{12}	$\bar{x}_3\bar{x}_2$	20

Fig. 5.12 Optimal state codes for Moore FSM S_{15}

T_1T_2 \ T_3T_4	00	01	11	10
00	a_1	a_3	a_6	*
01	a_{13}	a_{14}	*	a_5
11	a_{11}	a_9	a_8	a_{12}
10	a_7	a_2	a_4	a_{10}

Taking into account the codes from Fig. 5.12, the system (5.17) can be transformed and represented as the following:

$$\begin{aligned}
B_1 &= \bar{T}_1\bar{T}_2\bar{T}_4; & B_5 &= T_1\bar{T}_2T_4;\\
B_2 &= T_2T_3\bar{T}_4; & B_6 &= T_1\bar{T}_2\bar{T}_3\bar{T}_4 \vee T_2T_3T_4;\\
B_3 &= T_2\bar{T}_3\bar{T}_4 \vee \bar{T}_1T_2\bar{T}_3; & B_7 &= T_1T_2\bar{T}_3T_4 \vee \bar{T}_2T_3\bar{T}_4.\\
B_4 &= \bar{T}_1\bar{T}_2T_4;
\end{aligned} \tag{5.18}$$

As our analysis of system (5.18) shows, there are the following sets of classes $\Pi_{RG} = \{B_1, B_2, B_4, B_5\}$ and $\Pi_{TC} = \{B_3, B_6, B_7\}$ for the Moore FSM S_{15}. Let the system of its microoperations be the following:

$$\begin{aligned}
y_1 &= A_3 \vee A_5 \vee A_6 \vee A_{14}; & y_5 &= A_3 \vee A_6 \vee A_8 \vee A_9 \vee A_{14};\\
y_2 &= A_2 \vee A_7 \vee A_{11} \vee A_{13}; & y_6 &= A_2 \vee A_3 \vee A_4 \vee A_6 \vee A_7;\\
y_3 &= A_2 \vee A_4 \vee A_8 \vee A_9 \vee A_{10}; & y_7 &= A_2 \vee A_3 \vee A_{10} \vee A_{12} \vee A_{13}.\\
y_4 &= A_4 \vee A_5 \vee A_8 \vee A_{10} \vee A_{12};
\end{aligned} \tag{5.19}$$

It means that $N = 7$, whereas it is enough two variables ($R_{TC} = 2$) for encoding of classes $B_i \in \Pi_{TC}$. Let us use the blocks BRAM having fixed outputs from the set $T(\text{BRAM}) = \{1, 2, 4\}$ and $Q = 64$ for implementing logic circuit of the block BY. It determines the following fixed structures for a single BRAM: 64×1, 32×2, and 16×4. From formula (5.13), the value $t_0 = 4$ can be found for the Moore

FSM S_{15}, it means that $t_F = 4$. It follows from formula (5.14) that system (5.19) can be implemented using $n_{BY} = 2$ blocks BRAM with the configuration 64×1; totally these blocks have 8 outputs. It can be found that $R_{BY} = 8 - 7 = 1$, therefore, condition (5.16) is violated. Thus, it is necessary to choose the model of $P_{C4}Y$ Moore FSM. Let us point out that this model is characterized by the equality $|\tau| = |Z| = 1$.

One from the codes for classes $B_i \in \Pi_{TC}$ should be reserved to identify the classes$B_i \notin \Pi_{TC}$. Let the classes $B_i \in \Pi_{TC}$ have the following codes: $K(B_3) = 01$, $K(B_6) = 10$, and $K(B_7) = 11$, then the following equations can be obtained from system (5.18):

$$\begin{aligned} \tau_1 &= B_6 \vee B_7 = T_1 T_2 T_4 \vee T_1 \bar{T}_2 \bar{T}_4; \\ \tau_2 &= B_3 \vee B_7 = T_2 \bar{T}_3 \vee \bar{T}_2 T_3 \bar{T}_4. \end{aligned} \tag{5.20}$$

Obviously, the set τ should include those functions, which require fewer numbers of PAL macrocells for their implementation. For the Moore FSM S_{15} both functions τ_1 and τ_2 are equivalent from this point of view, because each of them includes two terms. Let us form the following set $\tau = \{\tau_1\}$. Let us point out that the code 00 is used to identify the classes $B_i \notin \Pi_{TC}$. Thus, we can use the Boolean equation τ_1 from system (5.20) to implement the logic circuit of block BTC.

For the $P_{C4}Y$ Moore FSM S_{15}, the transformed structure table includes $H_0 = 20$ rows (Table 5.7).

For $P_{C4}Y$ Moore FSM, the block BP is specified by system

$$\Phi = \Phi(T, \tau, X). \tag{5.21}$$

This system is derived from the transformed structure table. For example, the following SOP $D_4 = F_1 \vee F_2 \vee F_6 \vee F_8 \vee F_{10} \vee F_{11} \vee F_{12} \vee F_{13} \vee F_{16} \vee F_{18} = \bar{\tau}_1 \bar{\tau}_2 \bar{T}_1 \bar{T}_3 \bar{T}_4 x_1 \vee \ldots \vee \tau_1 \bar{\tau}_2 \bar{x}_4 \bar{x}_5 \bar{x}_6 \vee \tau_1 \tau_2 x_3 \bar{x}_6$ can be derived from Table 5.7.

If logic circuit of the block BY is implemented using embedded memory blocks, then system (5.19) should be represented by Table 5.8.

Let H_i be the number of transitions from the state $a_m \in B_i$, whereas M_i be the number of states in the class $B_i \in \Pi_A$. The following formula can be used to find the number of ST rows for PY Moore FSM:

$$H = \sum_{i=1}^{I} H_i M_i. \tag{5.22}$$

The logic circuit for $P_{C4}Y$ Moore FSMS_{15} is shown in Fig. 5.13.

For the Moore FSM S_{15}, the classes $B_i \in \Pi_A$ are characterized by the following values: $H_1 = 3$, $M_1 = 1$; $H_2 = 2$, $M_2 = 2$; $H_3 = 4$, $M_3 = 3$; $H_4 = 1$, $M_4 = 2$; $H_5 = 2$, $M_5 = 2$; $H_6 = H_7 = 4$, $M_6 = M_7 = 2$. Thus, the structure table of PY Moore FSM includes $H = 41$ rows, as follows from (5.22). In contrary, the transformed ST of $P_{C4}Y$ Moore FSM S_{15} (Table 5.7) includes only $H_0 = 20$ rows.

Let $H(f)$ be the number of terms for SOP of some function f, q is the number of terms for a PAL-based macrocell. Let $n(f,q)$ be the number of macrocells having q terms, necessary to implement the logic circuit for Boolean function f. This number can be determined using the following formula:

Table 5.7 Transformed structure table of $P_{C4}Y$ Moore FSM S_{15}

B_i	$K(B_i)$ $\tau_1\tau_2$	$T_1T_2T_3T_4$	a_s	$K(a_s)$	X_h	Φ_h	h
B_1	00	$0*00$	a_2	1001	x_1	D_1D_4	1
			a_3	0001	$\bar{x}_1x_2$	D_4	2
			a_5	0110	$\bar{x}_1\bar{x}_2$	D_2D_3	3
B_2	00	$*110$	a_1	0000	x_3	–	4
			a_{10}	1010	$\bar{x}_3$	D_1D_3	5
B_3	10	$****$	a_4	1011	x_2	$D_1D_3D_4$	6
			a_7	1000	$\bar{x}_2x_3$	D_1	7
			a_6	0011	$\bar{x}_2\bar{x}_3x_4$	D_3D_4	8
			a_{13}	0100	$\bar{x}_2\bar{x}_3\bar{x}_4$	D_2	9
B_4	00	$00*1$	a_8	1111	1	$D_1D_2D_3D_4$	10
B_5	00	$10*1$	a_3	0001	x_3	D_4	11
			a_9	1101	$\bar{x}_3$	$D_1D_2D_4$	12
B_6	10	$****$	a_2	1001	x_4	D_1D_4	13
			a_{11}	1100	$\bar{x}_4x_5$	D_1D_2	14
			a_{12}	1110	$\bar{x}_4\bar{x}_5x_6$	$D_1D_2D_3$	15
			a_{14}	0101	$\bar{x}_4\bar{x}_5\bar{x}_6$	D_2D_4	16
B_7	11	$****$	a_1	0000	x_3x_6	–	17
			a_3	0001	$x_3\bar{x}_6$	D_4	18
			a_{10}	1010	$\bar{x}_3x_2$	D_1D_3	19
			a_{12}	1110	$\bar{x}_3\bar{x}_2$	$D_1D_2D_3$	20

Table 5.8 Specification of block BY for $P_{C4}Y$ Moore FSM S_{15}

$T_1T_2T_3T_4$	y_1	y_2	y_3	y_4	y_5	y_6	y_7	τ_2	a_m
0 0 0 0	0	0	0	0	0	0	0	0	a_1
0 0 0 1	1	0	0	0	1	1	1	0	a_3
0 0 1 0	0	0	0	0	0	0	0	0	*
0 0 1 1	1	0	0	0	1	1	0	0	a_6
0 1 0 0	0	1	0	0	0	0	1	1	a_{13}
0 1 0 1	1	0	0	0	1	0	0	1	a_{14}
0 1 1 0	1	0	0	1	0	0	0	0	a_5
0 1 1 1	0	0	0	0	0	0	0	0	*
1 0 0 0	0	1	0	0	0	1	0	0	a_7
1 0 0 1	0	1	1	0	0	1	1	0	a_2
1 0 1 0	0	0	1	1	0	0	1	1	a_{10}
1 0 1 1	0	0	1	1	0	1	0	0	a_4
1 1 0 0	0	1	0	0	0	0	0	1	a_{11}
1 1 0 1	0	0	1	0	1	0	0	1	a_9
1 1 1 0	0	0	0	1	0	0	1	0	a_{12}
1 1 1 1	0	0	1	1	1	0	0	0	a_8

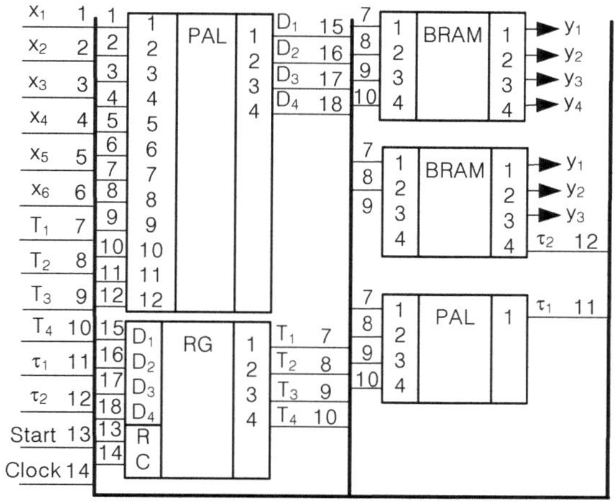

Fig. 5.13 Logic circuit of $P_{C4}Y$ Moore FSM S_{15}

$$n(f,q) = \left\lceil \frac{H(f)-q}{q-1} \right\rceil + 1. \tag{5.23}$$

Using formulae (5.14) and (5.23), it is possible to calculate the hardware amount for logic circuit of any Moore FSM model. If microoperation y_7 is eliminated from the set Y of the Moore FSM S_{15}, then we get $N = 6$, $R_{BY} = 2$. It means that all code bits for the classes $B_i \in \Pi_{TC}$ are generated by the block BY. In this case the block BTC is absent and the $P_{C2}Y$ model is used for the Moore FSM S_{15}. Transformed structure tables are identical for models $P_{C4}Y$ and $P_{C3}Y$.

If the set of microoperations for FSM S_{15} includes some additional microoperation y_8, then $N = 8$, $R_{BY} = 0$. It means that all code bits for classes $B_i \in \Pi_{TC}$ are generated by the block BTC. In this case the $P_{C1}Y$ model is used for the Moore FSM S_{15}. Logic circuit of the block BTC is implemented using equations from system (5.20). Transformed structure tables are identical for models $P_{C4}Y$ and $P_{C1}Y$.

Some comparative characteristics are shown in Table 5.9. They are used to compare logic circuit implementations for the Moore FSM S_{15} based on different models, such as PY, $P_{C1}Y$ $P_{C2}Y$, $P_{C4}Y$, when the circuit is implemented using PAL-based macrocells with $q = 3$.

Three columns are used in the table to characterize each FSM model, namely: the number of terms for FSM functions; the number of macrocells used to implement these functions; the number of levels for logic circuits. In the row BP, these numbers are summed up for input memory functions D_1 - D_4, whereas in the row BTC for τ_1 and τ_2. The row FSM contains final characteristics for the particular model. The number of levels $L(f,q)$ is determined as

$$L(f,q) = \lceil \log_q n(f,q) \rceil + 1, \tag{5.24}$$

while the total number of logic circuit levels are determined as the maximum from the numbers obtained for functions $D_r \in \Phi$ and $\tau_r \in \tau$.

Table 5.9 Specification of block BY for $P_{C4}Y$ Moore FSM S_{15}

Model	PY			$P_{C1}Y$			$P_{C2}Y$			$P_{C4}Y$			P_0Y		
D_1	17	8	3	11	5	3	11	5	3	11	5	3	18	9	3
D_2	14	7	3	8	4	2	8	4	2	8	4	2	11	5	3
D_3	19	9	3	8	4	2	8	4	2	8	4	2	13	6	3
D_4	19	9	3	10	5	3	10	5	3	10	5	3	15	7	3
τ_1	0	0	0	2	1	1	0	0	0	2	1	1	0	0	0
τ_2	0	0	0	2	1	1	0	0	0	0	0	0	0	0	0
BP	69	33	3	37	18	3	37	18	3	37	18	3	57	27	3
BTC	0	0	0	4	2	1	0	0	0	2	1	1	0	0	0
FSM	69	33	3	41	20	3	37	18	3	39	19	3	57	27	3

Analysis of Table 5.9 shows that all models have better characteristics than the model PY. All models have the same number of blocks BRAM; their performance is equal. Let us point out that the discussed approach can be applied only if the following condition takes place:

$$S \geq L(f) + R + R_{TC}. \tag{5.25}$$

In formula (5.25), the symbol S stands for the number of macrocell inputs, the symbol $L(f)$ for the number of logical conditions used in the SOP of function f. For the P_0Y Moore FSM S_{15}, where $H = 32$, the required number of macrocells is fewer, than for the PY Moore FSM S_{15}. But this number is always bigger than for any from P_CY models.

5.3 Synthesis of Moore FSM with Logical Condition Replacement

As in case of Mealy FSM, the hardware amount for Moore FSM logic circuit can be decreased if the method of logical condition replacement is used. Let us discuss the basic model of MPY Moore FSM (Fig. 5.14).

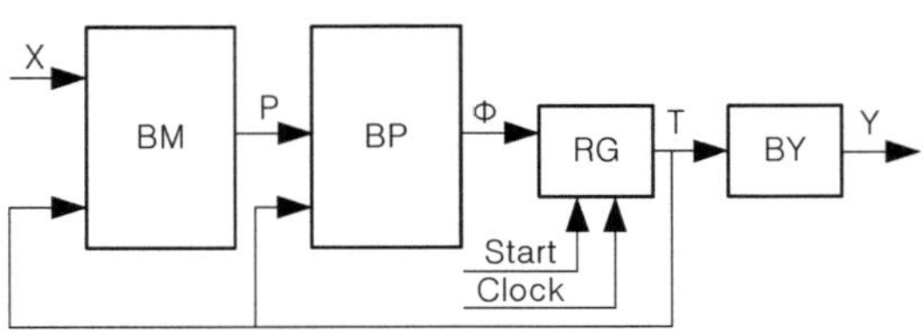

Fig. 5.14 Structural diagram of MPY Moore FSM

For MPY Moore FSM, a block BM generates system (4.3) including variables for logical condition replacement, a block BP generates functions (4.2), and a block

BY implements functions (5.2). Obviously, the logic circuit of block BM can be optimized using methods of state code transformation into either refined codes for classes of pseudoequivalent states, or codes of logical conditions. The first approach leads to MPCY Moore FSM, whereas the second method turns Moore FSM into MPLY Moore FSM. Decrease for the block BP hardware amount is possible due to application of the optimal state encoding, as well as various state code transformations (Table 5.10).

Table 5.10 Models of Moore FSM with logical condition replacement

LA	LB	LC
M MC ML	P P_{C1} P_{C4} P_0 P_{C2} P_{C3} P_C	Y

This table represents 15 different Moore FSM models with logical condition replacement. Each of them corresponds to a word $LA * LB * LC$. Let us discuss an example of the MP$_0$LY Moore FSM S_{16}, which is represented by the structure table of corresponding P$_0$Y Moore FSM (Table 5.11).

The method of optimal state encoding is used for encoding of states $a_m \in A$ of the Moore FSM S_{16}. The following partition $\Pi_A = \{B_1, \ldots, B_4\}$ can be constructed for the Moore FSM S_{16}, where $B_1 = \{a_1\}$, $B_2 = \{a_2, a_3\}$, $B_3 = \{a_4, a_5, a_6\}$, and $B_4 = \{a_7\}$. The class B_1 corresponds to the code $K(B_1) = *00$ (taking into account the unused input assignment 100), the class B_2 corresponds to the code $K(B_2) = 0*1$, the class B_3 corresponds to the code $K(B_3) = 1**$, and the class B_4 corresponds to the code $K(B_4) = 010$.

The structural diagram of MP$_0$LY Moore FSM is shown in Fig. 5.15.

Fig. 5.15 Structural diagram of MP$_0$LY Moore FSM

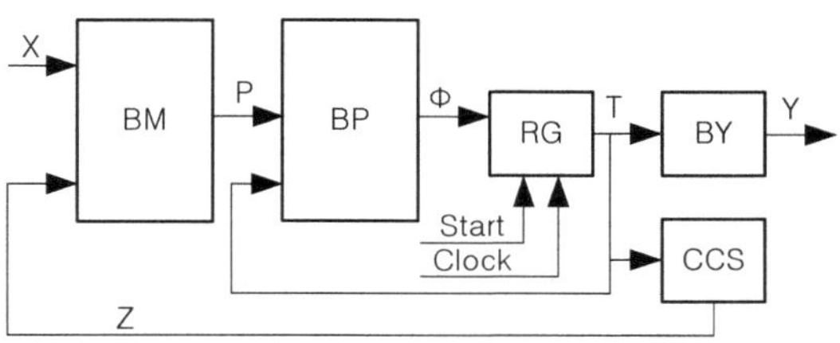

Functions of blocks for this model have been discussed already. Let us point out that a block CCS generates functions $Z(T)$. The following sets of logical conditions can be derived from Table 5.11: $X(a_1) = \{x_1\}$, $X(a_2) = \{x_2, x_3\}$, $X(a_3) = X(a_2)$, $X(a_4) = X(a_5) = X(a_6) = \{x_3, x_4\}$, and $X(a_7) = \emptyset$. It means that $G = 2$ and $P = \{p_1, p_2\}$. Let us distribute these logical conditions as it is shown in Table 5.12.

The following equation $p_2 = x_3$ can be derived from Table 5.12; it means that only logical conditions from the set $X(p_1) = \{x_1, x_2, x_3\}$ should be encoded. It is enough

Table 5.11 Structure table of P_0Y Moore FSM S_{16}

a_m	$K(a_m)$	a_s	$K(a_s)$	X_h	Φ_h	h
a_1 (–)	000	a_2	001	x_1	D_3	1
		a_3	011	$\bar{x}_1$	D_2D_3	2
a_2 (y_1y_2)	001	a_6	110	x_2	D_1D_2	3
		a_4	101	$\bar{x}_2x_3$	D_1D_3	4
		a_5	111	$\bar{x}_2\bar{x}_3$	$D_1D_2D_3$	5
a_3 (y_1y_3)	011	a_6	110	x_2	D_1D_2	6
		a_4	101	$\bar{x}_2x_3$	D_1D_3	7
		a_5	111	$\bar{x}_2\bar{x}_3$	$D_1D_2D_3$	8
a_4 (y_4y_5)	101	a_1	000	x_3	–	9
		a_7	010	$\bar{x}_3x_4$	D_2	10
		a_5	111	$\bar{x}_3\bar{x}_4$	$D_1D_2D_3$	11
a_5 (y_3y_5)	111	a_1	000	x_3	–	12
		a_7	010	$\bar{x}_3x_4$	D_2	13
		a_5	111	$\bar{x}_3\bar{x}_4$	$D_1D_2D_3$	14
a_6 (y_6y_2)	110	a_1	000	x_3	–	15
		a_7	010	$\bar{x}_3x_4$	D_2	16
		a_5	111	$\bar{x}_3\bar{x}_4$	$D_1D_2D_3$	17
a_7 (y_4y_5)	010	a_6	110	1	D_1D_2	18

Table 5.12 Logical condition replacement for P_0Y Moore FSM S_{16}

a_m	a_1	a_2	a_3	a_4	a_5	a_6	a_7
p_1	x_1	x_2	x_2	x_4	x_4	x_4	–
p_2	–	x_3	x_3	x_3	x_3	x_3	–

two variables for their encoding, thus we have the set $Z = \{z_1, z_2\}$. Let us encode the logical conditions by the following codes $K(x_1) = 00$, $K(x_2) = 01$, $K(x_4) = 10$, then the following equations represent these variables z_1 and z_2:

$$\begin{aligned} z_1 &= B_2 = \bar{T}_1 T_3; \\ z_2 &= B_3 = T_1 \end{aligned} \tag{5.26}$$

System (5.26) is used to implement the logic circuit of block CCS.

As an outcome of logical condition encoding, the following multiplexer function can be obtained:

$$p_1 = \bar{z}_1\bar{z}_2x_1 \vee \bar{z}_1z_2x_2 \vee z_1\bar{z}_2x_4. \tag{5.27}$$

System (5.27) is used to implement the logic circuit of block BM. Obviously, in our particular case this block is represented by a single multiplexer having two control inputs.

Table 5.13 Transformed structure table of MP_0LY Moore FSM S_{16}

B_i	$K(B_i)$	a_s	$K(a_s)$	P_h	Φ_h	h
B_1	$*00$	a_2	001	p_1	D_3	1
		a_3	011	$\bar{p}_1$	D_2D_3	2
B_2	$0*1$	a_6	110	p_1	D_1D_2	3
		a_4	101	$\bar{p}_1x_2$	D_1D_3	4
		a_5	111	$\bar{p}_1\bar{x}_3$	$D_1D_2D_3$	5
B_3	$1**$	a_1	000	x_3	–	6
		a_7	010	$\bar{x}_3p_1$	D_2	7
		a_5	111	$\bar{x}_3\bar{p}_1$	$D_1D_2D_3$	8
B_4	010	a_6	110	1	D_1D_2	9

System (4.2) can be derived from the transformed structure table. For the MP_0LY Moore FSM S_{16}, this table includes 9 rows (Table 5.13). Let us point out that the column P_h includes both the logical condition x_3 and the new variable p_1 determined by equation (5.27).

To specify the block BY, it is necessary to construct the table of microoperations. This step is executed in a trivial way. For the Moore FSM S_{16} block BY is specified by Table 5.14.

Table 5.14 Specification of block BY for MP_0LY Moore FSM S_{16}

$T_1T_2T_3$	y_1	y_2	y_3	y_4	y_5	y_6	m
0 0 0	0	0	0	0	0	0	1
0 0 1	1	1	0	0	0	0	2
0 1 1	1	0	1	0	0	0	3
1 0 1	0	0	1	1	0	0	4
1 1 1	0	0	1	0	1	0	5
1 1 0	0	1	0	0	0	1	6
0 1 0	0	0	1	1	0	0	7

The transformed ST is used to derive system (4.2); for example, the following equation can be found from Table 5.13: $D_1 = F_3 \vee \vee F_4 \vee F_5 \vee F_8 \vee F_9 = \bar{T}_1T_3 \vee T_1\bar{x}_3\bar{p}_1 \vee \bar{T}_1T_2\bar{T}_3$. The logic circuit of MP_0LY Moore FSM S_{16} is shown in Fig. 5.16.

This example shows that synthesis method for a three-level model of Moore FSM includes the following steps:

1. Construction of the partition Π_A of state set A by the classes of pseudoequivalent states.
2. Appropriate state encoding without logical condition replacement.
3. Logical condition replacement.
4. Construction of the transformed structure table.
5. Construction of tables specified FSM blocks.

Fig. 5.16 Logic circuit of MP_0LY Moore FSM S_{16}

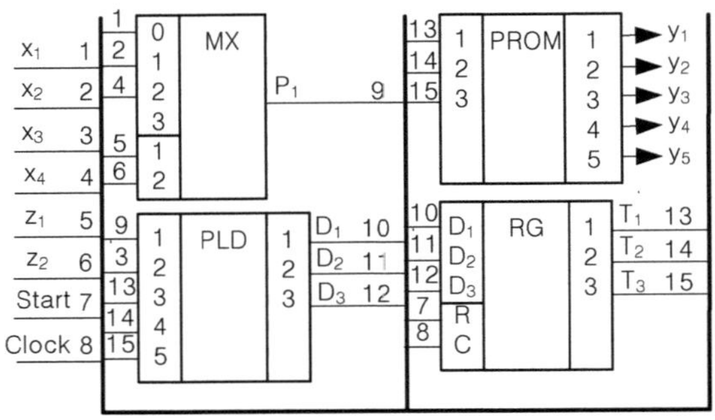

6. Construction of Boolean systems specified FSM blocks.
7. Logic circuit implementation using given logic elements.

Let us discuss an example of synthesis for the $MP_{C4}CY$ Moore FSM S_{15} represented by its transformed structure table (Table 5.6). Let us construct sets $X(B_i)$ including logical conditions $x_l \in X(a_m)$, where $a_m \in B_i$. The following sets can be found from Table 5.6: $X(B_1) = \{x_1, x_2\}$, $X(B_2) = \{x_3\}$, $X(B_3) = \{x_2, x_3, x_4\}$, $X(B_4) = \emptyset$, $X(B_5) = \{x_3\}$, $X(B_6) = \{x_4, x_5, x_6\}$, $X(B_7) = \{x_2, x_3, x_6\}$. Let us construct the table for logical condition replacement (Table 5.15), taking into account that $G = 3$.

Table 5.15 Specification of block BY for MP_0LY Moore FSM S_{16}

a_m	B_1	B_2	B_3	B_4	B_5	B_6	B_7
p_1	x_1	–	x_4	–	–	x_4	x_5
p_2	x_2	–	x_2	–	–	x_6	x_6
p_3	–	x_3	x_3	–	x_3	x_5	x_3

The structural diagram for model of $MP_{C4}LY$ Moore FSM includes four logic blocks and the register RG (Fig. 5.17).

Fig. 5.17 Structural diagram of $MP_{C4}LY$ Moore FSM

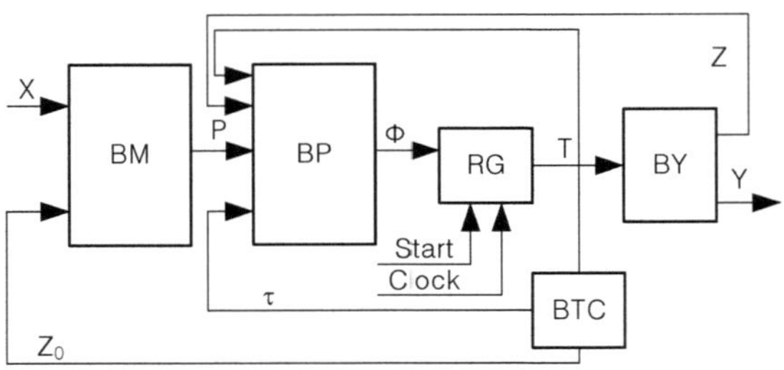

In this model, the block BCT generates functions $\tau(T)$, as well as functions $Z_0(T)$ used as control inputs for the multiplexers of block BM. The following sets of logical conditions can be derived from Table 5.15: $X(p_1) = \{x_1, x_2, x_4\}$, $X(p_2) = \{x_2, x_6\}$, $X(p_3) = \{x_3, x_5\}$. The logical conditions $x_l \in X(p_1)$ can be encoded using two variables (z_1 and z_2), whereas only single variable can be used for other logical conditions (z_3 for $x_l \in X(p_2)$ and z_4 for $x_l \in X(p_3)$). Let us encode the logical conditions in the following manner: $K(x_1) = 00$, $K(x_2) = 01$, $K(x_4) = 10$, $K(x_2) = 0$, $K(x_6) = 1$, $K(x_3) = 0$, $K(x_5) = 1$. In this case, the block BY is represented by the following system of Boolean functions:

$$\begin{aligned} p_1 &= \bar{z}_1\bar{z}_2x_1 \vee \bar{z}_1z_2x_2 \vee z_1\bar{z}_2x_4; \\ p_2 &= \bar{z}_3x_2 \vee z_3x_6; \\ p_3 &= \bar{z}_4x_3 \vee z_4x_5. \end{aligned} \tag{5.28}$$

To design the logic circuit of block BP, the transformed ST should be transformed once more. After this transformation, logical conditions $x_l \in X$ are replaced by variables $p_g \in P$ (as it follows from Table 5.15), and the column X_h is replaced by the column P_h (Table 5.16).

This table is used to construct system (5.12), which is the base for design of the logic circuit of block BP. The following Boolean equation $D_4 = = F_1 \vee F_2 \vee F_6 \vee F_8 \vee F_{10} \vee F_{11} \vee F_{12} \vee F_{13} \vee F_{16} \vee F_{18} = \bar{\tau}_1\bar{\tau}_2\bar{T}_1\bar{T}_3\bar{T}_4p_1 \vee \ldots \vee \tau_1\tau_2p_3\bar{p}_2$ can be derived from Table 5.16.

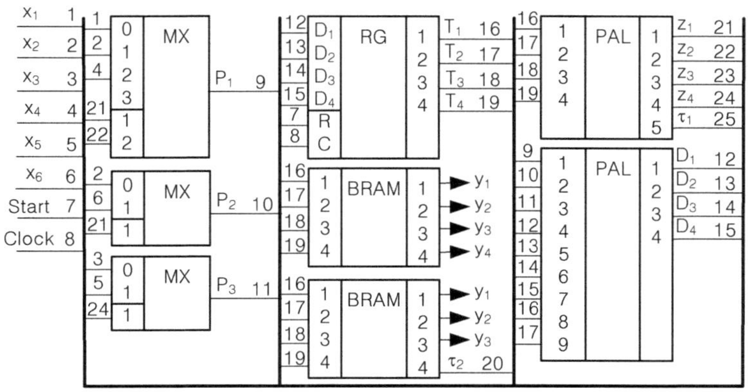

Fig. 5.18 Logic circuit of $\mathrm{MP_{C4}LY}$ Moore FSM S_{15}

Systems $\tau(T)$ and $Z_0(T)$should be constructed for implementation of the logic circuit of block BCT. In the discussed example, the system $\tau(T)$ has been derived already, whereas the system $Z_0(T)$ is the following one:

Table 5.16 Transformed structure table of $MP_{C4}CY$ Moore FSM S_{15}

B_i	$K(B_i)$ $\tau_1\tau_2$	$T_1T_2T_3T_4$	a_s	$K(a_s)$	P_h	Φ_h	h
B_1	00	0∗00	a_2	1001	p_1	D_1D_4	1
			a_3	0001	$\bar{p}_1p_2$	D_4	2
			a_5	0110	$\bar{p}_1\bar{p}_2$	D_2D_3	3
B_2	00	∗110	a_1	0000	p_3	–	4
			a_{10}	1010	$\bar{p}_3$	D_1D_3	5
B_3	10	∗∗∗∗	a_4	1011	p_2	$D_1D_3D_4$	6
			a_7	1000	$\bar{p}_2p_3$	D_1	7
			a_6	0011	$\bar{p}_2\bar{p}_3p_1$	D_3D_4	8
			a_{13}	0100	$\bar{p}_1\bar{p}_2\bar{p}_3$	D_2	9
B_4	00	00∗1	a_8	1111	1	$D_1D_2D_3D_4$	10
B_5	00	10∗1	a_3	0001	p_3	D_4	11
			a_9	1101	$\bar{p}_3$	$D_1D_2D_4$	12
B_6	10	∗∗∗∗	a_2	1001	p_1	D_1D_4	13
			a_{11}	1100	$\bar{p}_1p_3$	D_1D_2	14
			a_{12}	1110	$\bar{p}_1p_2\bar{p}_3$	$D_1D_2D_3$	15
			a_{14}	0101	$\bar{p}_1\bar{p}_2\bar{p}_3$	D_2D_4	16
B_7	11	∗∗∗∗	a_1	0000	p_2p_3	–	17
			a_3	0001	$p_3\bar{p}_2$	D_4	18
			a_{10}	1010	$\bar{p}_3p_1$	D_1D_3	19
			a_{12}	1110	$\bar{p}_3\bar{p}_1$	$D_1D_2D_3$	20

$$
\begin{aligned}
z_1 &= B_3 \vee B_6 = A_{11} \vee A_{13} \vee A_{14} \vee A_7 \vee A_8;\\
z_2 &= B_7 = A_9 \vee A_{10};\\
z_3 &= B_6 \vee B_7 = A_7 \vee A_8 \vee A_9 \vee A_{10};\\
z_4 &= B_6 = A_7 \vee A_8.
\end{aligned} \tag{5.29}
$$

Using the state codes from Karnaugh map (Fig. 5.12), we can get the final formulae for functions (5.29). For example, the following final equation $z_3 = T_1T_2T_3 \vee T_1\bar{T}_2\bar{T}_4$ can be obtained. To specify the block BY, Table 5.8 can be used for the $MP_{C4}CY$ Moore FSM S_{15}.The logic circuit of $MP_{C4}CY$ Moore FSM S_{15} is shown in Fig. 5.18.

These examples show the ways for design of logic circuits for multilevel Moore FSM models, represented by Table 5.10.

References

1. Barkalov, A., Titarenko, L.: Synthesis of Operational and Control Automata. UNITECH, Donetsk (2009)
2. Barkalov, A., Titarenko, L.: Moore fsm synthesis with coding of compatible microoperations fields. In: Proc. of IEEE East-West Design & Test Symposium - EWDTS 2007, Yerevan, Armenia, pp. 644–646. Kharkov National University of Radioelectronics (2007)

3. Barkalov, A., Titarenko, L., Chmielewski, S.: Optimization of logic circuit of moore fsm on CPLD. Pomiary Automatyka Kontrola 53(5), 18–20 (2007)
4. Barkalov, A., Titarenko, L., Chmielewski, S.: Optimization of moore fsm on cpld. International Journal of Applied Mathematics and Computer Science 17(4), 565–675 (2007)
5. Barkalov, A., Titarenko, L., Chmielewski, S.: Optimization of moore fsm on cpld. In: Proceedings of the Sixth Inter. conf. CAD DD 2007, Minsk, vol. 2, pp. 39–45 (2007)
6. Barkalov, A., Titarenko, L., Chmielewski, S.: Optimization of moore fsm on system-on chip. In: Proc. of IEEE East-West Design & Test Symposium – EWDTS 2007 (2007)
7. Barkalov, A., Titarenko, L., Chmielewski, S.: Decrease of hardware amount in logic circuit of moore FSM. Przegląd Telekomunikacyjny i Wiadomości Telokomunikacyjne (6), 750–752 (2008)
8. Barkalov, A., Titarenko, L., Chmielewski, S.: Optimization of moore control unit with refined state encoding. In: Proc. of the 15th Inter. Conf. MIXDES 2008, Poznań, Poland, pp. 417–420. Departament of Microeletronics and Computer Science, Technical University of Łódz (2008)
9. Barkalov, A., Titarenko, L., Chmielewski, S.: Optimization of moore fsm on system-on-chip using pal technology. In: Proc. of the International Conference TCSET 2008, Lviv-Slavsko, Ukraina, pp. 314–317. Ministry of Education and Science of Ukraine, Lviv Polytechnic National University, Publishing House of Lviv Polytechnic, Lviv (2008)
10. Barkalov, A., Wêgrzyn, A., Barkalov Jr., A.: Synthesis of control units with transformation of the codes of objects. In: Proc. of the IXth Inter. Conf. CADSM 2007, Lviv - Polyana, Ukraine, pp. 260–261. Lviv Polytechnic National University, Publishing House of Lviv Polytechnic National University, Lviv (2007)
11. Barkalov, A.A.: Principles of optimization of logic circuit of Moore FSM. Cybernetics and System Analysis (1), 65–72 (1998) (in Russian)
12. De Micheli, G.: Synthesis and Optimization of Digital Circuits. McGraw-Hill, New York (1994)

Chapter 6
FSM Synthesis with Transformation of GSA

Abstract. The chapter is devoted to design methods based on transformation of an interpreted graph-scheme of algorithm. The methods of decrease for the number of logical conditions per FSM state are discussed. In extreme case, all FSM transitions depend on single logical condition; it allows use of embedded memory blocks for implementation of FSM input memory functions. In this case all FSM blocks are implemented using standard library cells (not just macrocells of a particular FPLD chip). The second part of the chapter is devoted to hardware optimization for block of microoperations, based on verticalization of an interpreted GSA. It permits to decrease the number of decoders (up to 1) and bit capacity of microinstruction word, but this optimization is connected with increase for the number of cycles required for a control algorithm interpretation. At last, the models based on joint application of these methods are discussed.

6.1 Optimization of Logical Condition Replacement Block

As it was discussed before, the hardware amount of block BP logic circuit can be decreased due to replacement of logical conditions $x_l \in X$ by some additional variables $p_g \in P$, where $|P| = G$. The value of parameter G is determined by characteristics of a GSA $\Gamma$to be interpreted. This value can be diminished due to introducing some additional operator vertices into the GSA Γ [8,9].

It can be found that $G = 3$ for a subgraph Γ_0 (Fig. 6.1a). If the vertex b_2 is brought in (Fig. 6.1b), then the value of G is decreased up to 2, whereas bringing in the vertex b_3 (Fig. 6.1c) decreases G up to 1.

Thus, the introduction of additional operator vertices decreases the value of parameter G (the limit is equal to 1), but it increases the number of FSM states. Besides, introduction of additional operator vertices results in increase for the number of FSM models. Different Mealy FSM models are shown in Table 6.1.

In this table, the lower index g ($g = 1, \ldots, G$) stands for the number of variables used for the logical condition replacement. This table describes Mealy FSM models with different number of levels (1, 2, 3, and 4). All these models can be used to interpret the same GSA.

A. Barkalov and L. Titarenko: Logic Synthesis for FSM-Based Control Units, LNEE 53, pp. 129–154.
springerlink.com

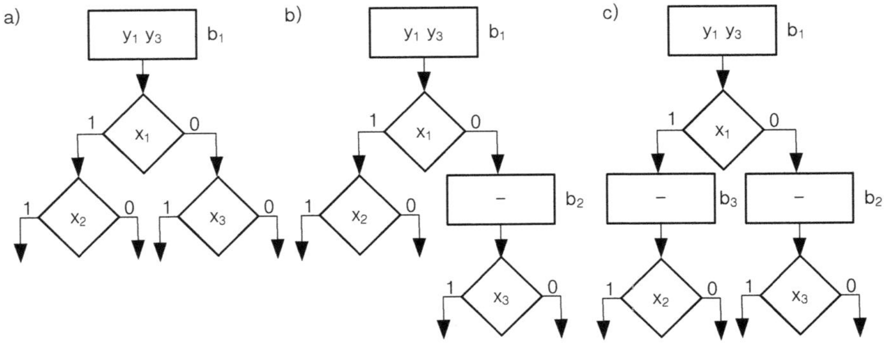

Fig. 6.1 Transformation of subgraph Γ_0

Table 6.1 Multilevel models of Mealy FSM

LA			LB	LC	LD
M_1	M_1C	M_1L	P	F	DY
	...			DY	
M_G	M_GC	M_GL			

Let the symbol NL_i stand for the number of models having i levels $(i = 1,\ldots,4)$. Table 6.1 determines the following numbers of different models:

1. $NL_1 = 1$ (There is only one single-level model, namely, P Mealy FSM).
2. $NL_2 = 3G + 3$ (The first term of equation corresponds to models with logical condition replacement, whereas the second determines PF-, PD- and PY Mealy FSMs).
3. $NL_3 = 3G * 3 + 2$ (The first term of equation corresponds to models determined by words $LA * LB * LC$, whereas the second determines the models described by words $LB * LC * LD$).
4. $NL_4 = 3G * 2$ (These Mealy FSM models are determined by the following words: $M_gPFD, M_gPFY, \ldots, M_gLPFY$).

Therefore, Table 6.1 represents totally $18G + 6$ different models for the same GSA. For an FSM with average complexity [4], where $G = 6$, there are 114 different models. Let us point out that each model can differ in its logic circuit hardware amount from other models. This statement is true for resulted FSM performance too.

Different Moore FSM models are shown in Table 6.2. This table determines the following numbers of different models:

1. $NL_1 = 7$ (This model corresponds to the word LB).
2. $NL_2 = 3G * 7 + 7$ (These models correspond either to words $LA * LB$, or $LB * LC$).
3. $NL_3 = 3G * 7$.

Table 6.2 Multilevel models of Moore FSM

LA			LB			LC
M_1	M_1C	M_1L	P	P_{C1}	P_{C4}	Y
	$\dots$		P_0	P_{C2}	P_{C3}	
M_G	M_GC	M_GL		P_C		

So, Table 6.2 represents $42G+14$ different models of Moore FSM. For FSMs with average complexity, we can get up to 266 different models.

Let us discuss an example of logic synthesis for the M_2PLFY Mealy FSM S_{19}, specified by the GSA Γ_3 (Fig. 6.2).

The FSM to be synthesized is represented by its model shown in Fig. 6.3. In this model, a block BM implements the logical condition replacement and generates the functions

$$\begin{aligned} p_1 &= p_1(Z_0, X); \\ p_2 &= p_2(Z_0, X). \end{aligned} \tag{6.1}$$

A block BP generates functions used for the encoding of transformed structure table rows, namely, functions

$$Z = Z(p_1, p_2, T). \tag{6.2}$$

A block BF generates variables used for encoding of the collections of microoperations, as well as input memory functions:

$$\begin{aligned} Z_1 &= Z_1(Z); \\ \Phi &= \Phi(Z). \end{aligned} \tag{6.3}$$

A block BY generates microoperations

$$Y = Y(Z_1). \tag{6.4}$$

At last, a code transformer CCS generates variables used as control inputs of multiplexers from the block BM:

$$Z_0 = Z_0(T). \tag{6.5}$$

Design of this FSM is reduced to construction of systems (6.1) – (6.5) and implementation of logic circuits for corresponding blocks using some macrocells.

As follows from Fig. 6.2, the states a_2 and a_5 are characterized by $|X(a_m)| > 2$. Thus, the corresponding subgraphs of GSA Γ_3 should be transformed. The transformation is reduced to introduction of three additional operator vertices into the GSA Γ_3 (Fig. 6.4).

In the FSM S_{17}, the number of states is increased from 5 to 7 after the transformation of the GSA Γ_3. But the transformation permits to get the value $G = 2$, that corresponds to the member M_2 in the formula M_2PLFY. In both cases, it is enough $R = 3$ variables $T_r \in T$ for the state encoding. Let us encode the states in a trivial

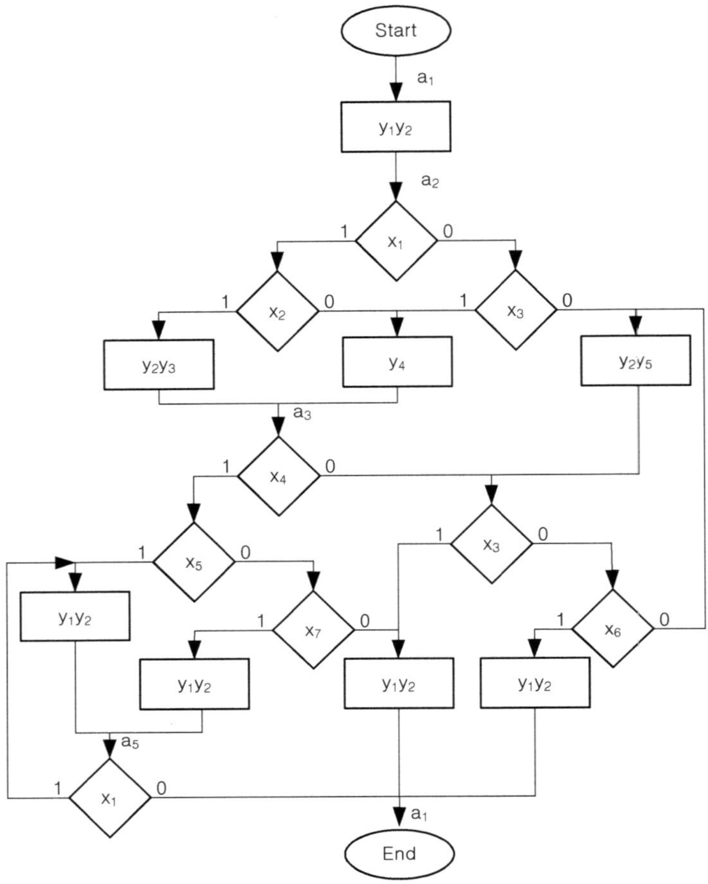

Fig. 6.2 Initial graph-scheme of algorithm Γ_3

Fig. 6.3 Structural diagram of M_2PLFY Mealy FSM

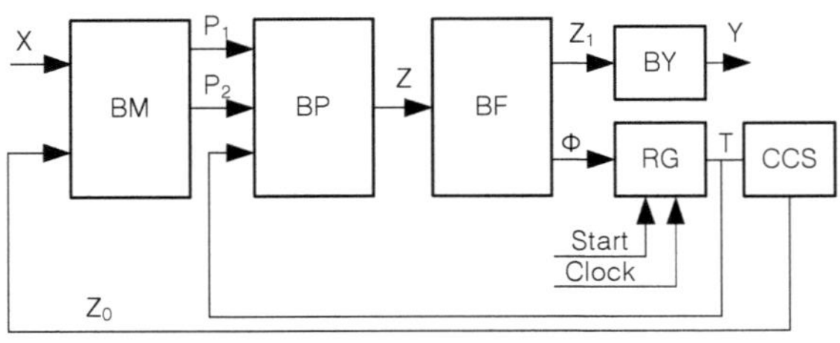

way, namely:$K(a_1) = 000, \ldots, K(a_7) = 110$. Using these codes, let us construct the structure table of Mealy FSM S_{17} (Table 6.3).

As it was mentioned above, it is enough $G = 2$ variables to replace the logical conditions, thus, there is a set $P = \{p_1, p_2\}$. Let us construct the table of logical condition replacement for FSM S_{17} (Table 6.4). As it can be found from the table, it is enough two variables $z_r \in Z_0$ to replace the logical conditions $x_l \in X(P_g)$. It is true, because there are only four pairs of logical conditions into Table 6.4, namely: $\langle x_1, x_2\rangle$, $\langle x_4, x_5\rangle$, $\langle x_3, x_6\rangle$, and $\langle \emptyset, x_7\rangle$. If the pair $\langle x_1, x_2\rangle$ corresponds to the code $K(x_1) = K(x_2) = 00$, then the equalities $p_1 = x_1$, $p_2 = x_2$ take place for the state a_5. Because the logical condition x_2 does not affect transitions from the state a_5, then a real value of this logical condition is not important. Therefore, the following set $Z_0 = \{z_1, z_2\}$ can be used in this particular case. Let us encode the remained pairs by the following codes:$K(x_4) = K(x_5) = 01$, $K(x_3) = K(x_4) = 10$, and $K(x_7) = 11$. As a result, the following system of equations can be constructed:

$$\begin{aligned} p_1 &= \bar{z}_1\bar{z}_2x_1 \vee \bar{z}_1z_2x_4 \vee z_1\bar{z}_2x_3; \\ p_2 &= \bar{z}_1\bar{z}_2x_2 \vee \bar{z}_1z_2x_5 \vee z_1\bar{z}_2x_4 \vee z_1z_2x_7. \end{aligned} \tag{6.6}$$

For the FSM S_{17}, the structure table includes $H = 16$ rows, therefore, it is enough $R_F = 4$ variables to encode the rows of this table. It gives the set of variables $Z = \{z_3, \ldots, z_6\}$. Let us encode the rows of Table 6.3 in the following way:

Table 6.3 Structure table of Mealy FSM S_{17}

a_m	$K(a_m)$	a_s	$K(a_s)$	X_h	Y_h	Φ_h	h
a_1	000	a_2	001	1	y_1y_2	D_3	1
a_2	001	a_3	010	x_1x_2	y_2y_3	D_2	2
		a_3	010	$x_1\bar{x}_2$	y_4	D_2	3
		a_6	101	$\bar{x}_1$	–	D_1D_3	4
a_3	010	a_5	100	x_4x_5	y_1y_2	D_1	5
		a_7	110	$x_4\bar{x}_5$	–	D_1D_2	6
		a_4	011	$\bar{x}_4$	–	D_2D_3	7
a_4	011	a_1	000	x_3	y_2y_5	–	8
		a_1	000	$\bar{x}_3x_6$	y_4	–	9
		a_4	011	$\bar{x}_3\bar{x}_6$	y_2y_5	D_2D_3	10
a_5	100	a_5	100	x_1	–	–	11
		a_1	000	$\bar{x}_1$	–	–	12
a_6	101	a_3	010	x_3	y_4	D_2	13
		a_4	011	$\bar{x}_3$	y_2y_5	D_2D_3	14
a_7	110	a_5	100	x_7	y_4y_6	D_1	15
		a_1	000	$\bar{x}_7$	y_2y_5	–	16

Table 6.4 Table of logical condition replacement for FSM S_{17}

a_m	a_1	a_2	a_3	a_4	a_5	a_6	a_7
p_1	–	x_1	x_4	x_3	x_1	x_3	–
p_2	–	x_2	x_5	x_6	–	–	x_7

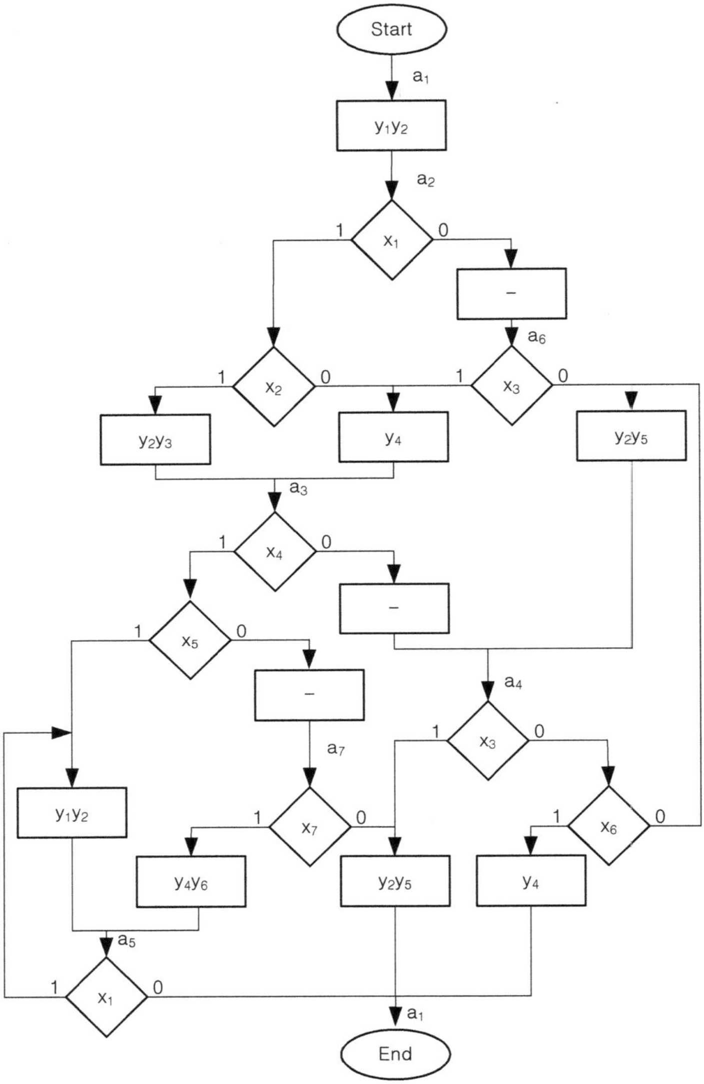

Fig. 6.4 Transformed GSA Γ_3

$K(F_1) = 0000, \ldots, K(F_{16}) = 1111$. Let us construct the transformed ST for Mealy FSM S_{17} (Table 6.5). The transformation is reduced to replacement of the column X_h by the column P_h, replacement of columns Y_h and Φ_h by the column Z_h, and deletion the columns a_s and $K(a_s)$ from the initial ST. This table is used to derive the equations of system (6.2), for example, the Boolean equation $z_3 = F_9 \vee \ldots \vee F_{16} = \bar{T}_1 T_2 T_3 \bar{p}_1 \vee T_1 \bar{T}_2 \vee T_1 \bar{T}_3$ can be derived from Table 6.5 (this equation is written after some minimization of initial expression extracted from the table).

For the FSM S_{17}, the structure table includes $T_0 = 6$ collections of microoperations, namely: $Y_1 = \emptyset$, $Y_2 = \{y_1, y_2\}$, $Y_3 = \{y_2, y_3\}$, $Y_4 = \{y_4\}$, $Y_5 = \{y_2, y_5\}$, $Y_6 = \{y_6\}$. These collections can be encoded using $R_0 = 3$ variables; it gives the set $Z_1 = \{z_7, z_8, z_9\}$. Let us encode the collections in the following way: $K(Y_1) = 000$, ..., $K(Y_6) = 101$. To design the logic circuit for block BY, it is enough to construct the Karnaugh map (Fig. 6.5). This map differs from the classical one, because each its cell can include more than one microoperation.

Fig. 6.5 Codes for collections of microoperations of Mealy FSM S_{17}

z_7 \ z_8z_9	00	01	11	10
0	-	y_1y_2	y_4	y_2y_3
1	y_2y_5	y_6	*	*

Table 6.5 Transformed structure table of Mealy FSM S_{17}

a_m	$K(a_m)$	P_h	Z_h	h
a_1	000	1	–	1
a_2	001	p_1p_2	z_6	2
		$p_1\bar{p}_2$	z_5	3
		$\bar{p}_1$	z_5z_6	4
a_3	010	p_1p_2	z_4	5
		$p_1\bar{p}_2$	z_4z_6	6
		$\bar{p}_1$	z_4z_5	7
a_4	011	p_1	$z_4z_5z_6$	8
		$\bar{p}_1p_2$	z_3	9
		$\bar{p}_1\bar{p}_2$	z_3z_6	10
a_5	100	p_1	z_3z_5	11
		$\bar{p}_1$	$z_3z_5z_6$	12
a_6	101	p_1	z_3z_4	13
		$\bar{p}_1$	$z_3z_4z_6$	14
a_7	110	p_2	$z_3z_4z_5$	15
		$\bar{p}_2$	$z_3z_4z_5z_6$	16

If embedded memory blocks BRAM are used to implement the logic circuit of block BY, then an input assignment $\langle z_7 z_8 z_9 \rangle$ is treated as an address of corresponding memory word. Word contents are determined by contents of corresponding cells of the Karnaugh map. If some macrocells are used to implement the logic circuit of block BY, then collections $Y_t \subseteq Y$ should be encoded in the optimal way and minimization of functions $y_n \in Y$ should be executed. Different types of macrocells determine different approaches for the collection encoding. Obviously, all these approaches have the same aim, which is decrease for the number of macrocells in logic circuit of the block BY. For the Mealy FSM S_{17}, the following equations, for example, can be derived from Fig. 6.5: $y_1 = \bar{z}_7\bar{z}_8z_9$, $y_2 = z_7\bar{z}_9 \vee z_8\bar{z}_9 \vee \bar{z}_7\bar{z}_8z_9$, and so on. If the collections are recoded as it shown in Fig. 6.6, then each microoperation $y_n \in Y$ is represented by the SOP with only single term, namely:

$$
\begin{aligned}
y_1 &= \bar{z}_7\bar{z}_8z_9; & y_4 &= z_1\bar{z}_9; \\
y_2 &= z_9; & y_5 &= z_1z_9; \\
y_3 &= z_8z_9; & y_6 &= z_8\bar{z}_9.
\end{aligned}
\tag{6.7}
$$

Fig. 6.6 Optimal codes for collections of microoperations of Mealy FSM S_{17}

z_7 \ z_8z_9	00	01	11	10
0	-	y_1y_2	y_2y_3	y_6
1	y_4	y_2y_5	*	*

To implement the logic circuit of block BF, it is necessary to construct corresponding table with columns Z, Z_1, Φ_h, h. In case of the Mealy FSM S_{17}, this table includes 16 rows (Table 6.6). Obviously, the number of rows is equal for both the initial structure table and the table specified the block BF.

In Table 6.6, the column Z includes row codes $K(F_h)$, determined by vectors $\langle z_3z_4z_5z_6\rangle$; the column Φ_h is taken from the structure table (in our example, from

Table 6.6 Specification for block BF of Mealy FSM S_{17}

Z	Z_1	Φ_h	h
0000	z_9	D_3	1
0001	z_8	D_2	2
0010	z_8z_9	D_2	3
0011	–	D_1D_3	4
0100	z_9	D_1	5
0101	–	D_1D_2	6
0110	–	D_2D_3	7
0111	z_7	–	8
1000	z_8z_9	–	9
1001	z_7	D_2D_3	10
1010	z_9	D_1	11
1011	–	–	12
1100	z_8z_9	D_2	13
1101	z_7	D_2D_3	14
1110	z_8z_9	D_1	15
1111	z_7	–	16

Table 6.7 Specification of block CCS for Mealy FSM S_{17}

a_m	$K(a_m)$	x_l	$K(x_l)$	Z_0	m
a_1	000	–	–	–	1
a_2	001	x_1x_2	00	–	2
a_3	010	x_4x_5	01	z_2	3
a_4	011	x_3x_6	10	z_1	4
a_5	100	x_1	00	–	5
a_6	101	x_3	10	z_1	6
a_7	110	x_7	11	z_1z_2	7

Table 6.3). The following procedure is executed to fill the column Z_1: 1) to take a collection of microoperations Y_t from the row h of initial ST; 2) to find the code of collection of microoperations $K(Y_t)$; 3) to write in the row h of table for block BF the variables from vector $\langle z_7z_8z_9\rangle$, which are equal to 1 in the code $K(Y_t)$. For the Mealy FSM S_{17}, codes $K(Y_t)$ are taken from Fig. 6.5.

If the logic circuit of block BF is implemented using macrocells, then functions (6.3) and (6.4) are derived from the table specified the block BF. The following SOP $z_7 = F_8 \vee F_{10} \vee F_{14} \vee F_{16} = z_4z_5z_6 \vee z_3\bar{z}_5z_6$ can be derived, for example, from Table 6.6. Obviously, structure table rows can be encoded in such a way, that the logic circuit of block BF includes minimal number of corresponding macrocells.

To form system (6.6), a table specified the block CCS should be constructed. For the Mealy FSM S_{17}, this block is specified by Table 6.7.

In common case, this table includes columns a_m, $K(a_m)$, $X(p_1)$, $K(x_e)_1, \ldots,$ $X(p_G)$, $K(x_e)_G$, Z_0, m. The purposes of these columns are clear. If the logic circuit of block CCS is implemented using embedded memory blocks, then the code $K(a_m)$ is considered as the word address. This word contains data Z_0 and corresponds to the row m of the table. If the logic circuit of block CCS is implemented using some macrocells, then system (6.6) is represented as a SOP. For example, the following SOP $z_1 = A_4 \vee A_6 \vee A_7$ can be derived for block CCS from Table 6.7. Taking into account the unused input assignment 111, this equation can be transformed into $z_1 = T_2T_3 \vee T_1T_3 \vee T_1T_2$. Obviously, states $a_m \in A$ can be encoded in the optimal way to minimize system (6.6). For example, the encoding shown in Fig. 6.7 results in the following equations: $z_1 = T_3$, $z_2 = T_1T_2$.

These tables and equations (6.1) – (6.6) allow implementation of the logic circuit for M$_2$PLFY Mealy FSM S_{17} (Fig. 6.8).

T_1 \ T_2T_3	00	01	11	10
0	a_1	a_4	a_6	a_2
1	a_5	*	a_7	a_3

Fig. 6.7 Optimal state codes for Mealy FSM S_{17}

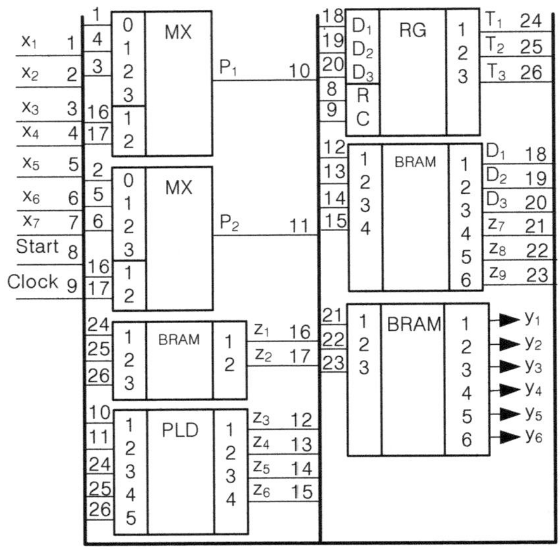

Fig. 6.8 Logic circuit of M_2PLFY Mealy FSM S_{17}

In Fig. 6.8, the logic circuits for blocks BF, BY, CCS are implemented using embedded memory blocks; the logic circuit of block BM is implemented using multiplexers; the logic circuit of block BP is implemented using macrocells. This example leads to the following general conclusion: *the structural decomposition of FSM logic circuit increases the number of regular functions*. It allows use of standard memory blocks for implementation of systems of Boolean functions represented a particular FSM. In turn, it simplifies the design process if FSM logic circuit is implemented using FPLD chips.

Next, the discussed example permits to formulate the general approach for design of FSM multilevel logic circuits, where FSM models are specified by Tables 6.1 and 6.2. To design a particular FSM circuit it is necessary:

1. To construct an FSM model, as well as general Boolean functions corresponding to each block of the model.
2. To transform an initial GSA (if necessary).
3. To construct tables corresponding to each block of the model. If some block is implemented using macrocells, then input variables of the block should be encoded in the optimal way. The criterion of optimality depends on the type of macrocells in use.
4. To implement the general logic circuit as a composition of the circuits for each block of FSM model to be designed.

6.2 Optimization for Block for Decoding of Microoperations

Because decoders are standard library elements, their use accelerates the process of a control unit design. Decoders are used for generation of microoperations in

PD Mealy FSM (Fig. 4.10). As we know, to design the PD Mealy FSM the set of microoperations Y should be divided by the classes of compatible microoperations $(Y^1,\ldots,Y^K)$. It means that a partition Π_Y should be found, such that each element of Π_Y corresponds to one class of compatible microoperations. Microoperations of each class are encoded using codes $K(y_n)$. The number of bits R_k for each microoperation encoding is determined by (4.12). The sum of these numbers gives the total number of bits R_D, required for microoperation encoding and determined by (4.13). Each class $Y^k \in \Pi_Y$ corresponds to decoder DC_k, the totality of these decoders forms the block BD (Fig. 4.10).

If all microoperations $y_n \in Y$ are compatible, then such a GSA Γ is named a *vertical GSA* (VGSA) [5]. The following condition

$$N_t \leq 1 \tag{6.8}$$

takes place for VGSA, where N_t is the number of microoperations in collection $Y_t \subseteq Y(\mathrm{t}=1,\ldots,\mathrm{T}_0)$. If this condition takes place, then PD Mealy FSM turns into PD_1 Mealy FSM, in which the block BD includes only one decoder having N outputs. The number of classes K in the partition Π_Y can be varied due to application of the procedure of *verticalization* to the initial GSA [5, 7]. The verticalization produces the family of PD Mealy FSMs with different amount of decoders implementing the block BD, namely PD_1-, $\mathrm{PD}_2-,\ldots$, and PD_k Mealy FSMs. The value of parameter K depends on characteristics of a GSA to be interpreted, namely on distribution of microoperations among GSA operator vertices. Let us discuss some problems connected with synthesis of PD Mealy FSM [1, 5]. Let us use for this discussion the fragment of GSA Γ_0 (Fig. 6.9a). Obviously, condition (6.8) is violated for this subgraph.

The sense of verticalization consists in presentation of each vertex $b_t \in B_1$ as a sequence of operator vertices $\langle b_t^1, b_t^2,\ldots\rangle$, including up to N_t elements. Two different approaches are possible for state marking:

1. The standard marking shown in Fig. 6.9b.
2. Saving of initial marks for FSM states (Fig. 6.9c, d).

The marked GSA shown in Fig. 6.9d corresponds to PD_2 Mealy FSM. One from its vertices includes an additional microoperation y_0, but we discuss it a bit latter. The drawback of PD_k FSM is decrease for digital system performance due to increase for the number of cycles needed to accomplish an algorithm to be interpreted. Besides, the successive execution of microoperations written in the same operator vertex is not always possible because of the data dependence among the microoperations $y_n \in Y_t$. Let, for example, microoperations written in the vertex b_5 (Fig. 6.9a) stand for the following actions: $y_1\#A := B+C; y_2\#B := A+B; y_3\#C := A+B$. Obviously, outcomes are different for parallel and successive modes of these microoperations execution. Let operands have the following values:$A=5$, $B=6$, and $C=8$. The outcomes of parallel execution are the values $A=14$, $B=11$, $C=11$. But if the microoperations are executed as a sequence $\langle y_1, y_2, y_3\rangle$, then the incorrect results $A=14$, $B=14+6=20$, and $C=14+20=34$ are obtained.

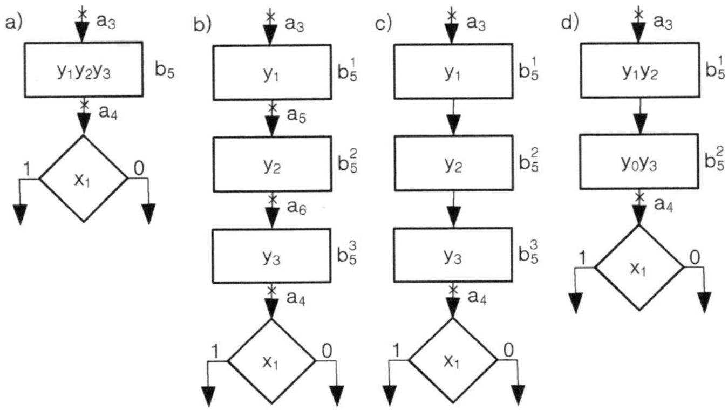

Fig. 6.9 Fragment of GSA Γ_0 before (a) and after (b, c, d) verticalization

To eliminate these drawbacks, a special variable y_0 is introduced, as well as a register RY. This variable controls the data-path synchronization, and the register is used to transform the sequence of microoperation into a parallel code corresponding to initial collection of microoperations. Let the symbol $T(y_n)$ stand for a flip-flop corresponding to microoperation$y_n \in Y$. The flip-flop $T(y_n)$ is set up, iff $y_n = 1$. If all microoperations $y_n \in Y_t$ are written into the register RY, then the variable y_0 is generated and the system data-path executes the collection $Y_t \subseteq Y(\mathrm{t} = 1, \ldots, \mathrm{T}_0)$. This mode can be organized quite easily, because all modern FPLD have a property of independent synchronization for flip-flops, which are registered outputs of macrocells [2, 10].

If states $a_m \in A$ are marked in the standard way [3,4], then the verticalization of GSA results in PD_V^1 Mealy FSM (Fig. 6.10).

In Fig. 6.10, outputs of the register RY are denoted as Y_R, where the symbol Y_R stands for a set of registered microoperations. Let us discuss an example of synthesis for the PD_V^1 Mealy FSM S_{18} specified by an initial GSA Γ_4 (Fig. 6.11).

1. **Verticalization of initial GSA.** This step is reduced to the successive splitting of operator vertices, if condition (6.8) for them is violated. The execution of verticalization causes no difficulties. The vertical GSA $V(\Gamma_4)$ with a standard

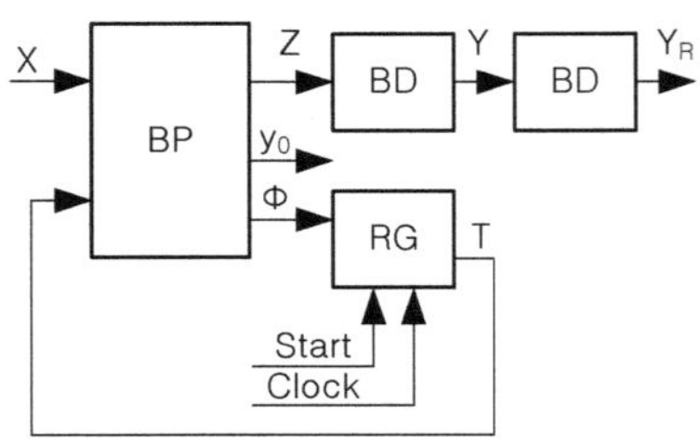

Fig. 6.10 Structural diagram of PD_V^1 Mealy FSM

state marking is shown in Fig. 6.12. To simplify the symbols, the vertices of VGSA $V(\Gamma_4)$ are renumbered.

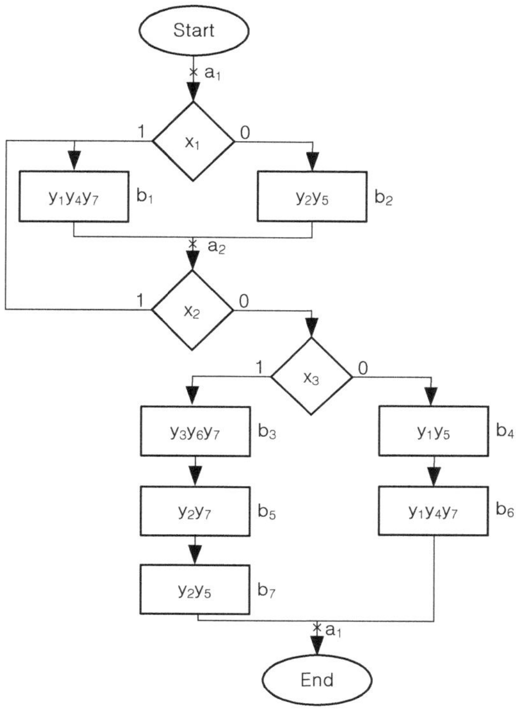

Fig. 6.11 Initial graph-scheme of algorithm Γ_4

Comparison of Fig. 6.11 and Fig. 6.12 shows that the FSM $S_{18}(\Gamma_4)$ has three state variables, whereas its counterpart $S_{18}(V(\Gamma_4))$ has four state variables.

2. **Microoperation encoding.** For the PD_V^1 Mealy FSM S_{18}, there are seven microoperations ($N = 7$), thus it is enough three variables for their encoding. It means that $R_D = 3$ and $Z = \{z_1, z_2, z_3\}$.
   In the common case, the number of encoding variables is determined as

$$R_D = \lceil \log_2(N+1) \rceil . \tag{6.9}$$

In (6.9), the number of microoperations is increased by 1 iff there is a collection $Y_t = \emptyset$ in the GSA to be interpreted. Let us use the following approach for microoperation coding: the more times microoperation $y_n \in Y$ appears in operator vertices of initial GSA, the more zeros its code includes. This approach is adaptation of the well-known algorithm for state assignment according with a frequency of their appearance in the FSM structure table [3, 4]. Let us name this approach as a *frequency encoding*. Let the symbol f_n stand for frequency of appearance for microoperation $y_n \in Y$. For the Mealy FSM S_{18}, there are

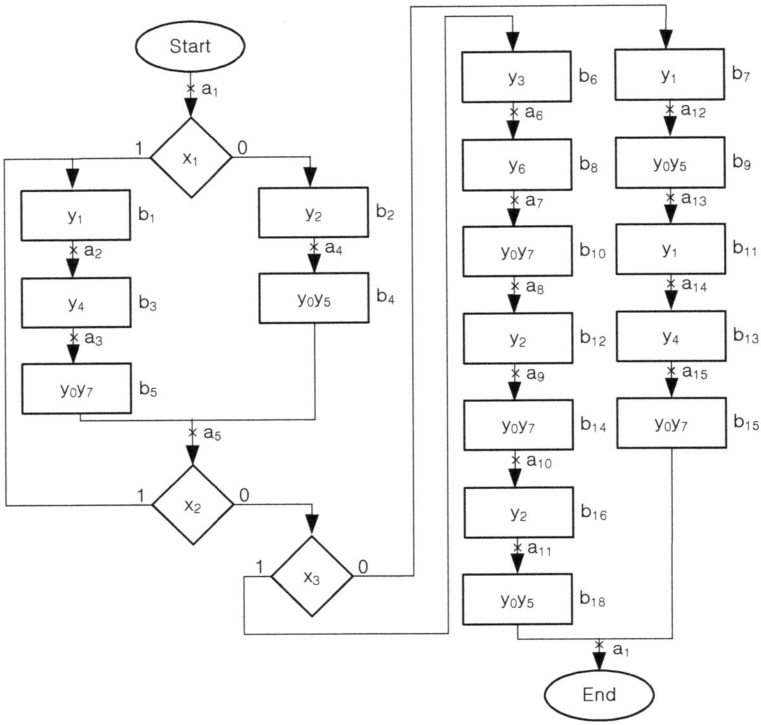

Fig. 6.12 Vertical graph scheme of algorithm $V(\Gamma_4)$

frequencies $f_1 = f_2 = f_5 = 3$, $f_3 = f_6 = 1$, $f_4 = 2$, $f_7 = 4$. It leads to the outcome of frequency encoding shown in Fig. 6.13.

Fig. 6.13 Frequency microoperation codes for Mealy FSM S_{18}

z_1 \ z_2z_3	00	01	11	10
0	y_7	y_1	y_4	y_2
1	y_5	y_6	*	y_3

The system of equations $y_1 = \bar{z}_1\bar{z}_2z_3, \ldots, y_7 = \bar{z}_1\bar{z}_2\bar{z}_3$ can be derived from the Karnaugh map (Fig.6.13). This system specifies the block BD.

3. **Construction of transformed structure table.** The table is constructed after the state encoding stage. Let the states $a_m \in A$ be encoded in a trivial way, namely: $K(a_1) = 0000$, …, $K(a_{15}) = 1110$. The transformed ST is constructed using VGSA. It includes the following columns: a_m, $K(a_m)$, a_s, $K(a_s)$, X_h, Z_h, Φ_h, h.

Table 6.8 Transformed structure table of PD_V^1 Mealy FSM S_{18}

a_m	$K(a_m)$	a_s	$K(a_s)$	X_h	Z_h	Φ_h	h
a_1	0000	a_2	0001	x_1	z_3	D_4	1
		a_4	0011	$\bar{x}_1$	z_2	D_3D_4	2
a_2	0001	a_3	0010	1	z_2z_3	D_3	3
a_3	0010	a_5	0100	1	y_0	D_2	4
a_4	0011	a_5	0100	1	y_0z_1	D_2	5
a_5	0100	a_2	0001	x_2	z_3	D_4	6
		a_6	0101	$\bar{x}_2x_3$	z_1z_2	D_2D_4	7
		a_{12}	1011	$\bar{x}_2\bar{x}_3$	z_3	$D_1D_2D_3$	8
a_6	0101	a_7	0110	1	z_1z_3	D_2D_3	9
a_7	0110	a_8	0111	1	y_0	$D_2D_3D_4$	10
a_8	0111	a_9	1000	1	z_1	D_1	11
a_9	1000	a_{10}	1001	1	y_0	D_1D_4	12
a_{10}	1001	a_{11}	1010	1	z_2	D_1D_3	13
a_{11}	1010	a_1	0000	1	y_0z_1	–	14
a_{12}	1011	a_{13}	1100	1	y_0z_1	D_1D_2	15
a_{13}	1100	a_{14}	1101	1	z_3	$D_1D_2D_4$	16
a_{14}	1101	a_{15}	1110	1	z_2z_3	$D_1D_2D_3$	17
a_{15}	1110	a_1	0000	1	y_0	–	18

The column Z_h includes variables $z_r \in Z$ equal to 1 for the code $K(y_n)$ of microoperation formed for transition $\langle a_m, a_s \rangle$. Besides, this column includes the variable y_0. In the discussed example, the transformed ST includes $H = 18$ rows (Table 6.8). The column Z_h is filled in the following way. For example, the microoperation y_6 is generated for transition from a_6 into a_7 (the row 9). Because $K(y_6) = 101$, then the row 9 of column Z_h contains the variables z_1 and z_3. Next, both y_0 and y_7 are generated for transition from a_7 into a_8 (the row 10). Because $K(y_7) = 000$ (Fig. 6.13), then the row 10 of column Z_h contains only y_0. Using the same approach, all rows of transformed ST (Table 6.8) are filled.

This table is used to derive systems $Z(X,T)$, $y_0(X,T)$, and $\Phi(X,T)$. For example, the following functions $z_1 = F_5 \vee F_7 \vee F_9 \vee F_{11} \vee F_{14} \vee F_{15} = \bar{T}_1 T_3 T_4 \vee$ $\vee \bar{T}_1 T_2 \bar{T}_3 \bar{T}_4 \bar{x}_2 x_3 \vee \bar{T}_1 T_2 \bar{T}_3 \bar{T}_4 \vee T_1 \bar{T}_2 T_3$; $y_0 = F_4 \vee F_5 \vee F_{10} \vee F_{12} \vee F_{14} \vee F_{15} \vee \vee F_{18} =$ $\bar{T}_1 \bar{T}_2 T_3 \vee T_2 T_3 \bar{T}_4 \vee T_1 \bar{T}_2 \bar{T}_4 \vee T_1 \bar{T}_2 T_3$; $D_4 = F_1 \vee F_2 \vee F_6 \vee F_7 \vee F_8 \vee \vee F_{10} \vee F_{12} \vee F_{16} =$ $\bar{T}_2 \bar{T}_3 \bar{T}_4 \vee \bar{T}_1 T_2 \bar{T}_4 \vee T_1 \bar{T}_3 \bar{T}_4$ can be derived from Table 6.8.

Obviously, states $a_m \in A$ can be encoded in the optimal way to minimize the number of macrocells (or LUT elements) in the logic circuit of block BP.

4. **Design of FSM logic circuit.** Logic circuit of FSM is designed using systems of equations obtained previously. For our example, the logic circuit is shown in Fig. 6.14.

In this circuit, the additional block PLD generates signals R_0 (to clear the register RY) and C_0 (to synchronization of the register RY). The outputs of register RY are denoted as y_{Rn}, because they correspond to registered outputs of microoperations

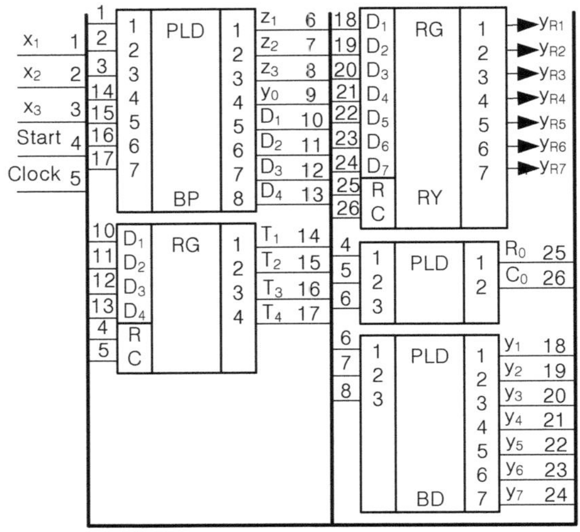

Fig. 6.14 Logic circuit of PD_V^1 Mealy FSM S_{18}

$y_n \in Y$. A practical circuit of the register RY , as well as the block of its control, depends on microchips in use; we do not discuss this step in the book.

The number of inputs for block BP can be decreased, if both an initial GSA and an equivalent VGSA have the same marks of the states. It is possible due to use of some method, which is an analogue of the code sharing method used in compositional microprogram control units [6]. These methods can be found in [1, 5], so we do not discuss them here. In our book we assume that PD_k Mealy FSM is synthesized using the standard marking of the interpreted GSA. To minimize the number of states for PD_k Mealy FSM ($1 < k < K$), it is necessary to find the best distribution of microoperations $y_n \in Y$ among the classes of compatible microoperations Y^k . For example, there is no changing in the fragment Γ_0 (Fig. 6.15a) after its verticalization, if PD_2 Mealy FSM is synthesized and $y_1 \in Y^1$, $y_2 \in Y^2$.

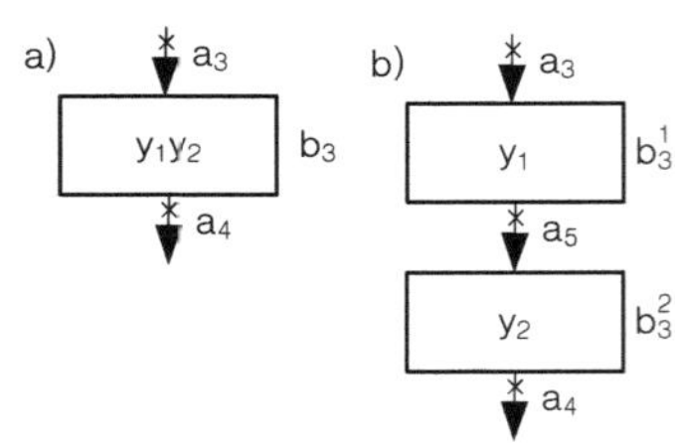

Fig. 6.15 Influence of compatibility of microoperations on the number of states

But if both microoperations belong to the same class of compatibility (for example, if $y_1, y_2 \in Y^1$), then the vertex b_3 is transformed into the vertices b_3^1 and b_3^2 (as shown in Fig. 6.15b). Obviously, the second case is connected with increase for the state number of FSM to be designed in comparison with initial P Mealy FSM.

6.3 Synthesis of Multilevel FSM Models

Jointed use of previously discussed methods results in multilevel models of Mealy and Moore FSMs. The multilevel models of Mealy FSM are represented in Table 6.9. In this table, the symbol D_k informs that the procedure of verticalization was applied to the initial GSA and the final partition Π_Y includes k classes$(k = 1, \ldots, K)$. Obviously, the value K is determined by characteristics of GSA to be interpreted.

Table 6.9 Multilevel models of Mealy FSM

LA			LB	LC	LD
M_1	M_1C	M_1L	P	F	Y
	$\ldots$			Y	D_1
M_G	M_GC	M_GL		D_1	$\vdots$
				$\vdots$	D_K
				D_K	

The following numbers of Mealy FSM models with different number of levels can be derived from Table 6.9:

1. $NL_1 = 1$ (It is P Mealy FSM).
2. $NL_2 = 3G + K + 2$ (The first member of this formula determines the models with logical condition replacement, the second member corresponds to models with verticalization of initial GSA, and the third member determines PF and PY Mealy FSM).
3. $NL_3 = 3G * (K+2) + K + 1$ (The first member of this formula determines the models with logical condition replacement and encoding of collections of microoperations, the second member corresponds to PFD Mealy FSM with verticalization of initial GSA, and the third member determines PFY Mealy FSM).
4. $NL_4 = 3G * (K+1)$ (This formula determines the number of FSM including the block BF used for encoding of structure table rows).

Totally, Table 6.9 determines $6GK + 12G + 2K + 4$ different models of Mealy FSM. For FSM with average complexity ($G = K = 6$) [4], there are 304 different models. If $G = K = 8$, then the number of models is increased up to 628. A synthesis method for multilevel model is a collection of methods for designing corresponding models with less number of levels. Let us discuss an example of the M_2PFD_2 Mealy FSM S_{19}, represented by a GSA Γ_5 (Fig. 6.16).

For this GSA, it is necessary $G = 3$ variables for the logical condition replacement. But according to the task, G should be equal 2. Thus, some additional operator vertices should be introduced into the GSA Γ_5 to replace the logical conditions $x_l \in X = \{x_1, \ldots, x_7\}$ by new variables $p_g \in P = \{p_1, p_2\}$. Next, some operator vertices of the GSA Γ_5 include more, than two microoperations. But according to

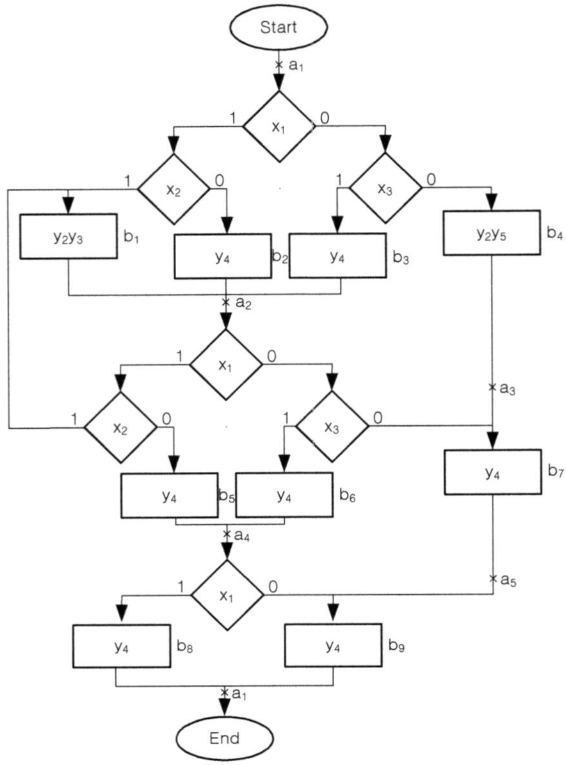

Fig. 6.16 Initial graph-scheme of algorithm Γ_5

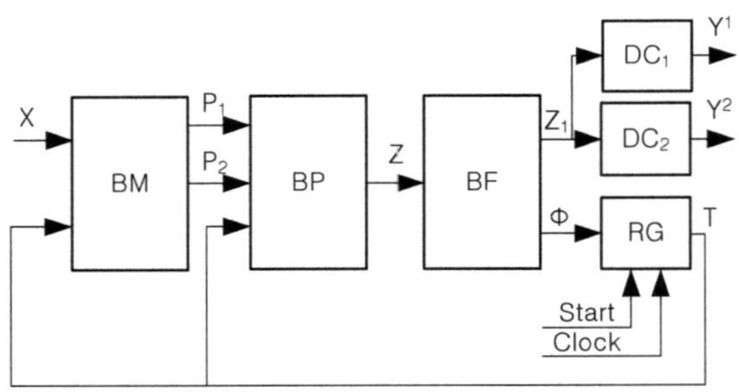

Fig. 6.17 Structural diagram of M_2PFD_2 Mealy FSM

the task, K should be equal 2. Thus, the procedure of verticalization should be applied for the GSA Γ_5.The structural diagram of M_2PFD_2 Mealy FSM is shown in Fig. 6.17.

In this model, the block BM generates functions

$$\begin{aligned} p_1 &= p_1(X,T), \\ p_2 &= p_2(X,T), \end{aligned} \tag{6.10}$$

Table 6.10 Logical condition replacement for Mealy FSM S_{19}

a_m	a_1	a_2	a_3	a_4	a_5	a_6	a_7	a_8	a_9	a_{10}	a_{11}	a_{11}
p_1	x_1	–	x_3	–	–	x_4	x_6	–	–	–	–	–
p_2	x_2	–	–	–	–	x_5	–	–	x_7	–	–	–

the block BP generates functions (6.2), the block BF generates functions (6.3) and (6.4). The block BD consists from two decoders, DC_1 and DC_2, implementing microoperations $y_n \in Y^k$, where $k = 1,2$:

$$Y^k = Y^k(Z_1). \tag{6.11}$$

1. **Transformation of initial GSA.** To satisfy the condition $G = 2$, the operator vertices b_{10} and b_{11} should be introduced into the initial GSA Γ_5. It leads to transformed GSA $V(\Gamma_5)$ shown in Fig. 6.18. To satisfy the condition $K = 2$, it is necessary to distribute microoperations $y_n \in Y$ between two classes of compatibility, namely Y^1 and Y^2.

   The distribution should be executed in such a way, that transformed GSA includes minimal possible amount of new operator vertices. There is no any known algorithm for solution of this problem, because of it let us distribute the microoperations using some heuristics. Finally, we can get the distribution:$Y^1 = \{y_1,y_3,y_5,y_7\}$ and $Y^2 = \{y_2,y_4,y_6\}$. In this case, new vertices $b_{12} - b_{16}$ are introduced into the transformed GSA $V(\Gamma_5)$ (Fig. 6.18).
2. **Logical condition replacement.** This step is executed using the well-known procedure and the results are shown in in Table 6.10.

   According to the task (determined by the formula of FSM model), the states should be assigned using the approach of multiplexer encoding. Let us remind that this approach decreases the parameters of multiplexers (the number of their control and data inputs) from the block BM. The multiplexer codes for Mealy FSM S_{19} are shown in Fig. 6.19.

   Analysis of Fig. 6.19 shows that the logical conditions $x_l \in X(p_1)$ are identified by state variables T_3 and T_4, whereas the logical conditions $x_l \in X(p_2)$ are identified by state variables T_2 and T_3. The following system can be constructed using the codes from Fig. 6.19:

$$\begin{aligned} p_1 &= \bar{T}_3\bar{T}_4x_1 \vee \bar{T}_3T_4x_3 \vee T_3\bar{T}_4x_4 \vee T_3T_4x_6; \\ p_2 &= \bar{T}_2\bar{T}_3x_2 \vee \bar{T}_2T_3x_5 \vee T_2\bar{T}_3x_7. \end{aligned} \tag{6.12}$$

   Let us point out that the authors do not know an algorithm of multiplexer state encoding resulted in the minimal hardware amount for the block BM, so it can be a subject for further research.
3. **Construction of FSM structure table.** This step is executed using the standard approach [3, 4]. In the discussed case, the structure table includes $H = 19$ rows (Table 6.11).

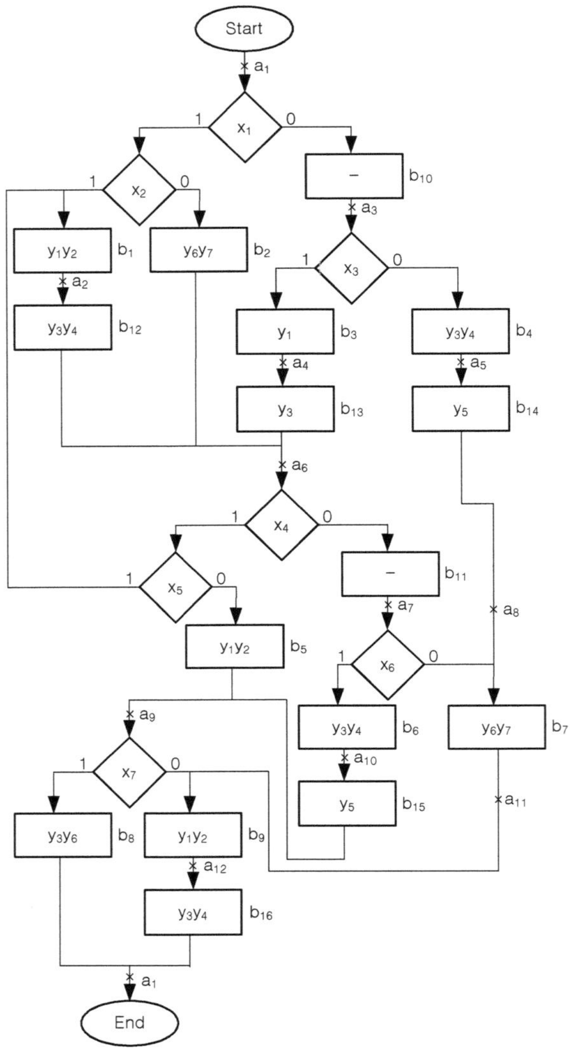

Fig. 6.18 Transformed GSA $V(\Gamma_5)$

4. **Encoding of structure table rows.** Obviously, it is enough $R_F = 5$ variables from the set $Z = \{z_1, \ldots, z_5\}$ to encode $H = 19$ rows of the structure table. Let us encode the rows in a trivial way: $K(F_1) = 00000, \ldots, K(F_{19}) = 10100$.
5. **Construction of transformed structure table.** This step is reduced to replacement of logical conditions $x_l \in X$ by the variables p_1 and p_2, as well as replacement of columns $K(a_s)$ - Φ_h of initial ST by the column Z_h. As in the case of PF Mealy FSM, this column contains only the variables $z_r \in Z$ equal to 1 in the code $K(F_h)$ $(h = 1, \ldots, 19)$. The transformed structure table of M_2PFD_2 Mealy FSM S_{19} includes 19 rows too (Table 6.12).

Fig. 6.19 Multiplexer state codes of Mealy FSM S_{19}

T_1T_2 \ T_3T_4	00	01	11	10
00	a_1	a_3	a_7	a_6
01	a_9	*	*	*
11	a_{10}	a_{11}	*	a_{12}
10	a_2	a_4	a_8	a_5

Table 6.11 Structure table of M_2PFD_2 Mealy FSM S_{19}

a_m	$K(a_m)$	a_s	$K(a_s)$	X_h	Y_h	Φ_h	h
a_1	0000	a_2	1000	x_1x_2	y_1y_2	D_1	1
		a_6	0010	$x_1\bar{x}_2$	y_6y_7	D_3	2
		a_3	0001	$\bar{x}_1$	–	D_4	3
a_2	1000	a_6	0010	1	y_3y_4	D_3	4
a_3	0001	a_4	1001	x_3	y_1	D_1D_4	5
		a_5	1010	$\bar{x}_3$	y_3y_4	D_1D_3	6
a_4	1001	a_6	0010	1	y_3	D_3	7
a_5	1010	a_8	1011	1	y_5	$D_1D_3D_4$	8
a_6	0010	a_2	1000	x_4x_5	y_1y_2	D_1	9
		a_9	0100	$x_4\bar{x}_5$	y_1y_2	D_2	10
		a_7	0011	$\bar{x}_4$	–	D_3D_4	11
a_7	0011	a_{10}	1100	x_6	y_3y_4	D_1D_2	12
		a_{11}	1101	$\bar{x}_6$	y_6y_7	$D_1D_2D_4$	13
a_8	1011	a_{11}	1101	1	y_6y_7	$D_1D_2D_4$	14
a_9	0100	a_1	0000	x_7	y_3y_6	–	15
		a_{12}	1110	$\bar{x}_7$	y_1y_2	$D_1D_2D_3$	16
a_{10}	1100	a_9	0100	1	y_5	D_2	17
a_{11}	1101	a_{12}	1110	1	y_1y_2	$D_1D_2D_3$	18
a_{12}	1110	a_1	0000	1	y_3y_4	–	19

This table is used to derive system (6.2). For example, the following functions $z_1 = F_{17} \vee F_{18} \vee F_{19} = T_1T_2$, $z_2 = F_9 \vee \ldots \vee F_{16} = \bar{T}_1T_3 \vee T_3T_4 \vee \bar{T}_1T_2$ can be derived from Table 6.12 (after minimization).

6. **Encoding of the classes of compatible microoperations.** This step is executed using the rules applied for PD Mealy FSM. Obviously, it is enough three variables (z_6, z_7, z_8) for encoding of the microoperations $y_n \in Y^1$, whereas the microoperations $y_n \in Y^2$ are encoded using only two variables (z_9, z_{10}).

   Therefore, the set of encoding variables $Z_1 = \{z_6, \ldots, z_{10}\}$ is constructed. Let zero codes be used to represent a case when microoperations from a particular class (Y^1, or Y^2, or both) are absent in some collection of microoperations. Let us encode microoperations $y_n \in Y$ as it is shown in Table 6.13.

Table 6.12 Transformed structure table of M_2PFD_2 Mealy FSM S_{19}

a_m	$K(a_m)$	P_h	Z_h	h
a_1	0000	p_1p_2	–	1
		$p_1\bar{p}_2$	z_5	2
		$\bar{p}_1$	z_4	3
a_2	1000	1	z_4z_5	4
a_3	0001	p_1	z_3	5
		$\bar{p}_1$	z_3z_5	6
a_4	1001	1	z_3z_4	7
a_5	1010	1	$z_3z_4z_5$	8
a_6	0010	p_1p_2	z_2	9
		$p_1\bar{p}_2$	z_2z_5	10
		$\bar{p}_1$	z_2z_4	11
a_7	0011	p_1	$z_2z_4z_5$	12
		$\bar{p}_1$	z_2z_3	13
a_8	1011	1	$z_2z_3z_5$	14
a_9	0100	p_2	$z_2z_3z_4$	15
		$\bar{p}_2$	$z_2z_3z_4z_5$	16
a_{10}	1100	1	z_1	17
a_{11}	1101	1	z_1z_5	18
a_{12}	1110	1	z_1z_4	19

Table 6.13 Transformed structure table of M_2PFD_2 Mealy FSM S_{19}

Y^1	$K(y_n)$	Y^2	$K(y_n)$
$\emptyset$	000	$\emptyset$	00
y_1	001	y_2	01
y_3	010	y_4	10
y_5	011	y_6	11
y_7	100	–	–

This table defines unambiguously the block BD, for example, $y_1 = \bar{z}_6\bar{z}_7z_8$, $y_2 = \bar{z}_9z_{10}$ and so on. Functions (6.11) can be minimized using "don't care" input assignments. For example, the function y_7 can be minimized up to the term $y_7 = z_6\bar{z}_8$ taking into account the unused input assignment 110.

7. **Specification of block BF.** This block is specified by the table with columns $K(F_h)$, Z_h, Φ_h, h. The column $K(F_h)$ contains code of the row h, the column Φ_h includes input memory functions $D_r \in \Phi$ from the row h of initial ST. To fill the column Z_h, it is necessary to take the codes of microoperations $y_n \in Y^k$ from corresponding table and to write in the row h of column Z_h the variables $z_r \in Z_1$ corresponding to the bits equal to 1 in the codes of microoperations. Obviously, the table specifying the block BF includes H rows (Table 6.14 for our

example). Embedded memory blocks are the best elements for implementing the logic circuit of block BF.

8. **Implementation of FSM logic circuit.** This step is reduced to implementation of the circuit on the base of tables and systems of equations obtained from the previous steps. For the M_2PFD_2 Mealy FSM S_{19}, the logic circuit is shown in Fig. 6.20.

Acting in the same way, the logic circuit can be designed for any model from Table 6.9. For Moore FSM, there is no sense in verticalization. Because of it,

Table 6.14 Specification of block BF of M_2PFD_2 Mealy FSM S_{19}

$K(F_h)$	Z_h	Φ_h	h	$K(F_h)$	Z_h	Φ_h	h
00000	z_8z_{10}	D_1	1	01010	–	D_3D_4	11
00001	$z_6z_9z_{10}$	D_3	2	01011	z_7z_9	D_1D_2	12
00010	–	D_4	3	01100	$z_6z_9z_{10}$	$D_1D_2D_4$	13
00011	z_7z_9	D_3	4	01101	$z_6z_9z_{10}$	$D_1D_2D_4$	14
00100	z_8	D_1D_4	5	01110	$z_7z_9z_{10}$	–	15
00101	z_7z_9	D_1D_3	6	01111	z_8z_{10}	$D_1D_2D_3$	16
00110	z_7	D_3	7	10000	z_7z_8	D_2	17
00111	z_7z_8	$D_1D_3D_4$	8	10001	z_8z_{10}	$D_1D_2D_3$	18
01000	z_8z_{10}	D_1	9	10010	z_7z_9	–	19
01001	z_8z_{10}	D_2	10	–	–	–	20

multilevel models of Moore FSM are still represented by Table 6.2. There is a very important particular case with $G = 1$. In this case all model blocks can be implemented using only standard library cells, because the systems of equations for FSM logic circuit belong to classes of multiplexer and regular functions [7]. Let us discuss an example of the M_1PY Moore FSM S_{20} logic circuit design, where the FSM is specified by its structure table (Table 6.15). For the Moore FSM S_{20}, all transitions depend on $G = 1$ variable, it determines the set $P = \{p_1\}$. Thus, there is no need in distribution of logical conditions. The following equation can be found from Table. 6.15:

$$p_1 = \bar{T}_1\bar{T}_2x_1 \vee \bar{T}_1T_2x_2 \vee T_1T_2x_3. \tag{6.13}$$

If the logic circuit of block BY is implemented with some macrocells, then the corresponding equations are derived from Table 6.15. But it is enough to use their truth table, if the logic circuit is implemented with embedded memory blocks BRAM (Table 6.16 in the example).

To specify the block BP, it is necessary to construct the table with columns $K(a_m)$, p_1, Φ_h, h, having $H = 2M$ rows. Let a transition from state $a_m \in A$ be executed unconditionally and let this transition be represented by the row h of the initial structure table. In this case the row is duplicated for both $p_1 = 0$ and $p_1 = 1$. After such duplication, the transition does not depend on values of logical conditions

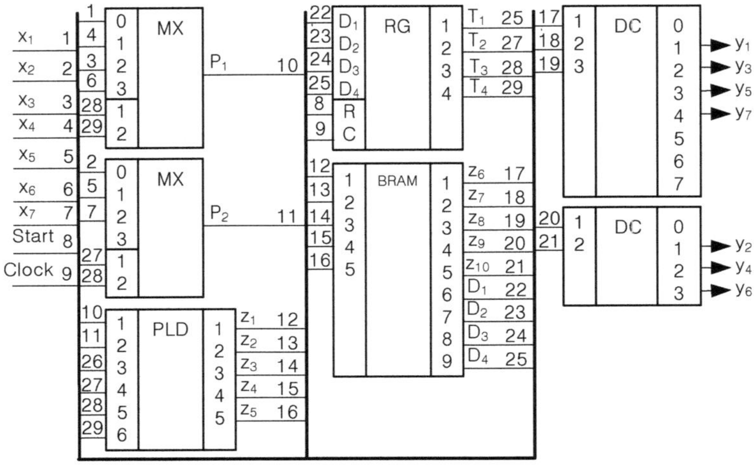

Fig. 6.20 Logic circuit of M_2PFD_2 Mealy FSM S_{19}

Table 6.15 Structure table of Moore FSM S_{20}

a_m	$K(a_m)$	a_s	$K(a_s)$	X_h	Φ_h	h
a_1 (–)	00	a_2	01	x_1	D_2	1
		a_3	10	$\bar{x}_1$	D_1	2
a_2 (y_1y_2)	01	a_2	01	x_2	D_2	3
		a_4	11	$\bar{x}_2$	D_1D_2	4
a_3 ($y_2y_3y_4$)	10	a_1	00	1	–	5
a_4 ($y_1y_3y_5$)	11	a_3	10	x_3	D_1	6
		a_2	01	$\bar{x}_3$	D_2	7
a_7 (y_3y_6)	11	a_1	000	1	–	8

Table 6.16 Specification of block BY for Moore FSMS_{20}

T_1T_2	y_1	y_2	y_3	y_4	y_5
0 0	0	0	0	0	0
0 1	1	1	0	0	0
1 0	0	1	1	1	0
1 1	1	0	1	0	1

$x_l \in X$. For the Moore FSM S_{20}, the transitions are duplicated for the rows 5 and 6 of Table 6.17.

For an FSM with $G = 1$, the vector $\langle K(a_m), p_1 \rangle$ is treated as a memory address, where a word content is taken from the corresponding row of initial ST. For the Moore FSM S_{20}, the second row of Table 6.17 corresponds to the first row of Table 6.15, the first row of Table 6.17 corresponds to the second row of Table 6.15, rows 5 and 6 correspond to the row 5 of the initial Moore FSM structure table (Table 6.15).

The logic circuit of M_1PY Moore FSM (Fig. 6.21 in the discussed case) is constructed in a trivial way. The circuit of block BM is determined by equation (6.13); the logic circuit of block BP is implemented using BRAMs and it is determined by Table 6.17; the logic circuit of block BY is implemented using BRAMs too and it is determined by Table 6.16.

Table 6.17 Specification of block BP for M_1PY Moore FSM S_{20}

$K(a_m)$	p_1	Φ_h	h
00	0	D_1	1
00	1	D_2	2
01	0	D_1D_2	3
01	1	D_2	4
10	0	–	5
10	1	–	6
11	0	D_2	7
11	1	D_1	8

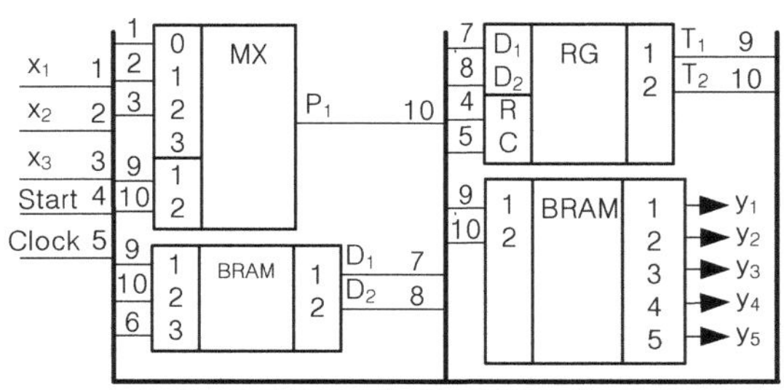

Fig. 6.21 Logic circuit of M_1PY Moore FSM S_{20}

References

1. Adamski, M., Barkalov, A., Bukowiec, A.: Structures of mealy fsm logic circuits under implementation of verticalized flow-chart. In: Proceedings of the IEEE East-West Design & Test Workshop, EWDTW 2005. Kharkov National University of Radioelectronics, Kharkov (2005)
2. Altera Corporation Webpage, http://www.altera.com

3. Baranov, S.: Logic and System Design of Digital Systems. TUT Press, Tallinn (2008)
4. Baranov, S.I.: Logic Synthesis of Control Automata. Kluwer Academic Publishers, Dordrecht (1994)
5. Barkalov, A., Bukowiec, A.: Synthesis of mealy finite states machines for interpretation of verticalized flow-charts. Informatyka Teoretyczna i Stosowana 5(8), 39–51 (2005)
6. Barkalov, A., Titarenko, L.: Logic Synthesis for Compositional Microprogram Control Units. Springer, Berlin (2008)
7. Barkalov, A., Titarenko, L.: Synthesis of Operational and Control Automata. UNITECH, Donetsk (2009)
8. Barkalov, A., Węgrzyn, M.: Design of Control Units With Programmable Logic. University of Zielona Góra Press, Zielona Góra (2006)
9. Minns, P., Elliot, I.: FSM-based digital design using Verilog HDL. John Wiley and Sons, Chichester (2008)
10. Xilinx Corporation Webpage, http://www.xilinx.com

Chapter 7
FSM Synthesis with Object Code Transformation

Abstract. The chapter is devoted to original optimization methods oriented on decrease of the number of outputs for FSM block generating input memory functions. These methods are based on the object code transformation. The FSM objects are either states or collections of microoperations. Sometimes, some additional identifiers are needed for one-to-one representation of different objects. Such optimization methods are discussed for both Mealy and Moore finite state machines. At last, the multilevel models of FSM with object code transformation, logical condition replacement and encoding of collections of microoperations are discussed. This chapter is written together with employee of "Nokia-Siemens Network" Alexander Barkalov (Ukraine).

7.1 Principle of Object Code Transformation

As it was mentioned before, the hardware reduction for FSM logic circuit is connected with the structural decomposition, which in turn is connected with increase for the number of levels in the FSM model. To optimize the hardware amount in block BY, it is necessary to generate some additional variables for encoding of microoperations (or collections of microoperations). The methods discussed in this Chapter are taken from [2–6]. These methods are based on one-to-one match among collections of microoperations and states. Let us name as *objects of FSM* its internal states $a_m \in A$ and collections of microoperations $Y_t \subseteq Y$. Let us point out that states and collections of microoperations are heterogeneous objects respectively each other, whereas different states, for example, are homogenous respectively each other. The optimization methods discussed in this Chapter are based on identification of one-to-one match among heterogeneous objects. If this match is found, then the block BP generates only codes for one object (which is a primary object), while a special code transformer generates the codes of another object (which is a secondary object).

Let us find a one-to-one match $A \to Y$among the states as primary objects and the microoperations as secondary objects. In this case, the block BP generates input

A. Barkalov and L. Titarenko: Logic Synthesis for FSM-Based Control Units, LNEE 53, pp. 155–191.
springerlink.com

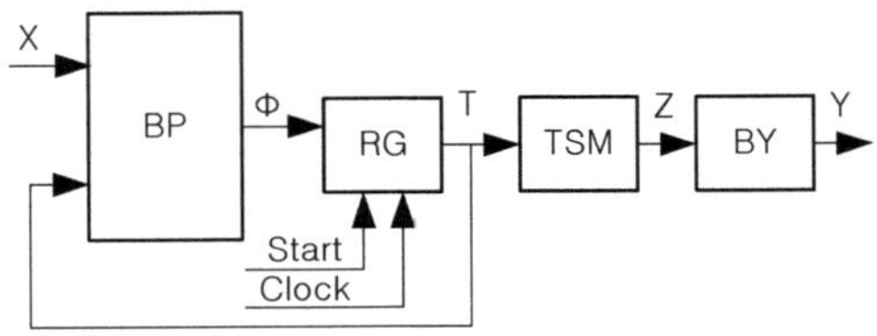

Fig. 7.1 Structural diagram of Mealy FSM1

memory functions $T_r \in T = \{T_1, \dots, T_R\}$ to encode the states, whereas a special state code transformer block TSM generates variables $z_r \in Z$ used for encoding of collections of microoperations. The structural diagram of Mealy FSM based on this principle is shown in Fig. 7.1. Let the symbol $\mathrm{PC_AY}$ stand for this model if collections of microoperations are encoded, whereas the symbol $\mathrm{PC_AD}$ stands for encoding of the classes of compatible microoperations. Let us name such models as FSM1.

Let us find a one-to-one match $Y \to A$ among the microoperations as primary objects and the states as secondary objects. In this case, the block BP generates variables $z_r \in Z$, whereas a special microoperation code transformer block TMS generates input memory functions $T_r \in T$. This approach results in the models of FSM2, denoted as $\mathrm{PC_YY}$ (if collections of microoperations are encoded) or as $\mathrm{PC_YD}$ (if classes of compatible microoperations are encoded). Their structural diagram is shown in Fig. 7.2.

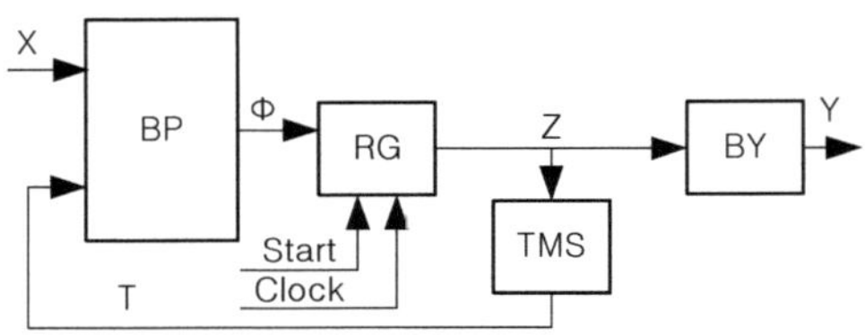

Fig. 7.2 Structural diagram of Mealy FSM2

These models correspond to cases when an FSM has the same numbers of states and collections of microoperations. If this condition is violated, then some additional identifiers should be used belonging to a set of identifiers V. In common case, the block BP generates variables T and V (Fig. 7.3) or variables Z and V (Fig. 7.4). All these variables are the outputs of the register RG.

Thus, in common case the number of bits in the register RG for Mealy FSM with object code transformation exceeds this number for equivalent PY or PD Mealy FSM. Obviously, the proposed approach can be applied iff the total hardware amount for blocks BP and TSM (TMS) is less, than the hardware amount for block BP of PY (PD) Mealy FSM. The same approach can be applied for Moore FSM. Let us point out that only application of the proposed approaches allows the economical implementation of PD Moore FSM.

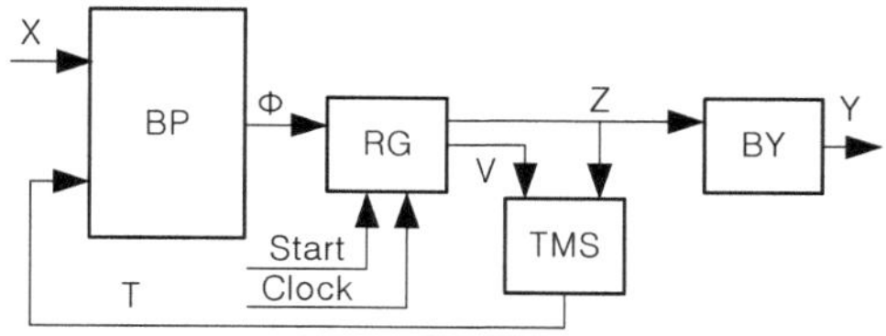

Fig. 7.3 Refined structural diagram of Mealy FSM1

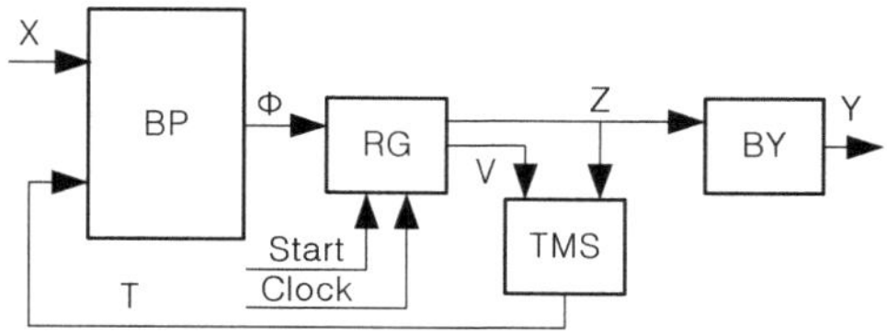

Fig. 7.4 Refined structural diagram of Mealy FSM2

7.2 Logic Synthesis for Mealy FSM with Object Code Transformation

Let the Mealy FSM S_{21} be specified by its structure table (Table 7.1). Consider logic synthesis for models of PC_AY, PC_AD, PC_YY and PC_YD Mealy FSM, based on the Mealy FSM S_{21}.

The following procedure is proposed for logic synthesis Mealy FSM1:

1. **One-to-one identification of collections of microoperations.** Let $T(a_s)$ be a set of collections of microoperations generated under transitions into the set $a_s \in A$, where $n_s = |Y(a_s)|$. In this case, it is necessary n_s identifiers for one-to-one identifications of collections $Y_t \subseteq Y(a_s)$. In common case, it is enough $K = \max(n_1, \ldots, n_M)$ identifiers for one-to-one identification of all collections $Y_t \subseteq Y$, these identifiers form the set $I = \{I_1, \ldots, I_K\}$. Let us encode each identifier $I_k \in I$ by a binary code $K(I_k)$ having $R_V = \lceil \log_2 K \rceil$ bits. Let us use the variables $v_r \in V = \{v_1, \ldots, v_{RV}\}$ for encoding of the identifiers.

   Let each collection $Y_t \in Y(a_s)$ correspond to the pair $\beta_{st} = \langle I_k, a_s \rangle$, where $I_k \in I$. Of course, an identifier $I_k \in I$should be different for different collections. In this case, a code $K(I_k)$ of set $Y_t \in Y(a_s)$ is determined by the following concatenation

$$K(Y_t) = K(I_k) * K(a_s). \tag{7.1}$$

   In (7.1) the symbol $*$ stands for concatenation of these codes.
2. **Encoding of collections of microoperations.** If the method of maximal encoding of collections of microoperations is applied, then let a collection $Y_t \subseteq Y$ be determined by a binary code $C(Y_t)$ having $Q = \lceil \log_2 T_0 \rceil$bits, where T_0 is the number of collections. If the method of encoding of the classes of compatible microoperations is used, then any collection Y_t is represented as the following vector [4]:

$$Y_t = \left\langle y_t^1, y_t^2, \ldots, y_t^J \right\rangle. \tag{7.2}$$

Table 7.1 Structure table of Mealy FSM S_{21}

a_m	$K(a_m)$	a_s	$K(a_s)$	X_h	Y_h	Φ_h	h
a_1	000	a_2	010	x_1	y_1y_2	D_1D_2	1
		a_3	011	$\bar{x}_1$	y_3	D_3	2
a_2	010	a_2	010	x_2	y_1y_2	D_2	3
		a_3	011	$\bar{x}_2x_3$	D_1	D_2	4
		a_4	100	$\bar{x}_2\bar{x}_3$	y_1y_2	D_1D_3	5
a_3	011	a_4	100	x_1	y_2y_5	D_1	6
		a_5	101	$\bar{x}_1$	y_6	D_1D_3	7
a_4	100	a_5	101	1	y_3y_7	D_1D_3	8
a_5	101	a_2	010	x_2x_3	y_1y_2	D_2	9
		a_3	011	$x_2\bar{x}_3$	y_3	D_2D_3	10
		a_5	101	$\bar{x}_2x_4$	y_3y_7	D_1D_3	11
		a_1	000	$\bar{x}_2\bar{x}_4$	–	–	12

In (7.2), the symbol J stands for the number of the classes of compatible microoperations, whereas the symbol y_t^j denotes microoperation $y_n \in Y_t$, belonged to the class j of compatible microoperations ($j = 1,\ldots,J$). Therefore, a code of collection Y_t is represented as a concatenation of microoperation codes.

3. **Construction of transformed structure table.** For FSM1, the transformed ST is used to generate input memory functions Φ and additional functions of identification V. These systems depend on the same terms and are represented as the following:

$$\phi_r = \bigvee_{h=1}^{H} C_{rh}A_m^hX_h \quad (r = 1,\ldots,R), \tag{7.3}$$

$$v_r = \bigvee_{h=1}^{H} C_{rh}A_m^hX_h \quad (r = 1,\ldots,R). \tag{7.4}$$

Obviously, to represent system (7.4) the column Y_h of initial ST should be replaced by columns: I_h is an identifier of the collection Y_h from pair $\beta_{s,h}$; $K(I_h)$ is a code of identifier I_k; V_h are variables $v_r \in V$, equal to 1 in the code $K(I_h)$.

4. **Specification of block TSM.** The block TSM generates variables $z_q \in Z$ represented as the following functions

$$Z = Z(V,T). \tag{7.5}$$

To construct system (7.5), it is necessary to built a table with columns a_s, $K(a_s)$, I_k, $K(I_k)$, Y_h, Z_h, h. The table includes all pairs $\beta_{t,s}$, determined the collection Y_1, next all pairs determined the collection Y_2, and so on. The number of their rows (H_0) is determined as a result of summation for numbers n_s(S $= 1,\ldots,$H). The column Z_h of the table includes variables $z_q \in Z$, equal to 1 in the code $K(Y_h)$. The system (7.5) can be represented as the following:

$$z_q = \bigvee_{h=1}^{H_0} C_{qh} X_h A_s^h \quad (q = 1, \ldots, Q). \tag{7.6}$$

In (7.6), the symbol V_h stands for conjunction of variables $v_r \in V$, corresponded to the code $K(I_k)$ of identifier from the row h of this table.

5. **Specification of block for generation microoperations.** This step is executed in the same manner, as it is done for PY or PD Mealy FSM.
6. **Synthesis of FSM logic circuit.** For the Mealy FSM1, there is no need in keeping codes of identifiers in the register RG. Therefore, these FSM models should be refined. The structural diagram of PC_AY Mealy FSM is shown in Fig. 7.5, whereas Fig. 7.6 shows the structural diagram of PC_AD Mealy FSM. In both cases, the block BP implements functions (7.3) – (7.4), the block TSM generates functions (7.6), blocks BY or BD implements microoperations $Y = Y(Z)$.

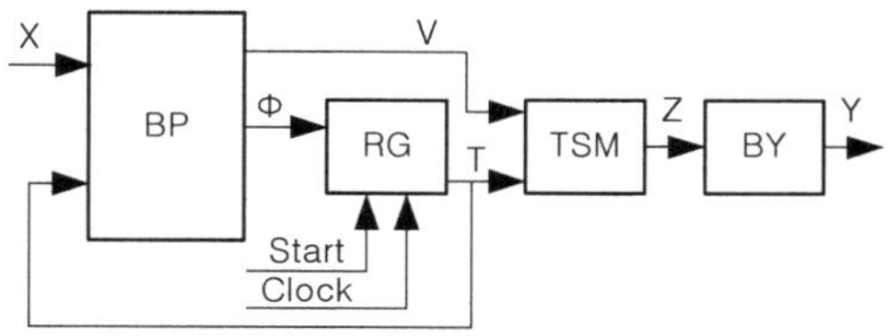

Fig. 7.5 Structural diagram of PC_AY Mealy FSM

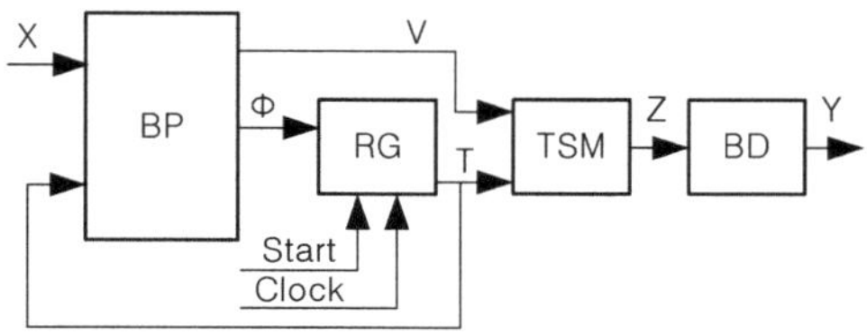

Fig. 7.6 Structural diagram of PC_AD Mealy FSM

In Table 7.1, there are the following collections of microoperations $Y_1 = \emptyset$, $Y_2 = \{y_1, y_2\}$, $Y_3 = \{y_3\}$, $Y_4 = \{y_4\}$, $Y_5 = \{y_5\}$, $Y_6 = \{y_6\}$, $Y_7 = \{y_7\}$, $T_0 = 7$.

Let us consider an example of logic synthesis for the PC_AY Mealy FSM S_{21}. The following sets can be derived from Table 7.1:$Y(a_1) = \{Y_1\}$,$Y(a_2) = \{Y_2\}$, $Y(a_3) = \{Y_3, Y_4\}$, $Y(a_4) = \{Y_2, Y_5\}$, $Y(a_5) = \{Y_6, Y_7\}$. It gives the value $K = 2$. Thus, it is enough two identifiers creating the set $I = \{I_1, I_2\}$; they can be encoded using $R_V = 1$ variables from the set $V = \{v_1\}$. Let $K(I_1) = 0$, $K(I_2) = 1$, then the following codes can be obtained using formula (7.1): $K(Y_1) = *000$, $K(Y_2^1) = *010$, $K(Y_2^2) = 0100$, $K(Y_3) = 0011$, $K(Y_4) = 1011$, $K(Y_5) = 1100$, $K(Y_6) = 0101$, and $K(Y_7) = 1101$. This example shows that there are m_t different codes determined a collection $Y_t \subseteq Y$ if this collection belongs to m_t different sets $Y(a_s)$. For example, for the collection $Y_2 \in Y(a_2) \cap Y(a_1)$ we have $m_2 = 2$, thus the collection Y_2 corresponds to codes $K(Y_2^1)$ and $K(Y_2^2)$.

There are $T_0 = 7$ different collections, thus $Q = 3$ and $Z = \{z_1, z_2, z_3\}$. Let the collections $Y_t \subseteq Y$ be encoded in the following way: $K(Y_1) = 000$,

Table 7.2 Transformed structure table of PC_AY Mealy FSM S_{21}

a_m	$K(a_m)$	a_s	$K(a_s)$	X_h	I_h	$K(I_k)$	V_h	Φ_h	h
a_1	000	a_2	010	x_1	–	–	–	D_1D_2	1
		a_3	011	$\bar{x}_1$	I_1	0	–	D_3	2
a_2	010	a_2	010	x_2	–	–	–	D_2	3
		a_3	011	$\bar{x}_2x_3$	I_2	1	v_1	D_2	4
		a_4	100	$\bar{x}_2\bar{x}_3$	I_1	0	–	D_1D_3	5
a_3	011	a_4	100	x_1	I_2	1	v_1	D_1	6
		a_5	101	$\bar{x}_1$	I_1	0	–	D_1D_3	7
a_4	100	a_5	101	1	I_2	1	v_1	D_1D_3	8
a_5	101	a_2	010	x_2x_3	–	–	–	D_2	9
		a_3	011	$x_2\bar{x}_3$	I_1	0	–	D_2D_3	10
		a_5	101	$\bar{x}_2x_4$	I_2	1	v_1	D_1D_3	11
		a_1	000	$\bar{x}_2\bar{x}_4$	–	–	–	–	12

$K(Y_2) = 001, \ldots, K(Y_7) = 110$. The transformed structure table (Table 7.2) should be constructed to find functions (7.3) – (7.4).

If the condition $n_s = 1$ takes place for some collection $Y_t \in Y(a_s)$, then there is no need in identifier code for this collection. This situation is marked as the symbol "–" in the corresponding row of transformed structure table. As it was mentioned, we can derive systems (7.3) – (7.4) from Table 7.2. For example, the following SOPs can be found: $v_1 = F_4 \vee F_5 \vee F_8 \vee F_{11} = A_2\bar{x}_2x_3 \vee A_3x_4 \vee \ldots = \bar{T}_1T_2\bar{T}_3\bar{x}_2x_3 \vee \bar{T}_1T_2T_3x_4 \vee \ldots, D_2 = F_2 \vee F_3 \vee F_4 \vee F_9 \vee F_{10} = A_1\bar{x}_1 \vee A_2x_2 \vee \ldots = \bar{T}_2\bar{T}_3x_2 \vee \bar{T}_1T_2\bar{T}_3x_2 \ldots$. Both systems are irregular, thus they are implemented using some macrocells.

Table 7.3 specifies the block TSM, it includes $H_0 = 8$ rows. This number is equal to the outcome of summation for the numbers n_s $(S = 1, \ldots, 5)$. System (7.6) is irregular and it is implemented using macrocells. For example, the SOP $z_1 = F_6 \vee F_7 \vee F_8 = A_4v_1 \vee A_5\bar{v}_1 = T_1\bar{T}_2\bar{T}_3v_1 \vee T_1\bar{T}_2T_3\bar{v}_1 \vee T_1T_2T_3\bar{v}_1 \vee T_1\bar{T}_2T_3v_1$ can be derived from the table specified the block TSM in our example.

The block BY is specified by the table of microoperations. For the PC_AY Mealy FSM S_{21}, this table includes $T_0 = 8$ rows (Table 7.4). Let us point out that codes $C(Y_t)$ are used as codes of collections Y_t.

The logic circuit of PC_AY Mealy FSM S_{21} is shown in Fig. 7.7. In this circuit, systems (7.3) and (7.4) are implemented using some macrocells creating the block BP; system (7.6) is implemented using some macrocells creating the block TSM; the system of microoperations $Y(Z)$ is implemented using embedded memory blocks PROM creating the block BY.

Let us consider an example of the logic synthesis for the PC_AD Mealy FSM S_{21}. Obviously, the outcome of one-to-one identification is the same for equivalent PC_AY and PC_AD Mealy FSM. To encode the collections of microoperations, it is necessary to find the partition Π_Y of the set of microoperations Y by the classes of pseudoequivalent microoperations [48]. For the FSM S_{21}, the following partition

Table 7.3 Specification of block TSM of PC$_A$Y Mealy FSM S_{21}

a_s	$K(a_s)$	I_h	$K(I_k)$	Y_h	Z_h	h
a_1	000	–	–	Y_1	–	1
a_2	010	–	–	Y_2	z_3	2
a_3	011	I_1	0	Y_3	z_2	3
a_3	011	I_2	1	Y_4	z_2z_3	4
a_4	100	I_1	0	Y_2	z_3	5
a_4	100	I_2	1	Y_5	z_1	6
a_5	101	I_1	0	Y_6	z_1z_3	7
a_5	101	I_2	1	Y_7	z_1z_2	8

Table 7.4 Table of microoperations for PC$_A$Y Mealy FSM S_{21}

Y_t	$C(Y_t)$	y_1	y_2	y_3	y_4	y_5	y_6	y_7
Y_1	000	0	0	0	0	0	0	0
Y_2	001	1	1	0	0	0	0	0
Y_3	010	0	0	1	0	0	0	0
Y_4	011	0	0	0	1	0	0	0
Y_5	100	0	1	0	0	1	0	0
Y_6	101	0	0	0	0	0	1	0
Y_7	110	0	0	1	0	0	0	1

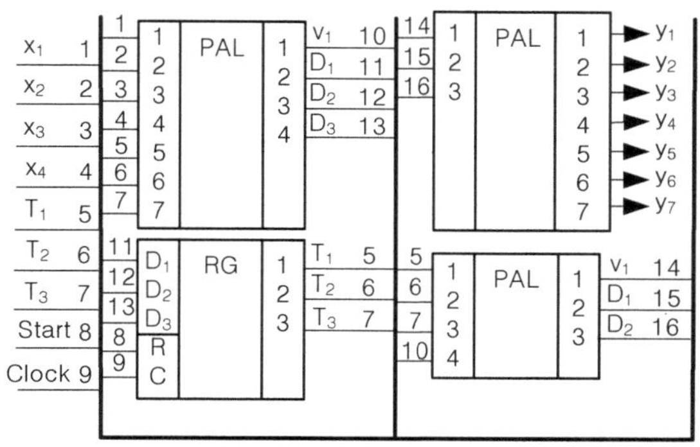

Fig. 7.7 Logic circuit of PC$_A$Y Mealy FSM S_{21}

$\Pi_Y = \{Y^1, Y^2\}$ with two classes can be found, where $Y^1 = \{y_1, y_3, y_4, y_5\}$, $Y^2 = \{y_2, y_6, y_7\}$. It is enough $Q_1 = 3$ variables to encode the microoperations $y_n \in Y^1$, and $Q_2 = 2$ variables for the microoperations $y_n \in Y^2$. It means that there is the set $Z = \{z_1, \ldots, z_5\}$, its cardinality is found as $Q = Q_1 + Q_2 = 5$. Let us encode microoperations $y_n \in Y$ in the way shown in Table 7.5. It leads to the codes $C(Y_t)$ of collections $Y_t \in Y$ shown in Table 7.6. Let us point out that if some microoperation $y_n^j \notin Y_t$, then the field j of code $C(Y_t)$ contains only zeros.

The transformed structure table of PC$_A$D Mealy FSM S_{21} is identical to the corresponding table of PC$_A$Y Mealy FSM S_{21} (Table 7.2). The table specifying the

Table 7.5 Codes of microoperations for PC_AYMealy FSM S_{21}

Y^1	$K(y_n^1)$ $z_1z_2z_3$	Y^2	$K(y_n^2)$ z_4z_5
y_1	0 0 1	y_2	0 1
y_3	0 1 0	y_6	1 0
y_4	0 1 1	y_7	1 1
y_5	1 0 0	–	–

Table 7.6 Codes of collections of microoperations for PC_AD Mealy FSM S_{21}

t	Y^1	$C(Y_t)$	t	Y^2	$C(Y_t)$
1	$\emptyset$	0 0 0 0 0	5	y_2y_5	1 0 0 0 1
2	y_1y_2	0 0 1 0 1	6	y_6	0 0 0 1 0
3	y_3	0 1 0 0 0	7	y_3y_7	0 1 0 1 1
4	y_4	0 1 1 0 0			

Table 7.7 Specification of block TSM for PC_ADMealy FSM S_{21}

a_s	$K(a_s)$	I_h	$K(I_k)$	Y_h	Z_h	h
a_1	000	–	–	Y_1	–	1
a_2	010	–	–	Y_2	z_3z_5	2
a_3	011	I_1	0	Y_3	z_2	3
a_3	011	I_2	1	Y_4	z_2z_3	4
a_4	100	I_1	0	Y_2	z_3z_5	5
a_4	100	I_2	1	Y_5	z_1z_5	6
a_5	101	I_1	0	Y_6	z_1z_4	7
a_5	101	I_2	1	Y_7	$z_1z_4z_5$	8

block TSM for both models is constructed in the same way. As a rule, this table for PC_AD Mealy FSM includes more variables $z_r \in Z$ (Table 7.7 in our example), than its counterpart for PC_AY Mealy FSM.

There is no need in a table specifying microoperations, because Table 7.5 contains inputs and outputs for decoders of the block BD. The logic circuit of PC_AD Mealy FSM S_{21} is shown in Fig. 7.8.

The following procedure is proposed to design a Mealy FSM2:

1. **One-to-one identification of states.** Let $A(Y_t)$ be a set of states, such that a collection $Y_t \subseteq Y$ is generated under some transitions in these states, and let $m_t = |A(Y_t)|$. In this case, it is enough m_t identifiers for one-to-one identifications of the states $a_m \in A(Y_t)$. It is necessary $K = \max(m_1, \ldots, m_T)$ variables for one-to-one identification of the states $a_m \in A$, let these identifiers form a set I. Let us encode an identifier $I_k \in I$ by a binary code $K(I_K)$ and let us construct a set of variables

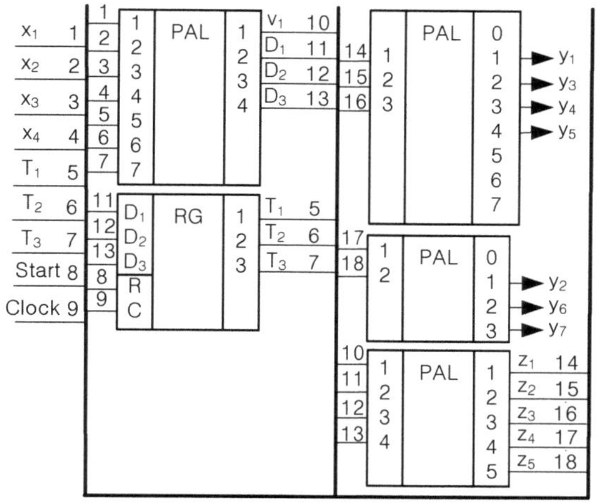

Fig. 7.8 Logic circuit of PC_AD Mealy FSM S_{21}

$V = \{v_1, \ldots, v_{R1}\}$ used for encoding of identifiers, where $R_1 = \lceil \log_2 K \rceil$. Let each state $a_s \in A(Y_t)$ correspond to a pair $\alpha_{t,s} = \langle I_k, Y_t \rangle$, then the code for state a_s is determined by the following concatenation:

$$C(a_s) = K(Y_t) * K(I_k). \tag{7.7}$$

2. **Encoding of collections of microoperations.** This step is executed using the approach discussed before.
3. **Construction of transformed structure table.** This table is used to derive functions (7.4) and $Z = Z(T,X)$. To construct it, the columns a_s, $K(a_s)$, Φ_h are eliminated from the initial structure table, in the same time the column Y_h is replaced by columns V_h and Z_h. The column Z_h contains variables $z_q \in Z$ equal to 1 in the code $K(Y_h)$. The system Z includes the following equations:

$$z_q = \bigvee_{h=1}^{H} C_{qh} A_m^h X_h \quad (q = 1, \ldots, Q). \tag{7.8}$$

4. **Specification of code transformer.** The code transformer TMS generates functions

$$\Phi = \Phi(V, Z). \tag{7.9}$$

This system can be specified by a table with the following columns: Y_t, $K(Y_t)$, I_k, $K(I_k)$, a_s, $K(a_s)$, Φ_h, h. The table includes all pairs $\langle I_k, Y_t \rangle$ for the state a_1, next, all pairs for a_2, and so on. The number of rows H_0 in this table is determined as a result of summation for the numbers m_t $(t = 1, \ldots, T)$. The system of input memory functions is represented as the following one:

$$\phi_r = \bigvee_{h=1}^{H_0} C_{rh} V_h Z_h \quad (r = 1, \ldots, R). \tag{7.10}$$

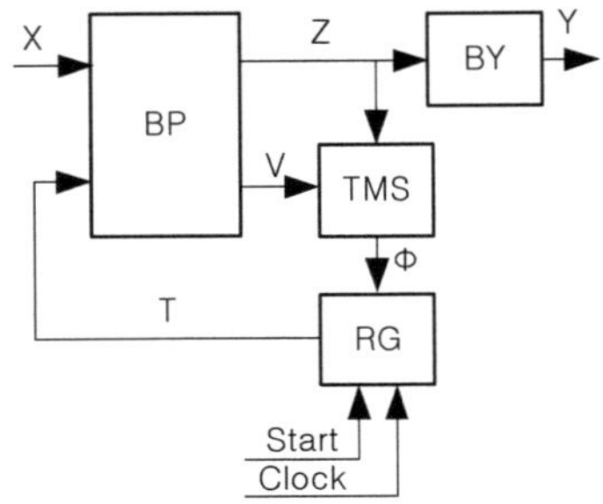

Fig. 7.9 Structural diagram of PC_AY Mealy FSM

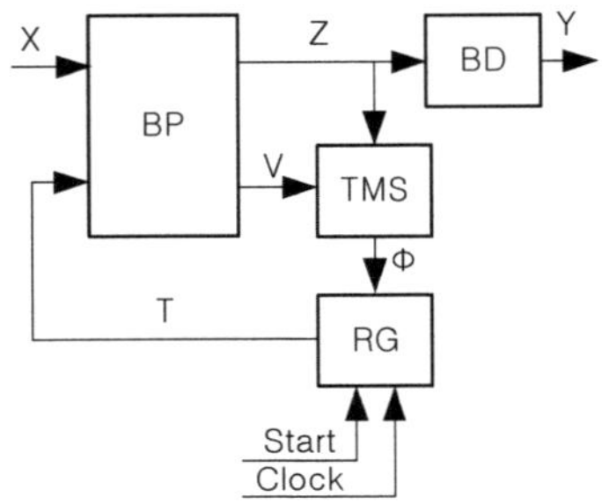

Fig. 7.10 Structural diagram of PC_AD Mealy FSM

In (7.10), the symbol Z_h stands for conjunction of variables $z_r \in Z$ corresponded to the collection of microoperations $Y_t \subseteq Y$ from the row h of the table specifying block TMS.

5. **Construction of the table of microoperations.** This step is executed using the same approach as the one applied for PC_AY Mealy FSM.
6. **Synthesis of FSM logic circuit.** For structural diagram shown in Fig. 7.2, the number of bits in the register RG is equal to $Q + R_V$. This number can be decreased up to R, using the structural diagrams shown in Fig. 7.9 and Fig. 7.10.

In both these cases, the block TMS generates input memory functions instead of state variables T. Due to such approach, it is enough R flip-flops in the register RG.

Let us discus an example of logic synthesis for the PC_AY Mealy FSM S_{21}. For FSM S_{21}, there are $T_0 = 7$ collections of microoperations, namely: $Y_1 = \emptyset$, $Y_2 = \{y_1, y_2\}$, $Y_3 = \{y_3\}$, $Y_4 = \{y_4\}$, $Y_5 = \{y_2, y_5\}$, $Y_6 = \{y_6\}$, $Y_7 = \{y_3, y_7\}$ (Table 7.1). Let us construct the sets $A(Y_t)$ and define their cardinality numbers: $A(Y_1) = \{a_1\}$, $m_1 = 1$; $A(Y_2) = \{a_2, a_4\}$, $m_2 = 2$; $A(Y_3) = \{a_3\}$, $m_3 = 1$; $A(Y_4) = \{a_3\}$, $m_4 = 1$; $A(Y_5) = \{a_4\}$, $m_5 = 1$; $A(Y_6) = \{a_5\}$, $m_6 = 1$, and $A(Y_7) = \{a_5\}$, $m_7 = 1$. Thus, it is enough $K = 2$ identifiers, that is $I = \{I_1, I_2\}$. The identifiers $I_k \in I$ can be encoded using $R_V = 1$ variable, that is $V = \{v_1\}$. Let the identifiers be encoded in the following way: $K(I_1) = 0$ and $K(I_2) = 1$. Let us find the pairs $\alpha_{t,s}$ for each element from the sets $A(Y_t)$. If $m_t = 1$, then the first component of corresponding pair is represented by the symbol $\emptyset$. This symbol corresponds to uncertainty in the code $C(a_s)^t$, where the superscript t means that the code of state a_s belongs to the pair $\alpha_{t,s}$. The following pairs can be constructed in the discussed

Table 7.8 State codes of PC$_A$Y Mealy FSM S_{21}

a_m	$C(a_s)^t$	$\alpha_{t,m}$	h	a_m	$C(a_s)^t$	$\alpha_{t,m}$	h
a_1	000*	$\alpha_{1,1}$	1	a_4	100*	$\alpha_{5,4}$	5
a_2	0010	$\alpha_{2,2}$	2	a_4	0011	$\alpha_{2,4}$	6
a_3	010*	$\alpha_{3,3}$	3	a_5	101*	$\alpha_{6,5}$	7
a_3	011*	$\alpha_{4,3}$	4	a_5	110*	$\alpha_{7,5}$	8

example: $\alpha_{1,1} = \langle \emptyset, Y_1 \rangle$, $\alpha_{2,2} = \langle I_1, Y_2 \rangle$, $\alpha_{2,4} = \langle I_2, Y_2 \rangle$, $\alpha_{3,3} = \langle \emptyset, Y_3 \rangle$, $\alpha_{4,3} = \langle \emptyset, Y_4 \rangle$, $\alpha_{5,4} = \langle \emptyset, Y_5 \rangle$, $\alpha_{6,5} = \langle \emptyset, Y_6 \rangle$, $\alpha_{7,5} = \langle \emptyset, Y_7 \rangle$. Using these pairs together with (7.7), we can get the codes $C(a_s)$ shown in Table 7.8.

This table includes $H_0 = m_1 + \ldots + m_T$ rows. As follows from Table 7.8, each from the states a_3, a_4, and a_5 have two different codes of the type (7.7). In common case, the number of codes $C(a_s)^t$ for some state $a_m \in A$ is equal to the number of different sets $A(Y_t)$, including this state a_m. The codes of collections of microoperations shown in Table 7.8 are the same as they were obtained before. The codes are placed in the three most significant positions of the column $C(a_m)$.

Using the known method, we can construct the transformed structure table of PC$_A$Y Mealy FSM S_{21} (Table 7.9) on the base of the initial structure table (Table 7.1). Using Table 7.9, we can derive systems (7.8) and (7.4), for example, $z_1 = F_6 \vee F_7 \vee F_8 \vee F_{11} = A_3x_4 \vee A_3\bar{x}_4 \vee A_4 \vee A_5\bar{x}_5x_4 = \bar{T}_1T_2\bar{T}_3x_4 \vee \ldots$; $v_1 = F_5 = A_2\bar{x}_2\bar{x}_3 = \bar{T}_1\bar{T}_2T_3\bar{x}_2\bar{x}_3$.

The table used for specification of the block TMS (Table 7.10) includes $H_2 = 2^{R_0} - H_1$ rows, where $R_0 = \lceil \log_2 H_0 \rceil$. It is necessary if the logic circuit of TMS is implemented with embedded memory blocks. In this case all possible addresses should be present. Let us point out that at least $H_1 = (2^Q - T)2^{R_1}$ rows contain zero output codes corresponded to unused collections of microoperations. For the FSM

Table 7.9 Transformed structure table of PC$_A$Y Mealy FSM S_{21}

a_m	$K(a_m)$	X_h	Z_h	V_h	h
a_1	000	x_1	z_3	–	1
		$\bar{x}_1$	z_2	–	2
a_2	010	x_2	z_3	–	3
		$\bar{x}_2x_3$	z_2z_3	–	4
		$\bar{x}_2\bar{x}_3$	z_3	v_1	5
a_3	011	x_1	z_1	–	6
		$\bar{x}_1$	z_1z_3	–	7
a_4	100	1	z_1z_2	–	8
a_5	101	x_2x_3	z_3	–	9
		$x_2\bar{x}_3$	z_2	–	10
		$\bar{x}_2x_4$	z_1z_2	–	11
		$\bar{x}_2\bar{x}_4$	–	–	12

Table 7.10 Specification of block TMS for PC$_A$Y Mealy FSM S_{21}

Y_t	$K(Y_t)$	I_k	$K(I_k)$	a_s	$K(a_s)$	Φ_h	h
Y_1	000	–	0	a_1	000	–	1
	000	–	1	a_1	000	–	2
Y_2	001	I_1	0	a_2	010	D_2	3
	001	I_2	1	a_4	100	D_1	4
Y_3	010	–	0	a_3	011	D_2D_3	5
	010	–	1	a_3	011	D_2D_3	6
Y_4	011	–	0	a_3	011	D_2D_3	7
	011	–	1	a_3	011	D_2D_3	8
Y_5	100	–	0	a_4	100	D_1	9
	100	–	1	a_4	100	D_1	10
Y_6	101	–	0	a_5	101	D_1D_3	11
	101	–	1	a_5	101	D_1D_3	12
Y_7	110	–	0	a_5	101	D_1D_3	13
	110	–	1	a_5	101	D_1D_3	14

S_{21}, there is $H_1 = 2$, it means that only 14 rows are in use, whereas there are totally $2^{R_0} = 16$ rows.

For the PC$_A$Y Mealy FSM S_{21}, table of microoperations is represented by Table 7.4; its logic circuit is shown in Fig. 7.11. This circuit corresponds to the model shown in Fig. 7.9. The logic circuit of block TMS is implemented using embedded memory blocks on the base of Table 7.10.

Let us point out that the logic circuit of block TMS can be implemented using some macrocells. In this case the following system of Boolean functions should be constructed:

$$D_r = \bigvee_{h=1}^{H_2} C_{rk} Z_h V_h \quad (r = 1, \dots, R). \tag{7.11}$$

If the column I_k contains the symbol "–" in the row h of the table specified block TMS, then $V_h = 1$. It allows minimizing system (7.11). For example, $D_1 = F_4 \vee F_9 \vee F_{10} \vee F_{11} \vee F_{12} \vee F_{15} \vee F_{14} = \bar{z}_1\bar{z}_2 z_3 v_1 \vee z_1\bar{z}_2\bar{z}_3 \vee z_1\bar{z}_2 z_3 \vee z_3\bar{z}_2 z_1$.

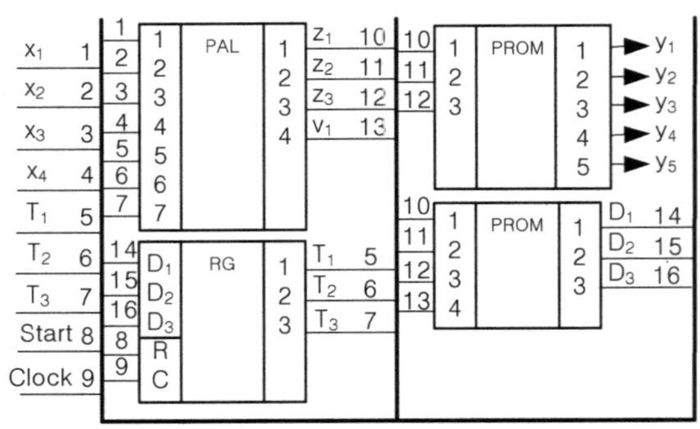

Fig. 7.11 Logic circuit of PC$_A$Y Mealy FSM S_{21}

Let us discuss an example of logic synthesis for the PC_AD Mealy FSM S_{21}, having the structural diagram shown in Fig. 7.10. The codes for its collections of microoperations are shown in Table 7.6. Using these codes of collections as well as the state codes from Table 7.8, it is possible to construct the table for state codes of PC_AD Mealy FSM S_{21} (Table 7.11).

Table 7.11 State codes for PC_AD Mealy FSM S_{21}

a_m	$C(a_s)^t$	$\alpha_{t,m}$	h	a_m	$C(a_s)^t$	$\alpha_{t,m}$	h
a_1	00000*	$\alpha_{1,1}$	1	a_4	10001*	$\alpha_{5,4}$	5
a_2	001010	$\alpha_{2,2}$	2	a_4	001011	$\alpha_{2,4}$	6
a_3	01000*	$\alpha_{3,3}$	3	a_5	00010*	$\alpha_{6,5}$	7
a_3	01100*	$\alpha_{4,3}$	4	a_5	01011*	$\alpha_{7,5}$	8

The transformed structure table of PC_AD Mealy FSM S_{21} (Table 7.12) is constructed in the same way, as it is done for PC_AY Mealy FSM.

Table 7.12 Transformed structure table of PC_AD Mealy FSM S_{21}

a_m	$K(a_m)$	X_h	Z_h	V_h	h
a_1	000	x_1	z_3z_5	–	1
		$\bar{x}_1$	z_2	–	2
a_2	010	x_2	z_3z_5	–	3
		$\bar{x}_2x_3$	z_2z_3	–	4
		$\bar{x}_2\bar{x}_3$	z_3z_5	v_1	5
a_3	011	x_4	z_1z_5	–	6
		$\bar{x}_1$	z_4	–	7
a_4	100	1	$z_2z_4z_5$	–	8
a_5	101	x_2x_3	z_3z_5	–	9
		$x_2\bar{x}_3$	z_2	–	10
		$\bar{x}_2x_4$	$z_2z_4z_5$	–	11
		$\bar{x}_2\bar{x}_4$	–	–	12

For PD Mealy FSM, the number of bits used in the code $K(Y_t)$ is much more than for equivalent PY Mealy FSM [8]. It means that the logic circuit of block TSM for PD Mealy FSM should be implemented using some macrocells. For the PC_AD Mealy FSM S_{21}, the table of block TSM includes $H_0 = 8$ rows (Table 7.13).

To implement the logic circuit of PC_AD Mealy FSM, its transformed ST is used to derive systems (7.4) and (7.8), whereas its table for block TSM is the base to derive system (7.11). For example, the following Boolean equation can be derived $D_1 = F_7 \vee F_8 = \bar{z}_1\bar{z}_2\bar{z}_3z_4\bar{z}_5 \vee \bar{z}_1z_2\bar{z}_3z_4z_5$ from Table 7.13. The logic circuit of PC_AD Mealy FSM S_{21} is shown in Fig. 7.12.

Table 7.13 Specification of block TSM for PC_AD Mealy FSM S_{21}

Y_t	$K(Y_t)$	I_k	$K(I_k)$	a_s	$K(a_s)$	Φ_h	h
Y_1	00000	–	0	a_1	000	–	1
Y_2	00101	I_1	0	a_2	001	D_3	2
		I_2	1	a_4	011	D_2D_3	3
Y_3	01000	–	0	a_3	010	D_2	4
Y_4	01100	–	0	a_3	010	D_2	5
Y_5	10001	–	0	a_4	011	D_2D_3	6
Y_6	00010	–	0	a_5	100	D_1	7
Y_7	01011	–	0	a_5	100	D_1	8

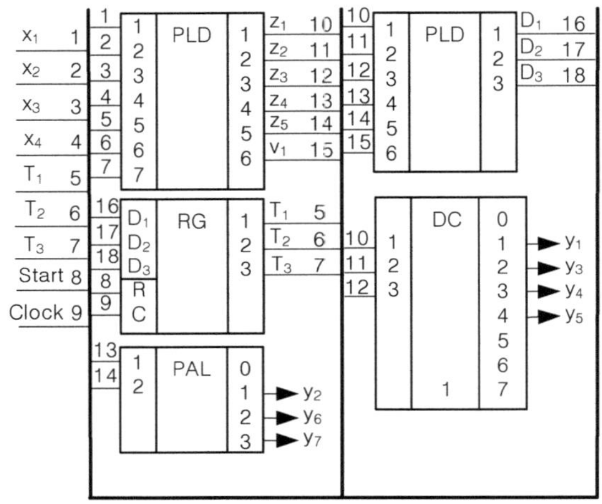

Fig. 7.12 Logic circuit of PC_AD Mealy FSM S_{21}

7.3 Logic Synthesis for Moore FSM with Object Code Transformation

Let us discuss some synthesis examples using the Moore FSM S_{22}, specified by its structure table (Table 7.14). There are $T = 4$ different collections of microoperations in this table, namely collections:$Y_1 = \emptyset$, $Y_2 = \{y_1, y_2\}$, $Y_3 = \{y_3\}$, $Y_4 = \{y_3, y_4\}$. These collections can be encoded using $Q = 2$ variables from the set $Z = \{z_1, z_2\}$. First of all, let us discuss an example of the PC_AY Moore FSM S_{22} synthesis.

The method for PC_AY Moore FSM synthesis includes the following steps:

1. **Encoding of collections of microoperations.** Each collection $Y_t \subseteq Y$ is encoded by a binary code $K(Y_t)$ having $Q = \lceil \log_2 T_0 \rceil$ bits, where T_0 is the number of different collections in GSA to be interpreted. The set of additional variables $Z = \{z_1, \ldots, z_Q\}$ should be built for encoding of collections $Y_t \subseteq Y$.
2. **Construction of table of microoperations.** This table includes the columns Y_t, $K(Y_t)$, $y_1, \ldots, y_N$, t; it specifies embedded memory blocks from the block BY.

Table 7.14 Specification of block TSM for PC_AD Mealy FSM S_{21}

a_m	$K(a_m)$	a_s	$K(a_s)$	X_h	Φ_h	h
a_1 (–)	000	a_2	001	x_1	D_3	1
		a_3	010	$\bar{x}_1 x_2$	D_2	2
		a_4	011	$\bar{x}_1 \bar{x}_2$	$D_2 D_3$	3
a_2 $(y_1 y_2)$	001	a_5	100	x_3	D_1	4
		a_6	101	$\bar{x}_3$	$D_1 D_3$	5
a_3 (y_3)	010	a_5	100	x_3	D_1	6
		a_6	101	$\bar{x}_3$	$D_1 D_3$	7
a_4 $(y_1 y_2)$	011	a_5	100	x_3	D_1	8
		a_6	101	$\bar{x}_3$	$D_1 D_3$	9
a_5 $(y_3 y_4)$	100	a_7	110	x_4	$D_1 D_2$	10
		a_1	000	$\bar{x}_4$	–	11
a_6 $(y_1 y_2)$	101	a_7	110	x_4	$D_1 D_2$	12
		a_1	000	$\bar{x}_4$	–	13
$a_7 (y_3 y_4)$	110	a_5	100	x_3	D_1	14
		a_6	101	$\bar{x}_3$	$D_1 D_3$	15

3. **Specification of block TSM.** This block is specified by the table having the following columns: a_m, Y_t, $K(Y_t)$, Z_m, m. The table is used for deriving functions $z_q \in Z$.
4. **Synthesis of FSM logic circuit.** The structural diagram of PC_AY Moore FSM is shown in Fig. 7.1. In this model, the block BP generates functions $\Phi = \Phi(T,X)$, derived from initial structure table, whereas the block BY generates output functions $Y = Y(Z)$, specified by the table of microoperations.

The collections of microoperations for the discussed example were found in the beginning of this Section. Let these collections of microoperations for the Moore FSM S_{22} be encoded in the following way: $K(Y_1) = 00$, $K(Y_2) = 01, \ldots, K(Y_4) = 11$. It allows to construct the tables for specification of microoperations (Table 7.15) and presentation of the block TSM (Table 7.16). The logic circuit of PC_AY Moore FSM S_{22} is shown in Fig. 7.13.

Obviously, the model of PC_AY Moore FSM can be applied if the following condition takes place:

$$R > Q. \tag{7.12}$$

Table 7.15 Table of microoperations for PC_AY Moore FSM S_{22}

Y_t	$K(Y_t)$	y_1	y_2	y_3	y_4	t
Y_1	00	0	0	0	0	1
Y_2	01	1	1	0	0	2
Y_3	10	0	0	1	0	3
Y_4	11	0	0	1	1	4

Table 7.16 Specification of block TSM for PC_AY Moore FSM S_{22}

a_m	$K(a_m)$	Y_t	$K(Y_t)$	Z_m	m
a_1	000	Y_1	00	–	1
a_2	001	Y_2	01	z_2	2
a_3	010	Y_3	10	z_1	3
a_4	011	Y_2	01	z_2	4
a_5	100	Y_4	11	z_1z_2	5
a_6	100	Y_2	01	z_2	6
a_7	101	Y_4	11	z_1z_2	7

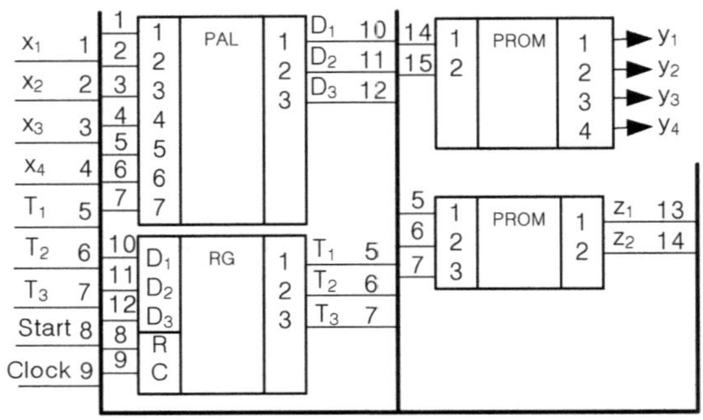

Fig. 7.13 Logic circuit of PC_AY Moore FSM S_{22}

If this condition is satisfied, then the complexity of block BY for PC_AY Moore FSM is 2^{R-Q} times less in comparison with this block complexity for equivalent PY Moore FSM. The main drawback of PC_AY Moore FSM is increase for the number of levels; it results in decrease for performance in comparison with equivalent PY Moore FSM.

The number of macrocells in logic circuit of PC_AY Moore FSM can be decreased due to taking into account existence of the pseudoequivalent states [7]. For example, the following partition $\Pi_A = \{B_1, B_2, B_3\}$ can be found for the Moore FSM S_{22}, where $B_1 = \{a_1\}$, $B_2 = \{a_2, a_3, a_4, a_7\}$, $B_3 = \{a_5, a_6\}$. Obviously, there are $I = 3$ classes of the pseudoequivalent states. If the method of optimal state encoding is used, then it results in the model P_EC_AY Moore FSM, having the same structural diagram as the one shown in Fig. 7.1. For the P_EC_AY Moore FSMS_{22}, the optimal state codes are shown in the Karnaugh map (Fig. 7.14). It allows obtaining the following codes for classes $B_i \in \Pi_A$:$K(B_1) = 0*0$, $K(B_2) = 1**$ and $K(B_3) = 0*1$.

T_1 \ T_2T_3	00	01	11	10
0	a_1	a_5	a_6	*
1	a_2	a_3	a_4	a_7

Fig. 7.14 Optimal state codes of Moore FSM S_{22}

Table 7.17 Transformed structure table of P_EC_AY Moore FSM S_{22}

B_i	$K(B_i)$	a_s	$K(a_s)$	X_h	Φ_h	h
B_1		a_2	100	x_1	D_1	1
		a_3	101	$\bar{x}_1x_2$	D_1D_3	2
		a_4	111	$\bar{x}_1\bar{x}_2$	$D_1D_2D_3$	3
B_2	$1{*}{*}$	a_5	001	x_3	D_3	4
		a_6	011	$\bar{x}_3$	D_2D_3	5
B_3	$0{*}1$	a_7	110	x_4	D_1D_2	6
		a_1	000	$\bar{x}_4$	–	7

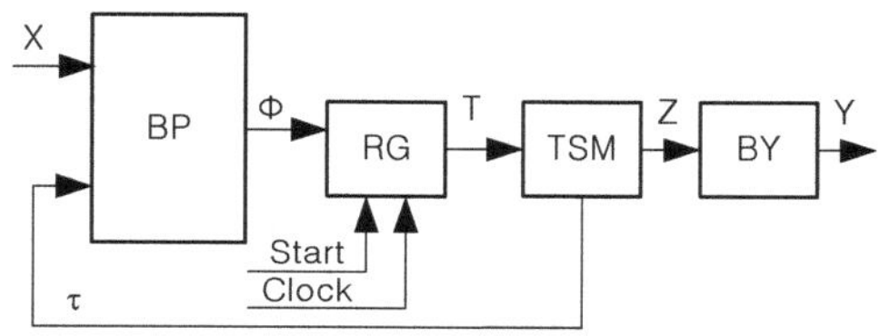

Fig. 7.15 Structural diagram of P_EC_AY Moore FSM

The transformed structure table of P_EC_AY Moore FSM S_{22} includes only $H_0 = 7$ rows (Table 7.17).

The logic circuit of P_EC_AY Moore FSM S_{22} differs from the circuit shown in Fig. 7.13 only in absence of the input T_2 for macrocells of the block BP. It is connected with the identity $T_2 \equiv *$ taking place for all codes of the classes $B_i \in \Pi_A$.

If the method of transformation of the state codes into the codes of the classes $B_i \in \Pi_A$ is applied, it leads to the model of P_EC_AY Moore FSM shown in Fig. 7.15.

The synthesis method for P_EC_AY Moore FSM includes the following steps:

1. Construction of the partition Π_A and encoding of the classes of pseudoequivalent states $B_i \in \Pi_A$.
2. Encoding of collections of microoperations.
3. Construction of the table of microoperations.
4. Construction of an expanded table specifying the block TSM.
5. Construction of transformed structure table.
6. Synthesis of FSM logic circuit.

For the Moore FSM S_{22}, there are $I = 3$ classes $B_i \in \Pi_A$, thus they can be encoded using $R_0 = 2$ variables from the set $\tau = \{\tau_1, \tau_2\}$. Let us encode the classes $B_i \in \Pi_A$ in the following way: $K(B_1) = 00$, $K(B_2) = 01$, $K(B_3) = 10$. Let us encode the collections of microoperations using the same codes as for the PC_AY Moore FSM S_{22}. In this case table of microoperations is represented by Table 7.17. The expanded table for block TSM (Table 7.18) includes additional columns B_i and τ_m. The column τ_m contains variables $\tau_r \in \tau$, used to encode states $a_m \in B_i$. The transformed structure table of P_CC_AY Moore FSM is constructed in a trivial way. In case of the P_CC_AY Moore FSM S_{22}, this table is represented by Table 7.19.

Table 7.18 Expanded table for block TSM of $P_C C_A Y$ Moore FSM S_{22}

a_m	$K(a_m)$	Y_t	$K(Y_t)$	Z_m	B_i	τ_m	m
a_1	000	Y_1	00	–	B_1	–	1
a_2	001	Y_2	01	z_2	B_2	τ_2	2
a_3	010	Y_3	10	z_1	B_2	τ_2	3
a_4	011	Y_2	01	z_2	B_2	τ_2	4
a_5	100	Y_4	11	z_1z_2	B_3	τ_1	5
a_6	101	Y_2	01	z_2	B_3	τ_1	6
a_7	110	Y_4	11	z_1z_2	B_3	τ_2	7

The transformed structure table is used to derive the system of input memory functions, represented in the following form:

$$\phi_r = \bigvee_{h=1}^{H_0} C_{rh} B_h X_h \quad (r = 1, \ldots, R). \tag{7.13}$$

The next SOP $D_3 = F_1 \vee F_3 \vee F_5 = B_1x_1 \vee B_1\bar{x}_1\bar{x}_2 \vee B_2\bar{x}_3 = \bar{\tau}_1\bar{\tau}_2x_1 \vee \bar{\tau}_1\bar{\tau}_2\bar{x}_1\bar{x}_2 \vee \bar{\tau}_1\tau_2\bar{x}_3$ can be derived, for example, from Table 7.19. The logic circuit of $P_C C_A$YMoore FSM S_{22} is shown in Fig. 7.16.

Table 7.19 Transformed structure table of the $P_C C_A Y$ Moore FSM S_{22}

B_i	$K(B_i)$	a_s	$K(a_s)$	X_h	Φ_h	h
B_1		a_2	001	x_1	D_3	1
		a_3	010	$\bar{x}_1x_2$	D_2	2
		a_4	011	$\bar{x}_1\bar{x}_2$	D_2D_3	3
B_2	$1{*}{*}$	a_5	100	x_3	D_1	4
		a_6	101	$\bar{x}_3$	D_1D_3	5
B_3	$0{*}1$	a_7	110	x_4	D_1D_2	6
		a_1	000	$\bar{x}_4$	–	7

The similar approaches are used to synthesize logic circuits of PC_AD, P_EC_AD, and P_CC_AD Moore FSMs. The only difference consists in the encoding method for collections of microoperations. For example, let us discuss a method for logic circuit design in case of the P_CC_AD Moore FSM S_{23} (Table 7.20).

For the Moore FSM S_{23}, the following partition $\Pi_A = \{B_1, B_2, B_3\}$ can be found, where $B_1 = \{a_1\}$, $B_2 = \{a_2, a_3\}$, $B_3 = \{a_4, a_5, a_6\}$. It means that $I = 3$, $R_0 = 2$, and $\tau = \{\tau_1, \tau_2\}$. Let us encode the classes $B_i \in \Pi_A$ in the following way: $K(B_1) = 00$, $K(B_2) = 01$, $K(B_3) = 10$. For the Moore FSM S_{23}, there are two classes of compatible microoperations, namely the class $Y^1 = \{y_1, y_3, y_5\}$ and the class $Y^2 = \{y_2, y_4, y_6\}$. Let us use variables z_1 and z_2 for encoding of microoperations $y_n \in Y^1$, whereas microoperations $y_n \in Y^2$ are encoded using variables z_3 and z_4. It determines the following set $Z = \{z_1, \ldots, z_4\}$, used to construct Table 7.21.

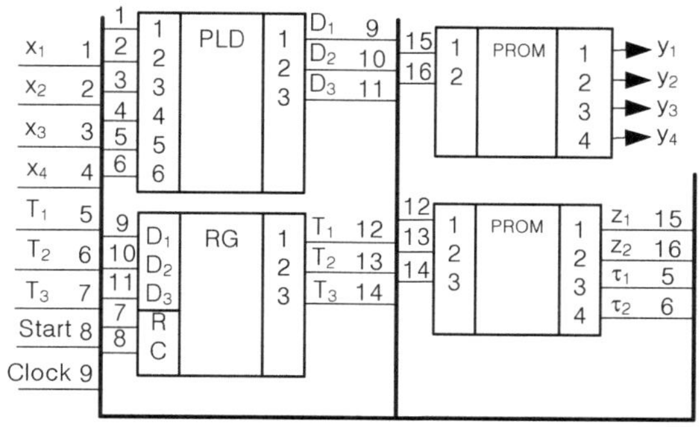

Fig. 7.16 Logic circuit of $P_C C_A Y$ Moore FSM S_{22}

Table 7.20 Structure table of Moore FSM S_{23}

a_m	$K(a_m)$	a_s	$K(a_s)$	X_h	Φ_h	h
a_1 (–)	000	a_2	001	x_1	D_3	1
		a_3	010	$\bar{x}_1$	D_2	2
a_2 (y_1y_2)	001	a_6	101	x_2	D_1D_3	3
		a_5	100	$\bar{x}_2x_3$	D_1	4
		a_4	011	$\bar{x}_2\bar{x}_3$	D_2D_3	5
a_3 (y_3y_4)	010	a_6	101	x_2	D_1D_3	6
		a_5	100	$\bar{x}_2x_3$	D_1	7
		a_4	011	$\bar{x}_2\bar{x}_3$	D_2D_3	8
a_4 (y_5y_6)	011	a_1	000	x_4	–	9
		a_5	100	$\bar{x}_4$	D_1	10
a_5 (y_1y_6)	100	a_1	000	x_4	–	11
		a_5	100	$\bar{x}_4$	D_1	12
a_6 (y_2y_3)	101	a_1	000	x_4	–	13
		a_5	100	$\bar{x}_4$	D_1	14

Table 7.21 Codes of microoperations for Moore FSM S_{23}

Y^1	$K(y_n^1)$ z_1z_2	Y^2	$K(y_n^2)$ z_3z_4
$\emptyset$	0 0	$\emptyset$	0 0
y_1	0 1	y_2	0 1
y_3	1 0	y_4	1 0
y_5	1 1	y_6	1 1

The block TSM of $P_C C_A Y$ Moore FSMS_{23} is represented by its expanded table (Table 7.22). This table is constructed using the same approach as the one used in design of $P_C C_A Y$ Moore FSM.

The system of input memory functions is constructed using the FSM transformed structure table. This table is constructed using the replacement of the states by corresponding classes of pseudoequivalent states. The transformed ST of $P_C C_A Y$

Table 7.22 Expanded table for block TSM of $P_C C_A Y$ Moore FSM S_{23}

a_m	$K(a_m)$	Y_t	$K(Y_t)$	Z_m	B_i	τ_m	m
a_1	000	Y_1	0000	–	B_1	–	1
a_2	001	Y_2	0101	z_2z_4	B_2	τ_2	2
a_3	010	Y_3	1010	z_1z_3	B_2	τ_2	3
a_4	011	Y_4	1111	$z_1z_2z_3z_4$	B_3	τ_1	4
a_5	100	Y_5	0111	$z_2z_3z_4$	B_3	τ_1	5
a_6	101	Y_6	1001	z_1z_4	B_3	τ_1	6

Table 7.23 Transformed structure table of $P_C C_A Y$ Moore FSM S_{23}

B_i	$K(B_i)$	a_s	$K(a_s)$	X_h	Φ_h	h
B_1	00	a_2	001	x_1	D_3	1
		a_3	010	$\bar{x}_1$	D_2	2
B_2	01	a_6	100	x_2	D_1D_3	3
		a_5	101	$\bar{x}_2x_3$	D_1	4
		a_4	011	$\bar{x}_2\bar{x}_3$	D_2D_3	5
B_3	10	a_1	110	x_4	–	6
		a_5	000	$\bar{x}_4$	D_1	7

Moore FSM S_{23} includes only $H_0 = 7$ rows (Table 7.23). It is twice less than in case of its initial ST (Table 7.20).

The logic circuit of $P_C C_A Y$ Moore FSM S_{23} is shown in Fig. 7.17.

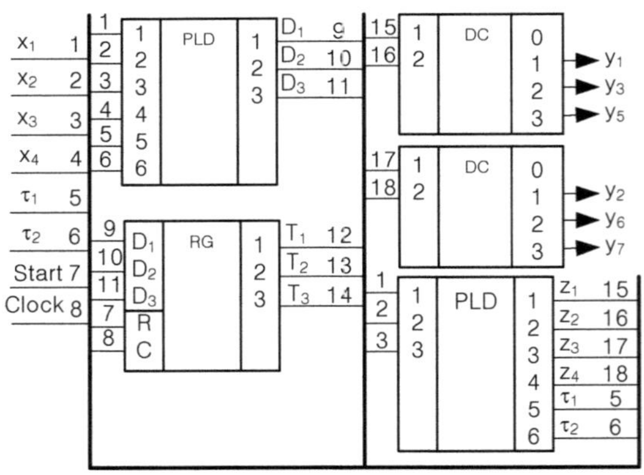

Fig. 7.17 Logic circuit of $P_C C_A Y$ Moore FSM S_{23}

Models of Moore FSM2 have structural diagrams similar to the one shown in Fig. 7.4. Synthesis methods for $PC_Y Y$ and $PC_Y D$ models include the following steps:

1. Encoding of collections of microoperations.
2. One-to-one identification of FSM states by collections of microoperations Y_t and identifiers $I_k \in I$.

3. Construction of transformed structure table.
4. Specification of code transformer for collections of microoperations.
5. Construction of table of microoperations.
6. Implementation of FSM logic circuit using some particular macrocells and embedded memory blocks.

Let us discus application of this method for logic synthesis of the PC_YY Moore FSM S_{22}, represented by its ST (Table 7.14).

Let us encode the collections of microoperations $Y_t \subseteq Y$ in the following way: $K(Y_1) = 00$, $K(Y_2) = 01$, $K(Y_3) = 10$, and $K(Y_4) = 11$. Remind, that $Y_1 = \emptyset$, $Y_2 = \{y_1, y_2\}$, $Y_3 = \{y_3\}$, and $Y_4 = \{y_3, y_4\}$. Let us find the sets of states $A(Y_t) \subseteq A$, such that they include the set $Y_t \subseteq Y$. For the Moore FSM S_{22}, there are the following sets: $A(Y_1) = \{a_1\}$, $A(Y_2) = \{a_2, a_4, a_6\}$, $A(Y_3) = \{a_3\}$, and $A(Y_4) = \{a_5, a_7\}$, having the following numbers of states $m_1 = m_3 = 1$, $m_2 = 3$, and $m_4 = 2$. It determines the set of identifiers $I = \{I_1, I_2, I_3\}$. It is enough $R_V = 2$ variables from the set $V = \{v_1, v_2\}$ for encoding of identifiers. Let us encode the identifiers $I_k \in I$ in the following way: $K(I_1) = 00$, $K(I_2) = 01$, $K(I_3) = 10$. Now the states $a_m \in A$ are one-to-one identified by pairs $\alpha_{1,1} = \langle Y_1, \emptyset \rangle$, $\alpha_{2,4} = \langle Y_2, I_2 \rangle$, $\alpha_{2,6} = \langle Y_2, I_3 \rangle$, $\alpha_{3,3} = \langle Y_3, \emptyset \rangle$, $\alpha_{4,5} = \langle Y_4, I_1 \rangle$, and $\alpha_{4,7} = \langle Y_4, I_2 \rangle$. The codes $C(a_m)^t$ of states $a_m \in A$ are shown in Table 7.24. In this table, two the most-significant bits of code $C(a_m)^t$ correspond to variables z_1 and z_2, whereas the variables v_1 and v_2 correspond to the least-significant bits of the state code.

The following two rules are used to construct the transformed ST of PC_YY Moore FSM. First, the column $K(a_s)$ is replaced by the column $C(a_s)^t$. Second, the column Φ_h contains $Q + R_V$ input memory functions, determined by the codes $C(a_s)^t$. For the PC_YY Moore FSM S_{22}, its transformed ST is represented by Table 7.25.

This table is used for deriving of input memory functions $\Phi = \Phi(T, X)$. The following Boolean equation $D_4 = F_3 \vee F_{10} \vee F_{12} = A_1\bar{x}_1\bar{x}_2 \vee A_5x_4 \vee A_6x_4 = \bar{T}_1\bar{T}_2\bar{T}_3\bar{x}_1\bar{x}_2 \vee T_1\bar{T}_2\bar{T}_3x_4 \vee T_1\bar{T}_2T_3x_4$,for example, is derived from Table 7.25.

The block TMS is specified by the table having columns α_t, m, $C(a_m)^t$, a_m, $K(a_m)$, T_m, m. The column T_m includes state variables $T_r \in T$equal to 1 in state codes $K(a_m)$. For the PC_YY Moore FSM S_{22}, this table is represented by Table 7.26.

The table of block TMS is used to derive state variables

$$T_r = \bigvee_{m=1}^{M} C_{rm} Z_t V_k \quad (r = 1, \ldots, R). \tag{7.14}$$

Table 7.24 State codes for PC_YY Moore FSM S_{22}

a_m	$C(a_m)^t$	$\alpha_{t,m}$	a_m	$C(a_m)^t$	$\alpha_{t,m}$
a_1	00∗∗	$\alpha_{1,1}$	a_5	1100	$\alpha_{4,5}$
a_2	0100	$\alpha_{2,2}$	a_6	0110	$\alpha_{2,6}$
a_3	10∗∗	$\alpha_{3,3}$	a_7	1101	$\alpha_{4,7}$
a_4	0101	$\alpha_{2,4}$			

Table 7.25 Transformed structure table of PC_YY Moore FSM S_{22}

a_m	$K(a_m)$	a_s	$C(a_s)^t$	X_h	Φ_h	h
a_1	000	a_2	0100	x_1	D_2	1
		a_3	10∗∗	$\bar{x}_1x_2$	D_1	2
		a_4	0101	$\bar{x}_1\bar{x}_2$	D_2D_4	3
a_2	001	a_5	1100	x_3	D_1D_2	4
		a_6	0110	$\bar{x}_3$	D_2D_3	5
a_3	010	a_5	1100	x_3	D_1D_2	6
		a_6	0110	$\bar{x}_3$	D_2D_3	7
a_4	011	a_5	1100	x_3	D_1D_2	8
		a_6	0110	$\bar{x}_3$	D_2D_3	9
a_5	100	a_7	1101	x_4	$D_1D_2D_4$	10
		a_1	00∗∗	$\bar{x}_4$	–	11
a_6	101	a_7	1101	x_4	$D_1D_2D_4$	12
		a_1	00∗∗	$\bar{x}_4$	–	13
a_7	110	a_5	1100	x_3	D_1D_2	14
		a_6	0110	$\bar{x}_3$	D_2D_3	15

Table 7.26 Specification of block TMS for PC_YY Moore FSM S_{22}

a_m	$C(a_m)^t$	$\alpha_{t,m}$	$K(a_m)$	T_m	m
a_1	00∗∗	$\alpha_{1,1}$	000	–	1
a_2	0100	$\alpha_{2,2}$	001	T_3	2
a_3	10∗∗	$\alpha_{3,3}$	010	T_2	3
a_4	0101	$\alpha_{2,4}$	011	T_2T_3	4
a_5	1100	$\alpha_{4,5}$	100	T_1	5
a_6	0110	$\alpha_{2,6}$	101	T_1T_3	6
a_7	1101	$\alpha_{4,7}$	110	T_1T_2	7

In (7.14), the symbol C_{rm} stands for a Boolean variable, equal to 1 iff the bit r of code $K(a_m)$ is equal to 1; the symbol Z_t determines a conjunction of variables $z_q \in Z$, corresponded to code $K(Y_t)$; the symbol V_k determines a conjunction of variables $v_r \in V$, corresponded to code $K(I_k)$. For the PC_YY Moore FSM S_{22}, we can find the SOP $T_3 = F_2 \vee F_4 \vee F_6 = \bar{z}_1z_2\bar{v}_1\bar{v}_2 \vee \bar{z}_1z_2\bar{v}_1v_2 \vee \bar{z}_1z_2v_1\bar{v}_2$, for example, from Table 7.26. The block of microoperations is specified by Table 7.15. The logic circuit of PC_YY Moore FSM S_{22} is shown in Fig. 7.18.

Obviously, the block used for generation of functions (7.14) can also generates variables $\tau_r \in \tau$, encoded the classes of pseudoequivalent states. Such an approach yields in P_CC_YY Moore FSM (Fig. 7.19). Obviously, there is no sense in optimal state encoding for this model.

Let us discuss an example of logic circuit design for the P_CC_YY Moore FSM S_{22}. The design method differs from the previous one because of necessity for finding partition Π_A and encoding the classes of pseudoequivalent states $B_i \in \Pi_A$. For the

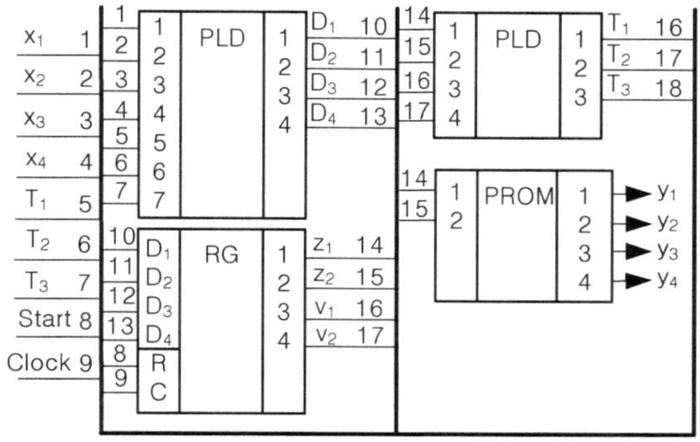

Fig. 7.18 Logic circuit of P_CC_YY Moore FSM S_{22}

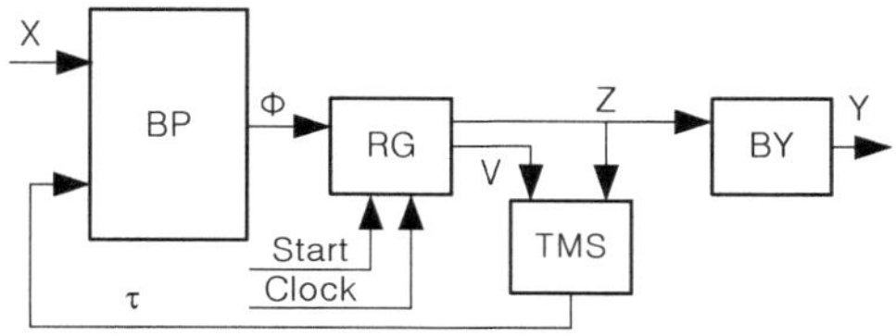

Fig. 7.19 Structural diagram of P_CC_YY Moore FSM

Table 7.27 Transformed structure table of P_CC_YY Moore FSM S_{22}

B_i	$K(B_i)$	a_s	$C(a_s)^t$	X_h	Φ_h	h
B_1	00	a_2	0100	x_1	D_3	1
		a_3	$10{*}{*}$	$\bar{x}_1x_2$	D_1	2
		a_4	0101	$\bar{x}_1\bar{x}_2$	D_2D_4	3
B_2	01	a_5	1100	x_3	D_1D_2	4
		a_6	0110	$\bar{x}_3$	D_2D_3	5
B_3	10	a_7	1101	x_4	$D_1D_2D_4$	6
		a_1	$00{*}{*}$	$\bar{x}_4$	–	7

P_CC_YY Moore FSM S_{22}, it could be constructed the partition $\Pi_A = \{B_1, B_2, B_3\}$ with classes $B_1 = \{a_1\}$, $B_2 = \{a_2, a_3, a_4, a_7\}$, and $B_3 = \{a_5, a_6\}$. Let us use the variables $\tau_r \in \tau = \{\tau_1, \tau_2\}$ for encoding of the classes $B_i \in \Pi_A$. Let us encode the classes $B_i \in \Pi_A$ in the following way:$K(B_1) = 00$, $K(B_2) = 01$, $K(B_3) = 10$. These codes allow to construct the transformed ST (Table 7.27) used to design the logic circuit of block BP for the P_CC_YY Moore FSM S_{22}.

The code transformer TMS is represented by Table 7.28. In this table, states $a_m \in B_i$ are replaced by corresponding classes B_i, whereas state codes $K(a_m)$ are replaced by corresponding code $K(B_i)$. Obviously, this replacement leads to replacement of the column T_m by the column τ_m.

The transformed ST is used to derive the functions $\Phi = \Phi(\tau, X)$. For example, the SOP $D_4 = B_1\bar{x}_1\bar{x}_2 \vee B_3x_4 = \bar{\tau}_1\bar{\tau}_2\bar{x}_1\bar{x}_2 \vee \tau_1\bar{\tau}_2x_4$ can be derived from Table 7.27. The system $\tau = \tau(Z, V)$ is derived from the table of block TMS. For example, the

Table 7.28 Specification of block TMS for P_CC_YY Moore FSM S_{22}

B_i	$C(a_m)^t$	$\alpha_{t,m}$	$K(B_i)$	τ_m	m
B_1	00∗∗	$\alpha_{1,1}$	00	–	1
B_2	0100	$\alpha_{2,2}$	01	τ_2	2
B_2	10∗∗	$\alpha_{3,3}$	01	τ_2	3
B_2	0101	$\alpha_{2,4}$	01	τ_2	4
B_3	1100	$\alpha_{4,5}$	10	τ_1	5
B_3	0110	$\alpha_{2,6}$	10	τ_1	6
B_2	1101	$\alpha_{4,7}$	01	τ_2	7

Fig. 7.20 Logic circuit of P_CC_YY Moore FSM S_{22}

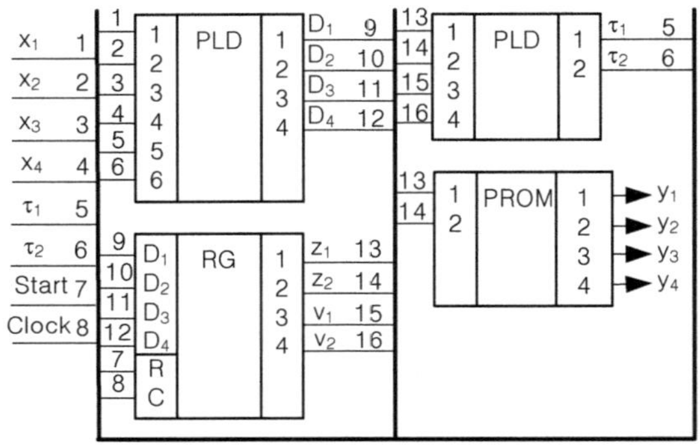

Fig. 7.21 Structural diagram of PC_YY Moore FSM

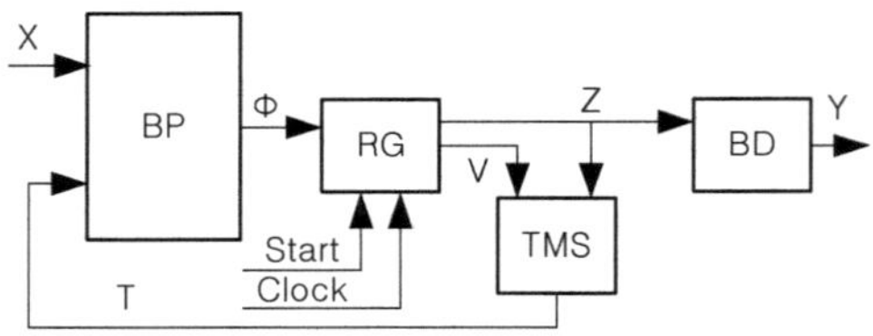

SOP $\tau_1 = F_5 \vee F_6 = z_1 z_2 \bar{v}_1 \bar{v}_2 \vee \bar{z}_1 z_2 v_1 \bar{v}_2$ can be derived from Table 7.28. The logic circuit of P_CC_YY Moore FSM S_{22} is shown in Fig. 7.20.

If the method of encoding of the classes of compatible microoperations is used, then it results in models of PC_YD Moore FSM (Fig.7.21). If this approach is used simultaneously with transformation of the classes of pseudoequivalent states, then it leads to models of P_CC_YD Moore FSM (Fig. 7.22).

The model of PC_YD Moore FSM can be viewed as a particular case of P_CC_YD Moore FSM. Let us discuss the method for model P_CC_YD Moore FSM synthesis.

There are the following steps in P_CC_YD Moore FSM logic synthesis:

1. Finding of the classes of compatible microoperations and their codes.
2. Identification of states $a_m \in A$ by pairs $\alpha_t, m = \langle Y_t, I_k \rangle$.
3. Finding of the partition of the state set A by the classes of pseudoequivalent states and encoding of classes $B_i \in \Pi_A$.

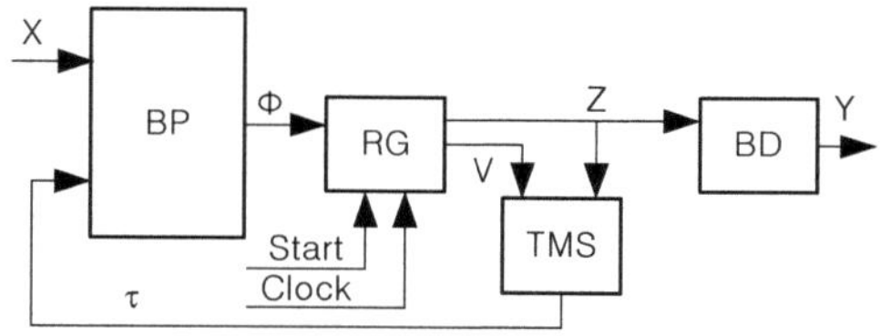

Fig. 7.22 Structural diagram of $P_C C_Y D$ Moore FSM

4. Construction of transformed structure table and functions $\Phi = \Phi(\tau, X)$.
5. Specification of block TMS and construction of system $\tau = \tau(Z, V)$.
6. Implementation of FSM logic circuit using given logic elements.

Let us discuss an example of this method application for logic synthesis of the PC_YY Moore FSM S_{24} (Table 7.29). In this case, the set of microoperations Y is divided by two classes, namely $Y^1 = \{y_1, y_3, y_5\}$ and $Y^2 = \{y_2, y_4, y_6\}$. The microoperations from the first class are encoded by variables z_1 and z_2, whereas variables z_3 and z_4 are used to encode microoperations from the second class. These four variables form the set Z. Let us encode microoperations $y_n \in Y^i$ in the way shown in Table 7.21, it leads to the following codes: $K(Y_1) = 0000$, where $Y_1 = \emptyset$; $K(Y_2) = 0101$, where $Y_2 = \{y_1, y_2\}$; $K(Y_3) = 1010$, where $Y_3 = \{y_3, y_4\}$; $K(Y_4) = 1111$, where $Y_4 = \{y_5, y_6\}$; $K(Y_5) = 1011$, where $Y_5 = \{y_3, y_6\}$.

The following sets of states $A(Y_t)$ can be found for the Moore FSM S_{24}: $A(Y_1) = \{a_1\}, A(Y_2) = \{a_2, a_5\}, A(Y_3) = \{a_3\}, A(Y_4) = \{a_4\}, A(Y_5) = \{a_6\}$. Therefore, there is a set of identifiers $I = \{I_1, I_2\}$ and its elements can be coded using only one variable from the set $V = \{v_1\}$. Let us construct the set of pairs $\alpha_{t,m}$ used for one-to-one identification of the FSM states. There are the following pairs for the Moore FSM S_{24}: $\alpha_{1,1} = \langle Y_1, \emptyset \rangle$, $\alpha_{2,2} = \langle Y_1, I_1 \rangle$, $\alpha_{2,5} = \langle Y_1, I_2 \rangle$, $\alpha_{3,3} = \langle Y_3, \emptyset \rangle$, $\alpha_{4,4} = \langle Y_4, \emptyset \rangle$,

Table 7.29 Structure table of Moore FSM S_{24}

a_m	$K(a_m)$	a_s	$K(a_s)$	X_h	Φ_h	h
a_1 (–)	000	a_2	001	x_1	D_3	1
		a_3	010	$\bar{x}_1$	D_2	2
a_2 (y_1y_2)	001	a_5	100	x_1	D_1	3
		a_4	011	$\bar{x}_1x_2$	D_1D_3	4
		a_1	000	$\bar{x}_1\bar{x}_2$	–	5
a_3 (y_3y_4)	010	a_5	100	x_1	D_1	6
		a_4	011	$\bar{x}_1x_2$	D_1D_3	7
		a_1	000	$\bar{x}_1\bar{x}_2$	–	8
a_4 (y_5y_6)	011	a_2	001	x_3	D_3	9
		a_6	101	$\bar{x}_3$	D_1D_3	10
a_5 (y_1y_2)	100	a_2	001	x_3	D_3	11
		a_6	101	$\bar{x}_3$	D_1D_3	12
a_6 (y_3y_6)	101	a_2	001	x_3	D_3	13
		a_6	101	$\bar{x}_3$	D_1D_3	14

Table 7.30 State codes for Moore FSM S_{24}

a_m	$C(a_m)^t$ $z_1z_2z_3z_4*$	m	a_m	$C(a_m)^t$ $z_1z_2z_3z_4v_1$	m
a_1	0 0 0 0 *	1	a_4	1 1 1 1 *	4
a_2	0 1 0 1 0	2	a_5	0 1 0 1 1	5
a_3	1 0 1 – *	2	a_6	1 0 1 1 *	6

and $\alpha_{5,6} = \langle Y_5, \emptyset \rangle$. Let the identifiers for the Moore FSM S_{24} have the following codes: $K(I_1) = 0$, $K(I_2) = 1$. It allows finding the codes $C(a_m)^t$ of the states shown in Table 7.30.

The state set A includes three classes of pseudoequivalent states, namely $B_1 = \{a_1\}$, $B_2 = \{a_2, a_3\}$, and $B_3 = \{a_4, a_5, a_6\}$. It is enough $R_0 = 2$ elements from the set $\tau = \{\tau_1, \tau_2\}$ for encoding of the classes $B_i \in \Pi_A$. Let the classes be encoded in a trivial way:$K(B_i) = 00, \ldots, K(B_3) = 10$. To construct the transformed structure table, it is necessary to replace the column a_m by the column B_i, the column $K(a_m)$ by the column $K(B_i)$, and the column $K(a_s)$ by the column $C(a_s)^t$. Besides, the column Φ_h includes input memory functions for loading of variables $z_q \in Z$ and $v_r \in V$ into the register RG. The block TMS is specified by its table having columns a_m, $C(a_m)^t$, B_i, $K(B_i)$, τ_m, and m. For both these tables, the codes $C(a_m)^t$ of states $a_m \in A$ are taken from a table similar to Table 7.30.

For the P_CC_YD Moore FSM S_{24}, its transformed ST and table specified the block TMS are shown in Table 7.31 and Table 7.32 respectively. Table 7.31 is used to derive the equations of system $\Phi = \Phi(\tau, X)$. For example, the following SOP $D_1 = F_2 \vee F_5 = B_1\bar{x}_1 \vee B_2\bar{x}_2\bar{x}_3 = \bar{\tau}_1\bar{\tau}_2\bar{x}_1 \vee \bar{\tau}_1\tau_2\bar{x}_2\bar{x}_3$ can be derived from the transformed structure table of the P_CC_YD Moore FSM S_{24}.

Table 7.32 is used to derive the system $\tau = \tau(Z, V)$, for example, the following SOP $\tau_2 = A_2 \vee A_3 = \bar{z}_1z_2\bar{z}_3z_4v_1 \vee z_1\bar{z}_2z_3\bar{z}_4$. The logic circuit of P_CC_YD Moore FSM S_{24} is shown in Fig. 7.23. In this circuit, the logic circuit of block BP is implemented using some macrocells, it implements input memory functions $\Phi = \Phi(\tau, X)$, loading variables $z_q \in Z$ and $v_r \in V$into the register RG. The logic circuit of block BD is implemented using standard decoders; it implements functions $Y = Y(Z)$, specified

Table 7.31 Transformed structure table of P_CC_YD Moore FSM S_{24}

B_i	$K(B_i)$	a_s	$C(a_s)^t$	X_h	Φ_h	h
B_1	00	a_2	01010	x_1	D_2D_4	1
		a_3	1010*	$\bar{x}_1$	D_1D_3	2
B_2	01	a_6	1011*	x_2	$D_1D_3D_4$	3
		a_5	01011	$\bar{x}_2x_3$	$D_2D_4D_5$	4
		a_4	1111*	$\bar{x}_2\bar{x}_3$	$D_1D_2D_3D_4$	5
B_3	10	a_7	0000*	x_4	–	6
		a_1	01011	$\bar{x}_4$	$D_2D_4D_5$	7

Table 7.32 Specification of block TMS for $P_C C_Y D$ Moore FSM S_{24}

a_m	$C(a_m)^t$	B_i	$K(B_i)$	τ_m	m
a_1	0000*	B_1	00	–	1
a_2	01010	B_2	01	τ_2	2
a_3	1010*	B_2	01	τ_2	3
a_4	1111*	B_3	10	τ_1	4
a_5	01011	B_3	10	τ_1	5
a_6	1011*	B_3	10	τ_1	6

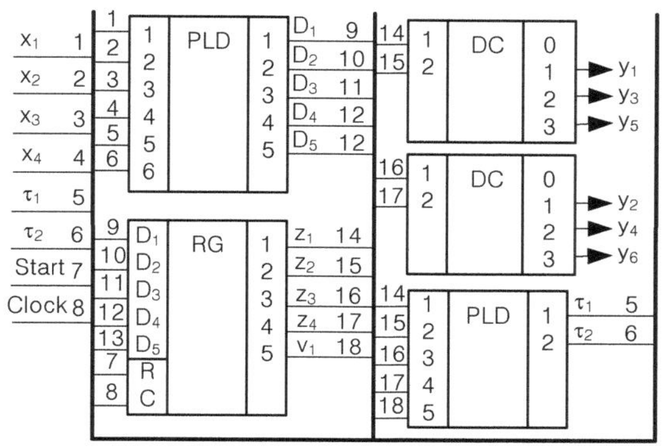

Fig. 7.23 Logic circuit of $P_C C_Y D$ Moore FSM S_{24}

by Table 7.21. The logic circuit of block TMS is implemented using macrocells, it generates functions $\tau = \tau(Z, V)$.

7.4 Multilevel Models of FSM with Object Code Transformation

The FSM models discussed in previous Sections are used to construct the multilevel models of FSM with object code transformation. Let the symbol G stand for the number of variables $p_g \in P$, used for replacement of logical conditions $x_l \in X$. In this case, the replacement of logical conditions leads to $3G$ different FSM models. Let the symbol K stand for the number of classes of compatible microoperations, then there are K models including the block D_k forming microoperations $y_n \in Y$. Let us point out that there is no sense in using the method of encoding of structure table rows, because the transformed table in this case contains both codes of states and microoperations. Therefore, multilevel models of Mealy FSM with object code transformation can include up to four levels. These models are represented by Table 7.33.

For Mealy FSM1, there are:

1. $n(BC) = (K+1)$ models with three levels;
2. $n(ABC) = 3G(K+1)$ models with four levels.

Table 7.33 Multilevel models of Mealy FSM with transformation of object codes

LA	LB	LC
		Y
M_1 M_1C M_1L	PC_A	D_1
$\vdots$	PC_Y	$\vdots$
M_G M_GC M_GL		D_K

For Mealy FSM2, there are:

1. $n(BC) = (K+1)$ models with two levels;
2. $n(ABC) = 3G(K+1)$ models with three levels.

In common case there are $n_1 = 6GK+6G+2K+2$ different models of Mealy FSM with object code transformation. If $G = K = 6$, then there are $n_1 = 268$ different models.

Let us discuss an example of logic synthesis for the MPLC$_Y$D$_K$ Mealy FSM S_{21} (Table 7.7). This model is shown in Fig. 7.24.

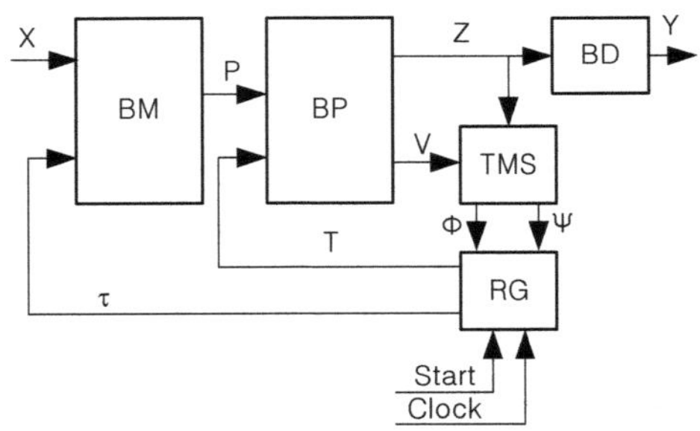

Fig. 7.24 Structural diagram of MPLC$_Y$D$_K$ Mealy FSM

In this model, the block TMS generates input memory functions loading codes of states $a_m \in A$ into the register RG, as well as functions $\Psi = \Psi(Z,V)$, used as variables for encoding of logical conditions $x_l \in X$. Synthesis method for MPLC$_Y$D$_K$ Mealy FSM includes the following steps:

1. Logical condition replacement and construction of the set P.
2. Logical condition encoding and construction of the set τ.
3. Finding of the classes of compatible microoperations, encoding of microoperations and construction of the set Z.
4. Identification of states $a_m \in A$ by pairs $\alpha_{t,m} = \langle Y_t, I_k \rangle$ and construction of the set V.
5. Construction of transformed structure table and systems $Z = Z(T,P)$ and $Z = Z(T,P)$.
6. Specification of the block TMS and deriving systems Φ and τ.
7. Implementation of the FSM logic circuit.

For the MPLC$_Y$D$_K$ Mealy FSM S_{21}, there are the following sets of logical conditions: $X(a_1) = \{x_1\}$, $X(a_2) = \{x_2,x_3\}$, $X(a_3) = \{x_4\}$, $X(a_4) = \emptyset$, and $X(a_5) = \{x_2,x_3,x_4\}$. It means that $G = 3$ and logical conditions are replaced by variables from the set $P = \{p_1, p_2, p_3\}$. The outcome of logical condition distribution is shown in Table 7.34.

Table 7.34 Replacement of logical conditions for FSM S_{21}

a_m	a_1	a_2	a_3	a_4	a_5
p_1	x_1	x_2	–	–	x_2
p_2	–	x_3	–	–	x_3
p_3	–	–	x_4	–	x_4

As follows from Table 7.34, the identities $p_2 = x_3$ and $p_3 = x_4$ take place. It means that there is only one multiplexer in the block BM. To encode the logical conditions $x_l \in X(p_1)$, it is enough only one variable from the set $\tau = \{\tau_1\}$. Let $K(x_1) = 0$ and $K(x_2) = 1$. There are two classes of compatible microoperations, namely $Y^1 = \{y_1,y_3,y_4,y_5\}$ and $Y^2 = \{y_2,y_6,y_7\}$. The codes of microoperations are shown in Table 7.5. From this table we can find the set of variables $Z = \{z_1,\dots,z_5\}$.

To identify the states, let us find sets $A(Y_t)$, where $Y_1 = \emptyset$, $Y_2 = \{y_1,y_2\}$, $Y_3 = \{y_3\}$, $Y_4 = \{y_4\}$, $Y_5 = \{y_2,y_5\}$, $Y_6 = \{y_6\}$, $Y_7 = \{y_3,y_7\}$. For the MPLC$_Y$D$_K$ Mealy FSM S_{21}, the following sets can be found: $A(Y_1) = \{a_1\}$, $A(Y_2) = \{a_2,a_4\}$, $A(Y_3) = \{a_3\}$, $A(Y_4) = \{a_3\}$, $A(Y_5) = \{a_4\}$, $A(Y_6) = \{a_5\}$, and $A(Y_7) = \{a_5\}$. Now we can find the pairs $\alpha_{t,m}$ for identification of states $a_m \in A$, namely: $\alpha_{1,1} = \langle Y_1,\emptyset\rangle$, $\alpha_{2,2} = \langle Y_2,I_1\rangle$, $\alpha_{2,4} = \langle Y_2,I_2\rangle$, $\alpha_{3,3} = \langle Y_3,\emptyset\rangle$, $\alpha_{4,3} = \langle Y_4,\emptyset\rangle$, $\alpha_{5,4} = \langle Y_5,\emptyset\rangle$, $\alpha_{6,5} = \langle Y_6,\emptyset\rangle$, and $\alpha_{7,5} = \langle Y_7,\emptyset\rangle$. Obviously, there are two identifiers included into the set $I = \{I_1,I_2\}$; it is enough only one variable v_1 for their encoding ($V = \{v_1\}$). Let $K(I_1) = 0$ and $K(I_2) = 1$.

The transformed structure table of MPLC$_Y$D$_K$ Mealy FSM includes the columns a_m, $K(a_m)$, P_h, Z_h, V_h, h. The column Z_h contains variables $z_q \in Z$, equal to 1 in the code $K(Y_h)$. The column V_h contains variables $v_r \in V$, equal to 1 in the identifier code for state a_s from the row h of initial ST. For the MPLC$_Y$D$_K$ Mealy FSM S_{21}, this table is shown in Table 7.35.

The systems $Z = Z(T,P)$ and $Z = Z(T,P)$ can be derived from the transformed ST. For example, the following systems $z_1 = F_6 = A_3 p_3 = \bar{T}_1 T_2 T_3 p_3$ and $v_1 = F_5 = A_2 \bar{p}_1 \bar{p}_2 = \bar{T}_1 T_2 \bar{T}_3 \bar{p}_1 \bar{p}_2$ can be derived from Table 7.35.

The code transformer is specified by a table having columns $\alpha_{t,m}$, $C(a_m)^t$, a_m, $K(a_m)$, Φ_m, $K(x_1)$, Ψ_m, m. The column $C(a_m)^t$ contains the code of state a_m, determined by the concatenation $K(Y_t) * K(I_k)$. The column Φ_m includes input memory functions D_r, equal to 1 for loading the code $K(a_m)$ into register RG . The column Ψ_m includes variables $\psi_r \in \Psi$, equal to 1 in the code $K(x_1)$ of logical condition determined the transition into state $a_m \in A$. For the MPLC$_Y$D$_K$ Mealy FSM S_{21}, the block TMS is specified by Table 7.36.

Table 7.35 Transformed structure table of $MPLC_YD_K$ Mealy FSM S_{21}

a_m	$K(a_m)$	P_h	Z_h	V_h	h
a_1	000	p_1	z_2z_5	–	1
		$\bar{p}_1$	z_2	–	2
a_2	010	p_1	z_3z_5	–	3
		$\bar{p}_1p_2$	z_2z_3	–	4
		$\bar{p}_1\bar{p}_2$	z_3z_5	v_1	5
a_3	011	p_3	z_1z_5	–	6
		$\bar{p}_3$	z_4	–	7
a_4	100	1	$z_2z_4z_5$	–	8
a_5	101	p_1p_2	z_3z_5	–	9
		$p_1\bar{p}_2$	z_2	–	10
		$\bar{p}_1p_3$	$z_2z_4z_5$	–	11
		$\bar{p}_1\bar{p}_3$	–	–	12

Table 7.36 Specification of block TMS for $MPLC_YD_K$ Mealy FSM S_{21}

a_m	$C(a_m)^t$	$\alpha_{t,m}$	$K(a_m)$	Φ_m	$K(x_1)$	Ψ_m	m
a_1	00000∗	$\alpha_{1,1}$	000	–	0	–	1
a_2	001010	$\alpha_{2,2}$	010	D_2	1	ψ_1	2
a_3	01000∗	$\alpha_{3,3}$	011	D_2D_3	–	–	3
a_3	01100∗	$\alpha_{4,3}$	011	D_2D_3	–	–	4
a_4	001011	$\alpha_{2,4}$	100	D_1	–	–	5
a_4	10001∗	$\alpha_{5,4}$	101	D_1	–	–	6
a_5	00010∗	$\alpha_{6,5}$	110	D_1D_3	–	–	7
a_5	01011∗	$\alpha_{7,5}$	110	D_1D_3	–	–	8

This table is used to derive the systems and Ψ. The SOPs $D_2 = F_2 \vee F_3 \vee F_4 = \bar{z}_1\bar{z}_2z_3\bar{z}_4z_5\bar{v}_1 \vee \bar{z}_1z_2z_3\bar{z}_4\bar{z}_5 \vee \bar{z}_1z_2\bar{z}_3\bar{z}_4\bar{z}_5$ and $\Psi_1 = F_2$ can be derived, for example, from Table 7.36. The logic circuit of $MPLC_YD_K$ Mealy FSM S_{21} is shown in Fig. 7.25.

Let us discuss an example of logic synthesis for the MPC_AY Mealy FSM S_{21} , represented by its structure table (Table 7.7). The structural diagram for this model is shown in Fig. 7.26. This model includes four combinational blocks (BM, BP, TSM, and BY) and the register RG. For this model, state codes are transformed into microoperations of FSM.

Synthesis method for MPC_AY Mealy FSM includes the following steps:

1. Logical condition replacement and construction of the set P.
2. Encoding of collections of microoperations and construction of set Z.
3. Construction of the set with pairs $\alpha_{t,m}$, identified collections of microoperations, and finding the set V of variables, encoding identifiers $I_k \in I$.
4. Construction of the transformed ST and deriving functions $\Phi = \Phi(Z,V)$ and $V = V(T,P)$.
5. Specification of the block TSM and construction of the system $Z = Z(T,P)$.

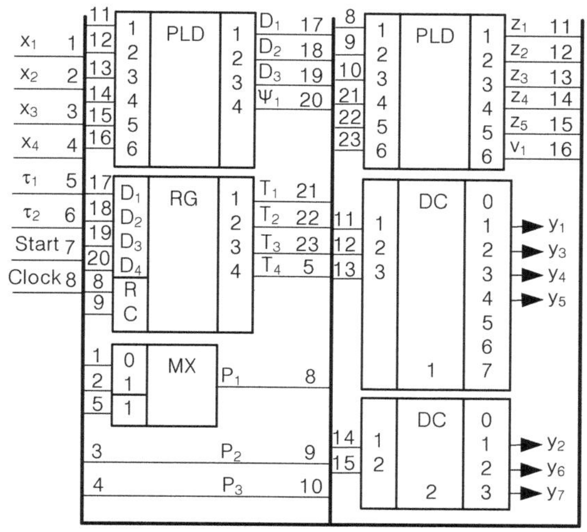

Fig. 7.25 Logic circuit of $MPLC_YD_K$ Mealy FSM S_{21}

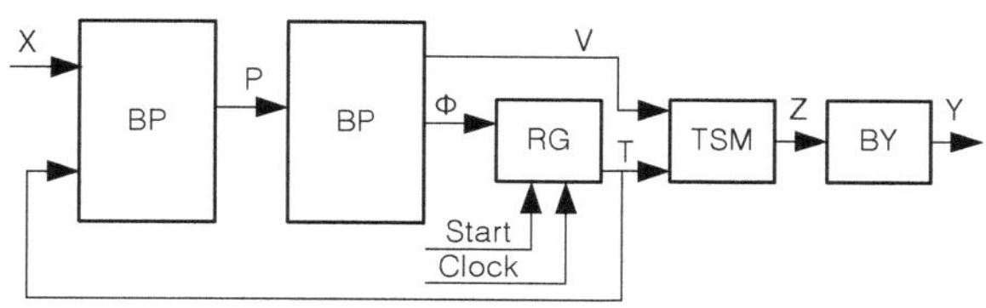

Fig. 7.26 Structural diagram of MPC_AY Mealy FSM

6. Construction of the table of microoperations and system $Y = Y(Z)$.
7. Implementation of FSM logic circuit.

As it was found before for the FSM S_{21}, the set P includes three elements, distribution of logical conditions $x_l \in X$ among elements of the set $P = \{p_1, p_2, p_3\}$ is shown in Table 7.34.

The initial ST includes $T_0 = 7$ collections of microoperations, namely: $Y_1 = \emptyset$, $Y_2 = \{y_1, y_2\}$, $Y_3 = \{y_3\}$, $Y_4 = \{y_4\}$, $Y_5 = \{y_2, y_5\}$, $Y_6 = \{y_6\}$, and $Y_7 = \{y_3, y_7\}$. It is enough $Q = 3$ variables from the set $Z = \{z_1, z_2, z_3\}$ for these collections encoding. Let the collections have the following codes:$K(Y_1) = 000, \ldots, K(Y_7) = 110$.

The initial ST determines pairs $\alpha_{1,1} = \langle a_1, \emptyset \rangle$ $\alpha_{2,2} = \langle a_2, I_1 \rangle$ $\alpha_{3,3} = \langle a_3, I_1 \rangle$ $\alpha_{3,4} = \langle a_3, I_2 \rangle$ $\alpha_{4,2} = \langle a_4, I_1 \rangle$, $\alpha_{4,5} = \langle a_4, I_2 \rangle$, $\alpha_{5,6} = \langle a_5, I_1 \rangle$, $\alpha_{5,7} = \langle a_5, I_2 \rangle$. Thus, there is a set $I = \{I_1, I_2\}$, such that its elements can be encoded using one variable $v_l \in V$. If $K(I_1) = 0$ and $K(I_2) = 1$, then the codes $C(Y_t)^m$ of collections $Y_t \subseteq Y$ are shown in Table 7.37.

There is only one difference between transformed structure tables of MPC_AY and MC_AYMealy FSMs. Namely, in the first case the column X_h is replaced by the column P_h (Table 7.38).

This table is used to derive functions $\Phi = \Phi(P,X)$ and $V = V(P,X)$. For example, the equations $D_3 = F_2 \vee F_4 \vee F_9 \vee F_{10} = A_1\bar{p}_1 \vee \ldots = \bar{T}_1\bar{T}_2\bar{T}_3 p_1 \vee \ldots$, $v_1 = F_4 \vee F_6 \vee$ $\vee F_8 \vee F_{11} = A_2\bar{p}_1 p_2 \vee \ldots = \bar{T}_1 T_2 \bar{T}_3 \bar{p}_1 p_2 \vee \ldots$can be derived from Table 7.38.

Table 7.37 Codes of collections of microoperations for MPC$_A$Y Mealy FSM S_{21}

Y_t	$\alpha_{t,m}$	$C(Y_t)^m$ $T_1T_2T_3T_4$	h	Y_t	$\alpha_{t,m}$	$C(Y_t)^m$ $T_1T_2T_3T_4$	h
Y_1	$\alpha_{1,1}$	0 0 0 *	1	Y_4	$\alpha_{3,4}$	1 1 0 *	5
Y_2	$\alpha_{2,2}$	0 1 0 *	2	Y_5	$\alpha_{4,5}$	1 0 0 1	6
	$\alpha_{4,2}$	1 0 0 0	3	Y_6	$\alpha_{5,6}$	1 0 1 0	7
Y_3	$\alpha_{3,3}$	0 1 1 0	4	Y_7	$\alpha_{5,7}$	1 0 1 1	8

Table 7.38 Transformed structure table of MPC$_A$Y Mealy FSM S_{21}

a_m	$K(a_m)$	a_s	$K(a_s)$	P_h	I_h	$K(I_h)$	V_h	Φ_h	h
a_1	000	a_2	010		*	*	–	D_2	1
		a_3	011	$\bar{p}_1$	I_1	0	–	D_2D_3	2
a_2	010	a_2	010		*	*	–	D_2	3
		a_3	011	$\bar{p}_1p_2$	I_2	1	v_1	D_2D_3	4
		a_4	100	$\bar{p}_1\bar{p}_2$	I_1	0	–	D_1	5
a_3	011	a_4	100	p_3	I_2	1	v_1	D_1	6
		a_5	101	$\bar{p}_3$	I_1	0	–	D_1D_3	7
a_4	100	a_5	101	1	I_2	1	v_1	D_1D_3	8
a_5	101	a_2	010	p_2	*	*	–	D_2	9
		a_3	011	$\bar{p}_2$	I_1	0	–	D_2D_3	10
		a_5	101	$\bar{p}_1p_3$	I_2	1	v_1	D_1D_3	11
		a_1	000	$\bar{p}_1\bar{p}_3$	*	*	–	–	12

For the MPC$_A$Y Mealy FSM S_{21}, the table of block TSM agrees with Table 7.3, whereas the table of microoperations with Table 7.4. The logic circuit of the MPC$_A$Y Mealy FSM S_{21} is shown in Fig. 7.27.

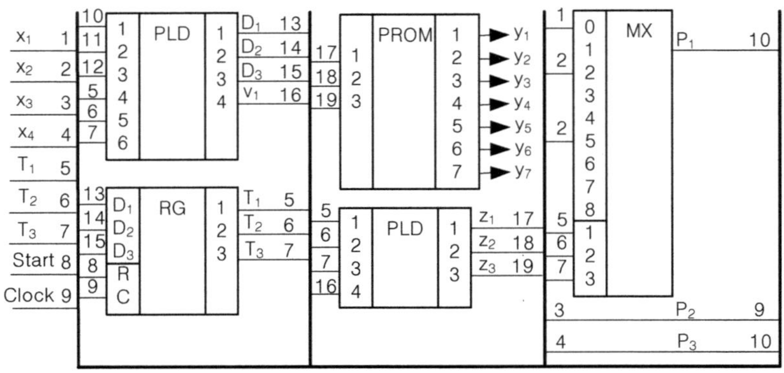

Fig. 7.27 Logic circuit of the MPC$_A$Y Mealy FSM S_{21}

Table 7.39 Multilevel models of Moore FSM with transformation of object codes

LA	LB	LC
M_1 M_1C M_1L	PC_A P_0C_A	Y
$\vdots$	$P_{C1}C_A \dots P_{C4}C_A$	D_1
	$P_{C1}C_Y \dots P_{C4}C_Y$	$\vdots$
M_G M_GC M_GL	PC_Y P_0C_Y	D_K

To reduce the hardware amount in logic circuit of Moore FSM with object code transformation, the following methods can be used: the transformation of initial GSA, the refined state encoding, the transformation of state codes into the codes of logical conditions, and the verticalization of initial GSA. All possible Moore FSM models are shown in Table 7.39.

The following numbers of Moore FSM multilevel models can be found from Table 7.39:

1. $NL_3 = 12(K+1)$. It is the number of models with encoding of collections of microoperations.
2. $NL_4 = 36G(K+1)$. It is the number of models with both encoding of collections of microoperations and logical condition replacement.

Thus, the total number of different models of Moore FSM with object code transformation is equal to $36G + 12K + 36GK + 12$. For interpretation of some GSA Γ, there are 1596 different models for Moore FSM with average complexity, that is $G = I = 6$ [1].

In common case, there are $n_1 = 268$ different models of Mealy FSM and $n_2 = 1596$ different models of Moore FSM for interpretation of the same GSA Γ having $G = I = 6$. It means that such a control algorithm can be implemented using at least $n = n_1 + n_2 = 1864$ different models of FSM with object code transformation. Such a huge plurality of possible solutions increases importance of the problem connected with a-priory choice of the optimal solutions. This problem can be solved using some rules and characteristics of both a GSA to be interpreted and logic elements to be used.

Let us discuss an example of logic synthesis for the MP$_0$LC$_A$Y Moore FSM S_{22} specified by its ST (Table 7.14). The structural diagram of MP$_0$LC$_A$Y Moore FSM is shown in Fig. 7.28.

In this model, the block BM implements functions $P = P(\tau, X)$ and it is used for the replacement of logical conditions $x_e \in X$ the block BP generates functions $\Phi = \Phi(T, P)$ used for loading of the code of collections of microoperations $C(Y_t)^m$ into the register RG; the block CCS generates variables $\tau = \tau(T)$ used for encoding of logical conditions $x_l \in X$; the block TSM generates variables $Z = Z(T)$ used for encoding of collections of microoperations; the block BY generates microoperations $Y = Y(Z)$. The synthesis method for MP$_0$LC$_A$Y Moore FSM includes the following steps:

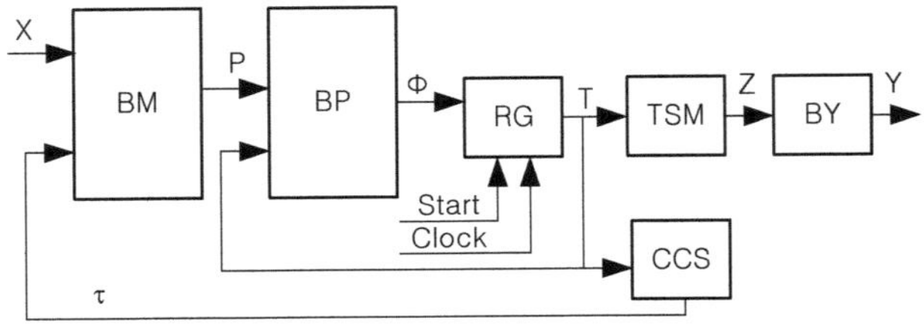

Fig. 7.28 Structural diagram of $\mathrm{MP_0LC_AY}$ Moore FSM

1. Optimal encoding for states $a_m \in A$.
2. Encoding of logical conditions $x_l \in X$.
3. Logical condition replacement and construction of functions P.
4. Identification of collections of microoperations by states $a_m \in A$.
5. Construction of transformed structure table and functions Φ.
6. Specification of block CCS and construction of system τ.
7. Encoding of collections and specification of block BY.
8. Specification of block TSM and construction of system Z.
9. Implementation of FSM logic circuit using some logic elements.

For the $\mathrm{MP_0LC_AY}$ Moore FSM S_{22}, there is the partition Π_A of the state set A by the three classes of pseudoequivalent states, namely: $B_1 = \{a_1\}$, $B_2 = \{a_2, a_3, a_4, a_7\}$, and $B_3 = \{a_5, a_6\}$. One of the possible variants for the optimal state encoding is shown in Fig. 7.29.

Fig. 7.29 Optimal state codes for Moore FSM S_{22}

T_1 \ T_2T_3	00	01	11	10
0	a_1	a_2	a_3	a_5
1	*	a_4	a_7	a_6

The following codes of the classes can be found from Fig. 7.29:$K(B_1) = *00$, $K(B_2) = **1$ and $K(B_3) = *10$. As it follows from these codes, there is no connection between the block BP and the state variable T_1. For the $\mathrm{MP_0LC_AY}$ Moore FSM S_{22}, the outcome of distribution of the logical conditions for their replacement is shown in Table 7.40.

As follows from Table 7.40, the block BM of $\mathrm{MP_0LC_AY}$ Moore FSM S_{22} is characterized by the sets $P = \{p_1, p_2\}$, $X(p_1) = \{x_1, x_4\}$, and $X(p_2) = \{x_2, x_3\}$. It is enough the variable τ_1 for encoding of logical conditions $x_l \in X(p_1)$, whereas the variable τ_2 can be used for encoding of logical conditions $x_l \in X(p_2)$. Thus, there is the set $\tau = \{\tau_1, \tau_2\}$. Let us encode the logical conditions in the following way: $K(x_1) = 0$, $K(x_4) = 1$, $K(x_2) = 0$, and $K(x_3) = 1$.

There are $T_0 = 4$ different collections of microoperations in Table 7.14, namely: $Y_1 = \emptyset$, $Y_2 = \{y_1, y_2\}$, $Y_3 = \{y_3\}$, and $Y_4 = \{y_3, y_4\}$. As it is found before, these

Table 7.40 Distribution of logical conditions for Moore FSM S_{22}

a_m	a_1	a_2	a_3	a_4	a_5	a_6	a_7
p_1	x_1	–	x_4	–	x_4	x_4	–
p_2	x_2	x_3	x_3	x_3	–	–	x_3

Table 7.41 Transformed ST of MP_0LC_AY Moore FSM S_{22}

B_i	$K(B_i)$	a_s	$K(a_s)$	P_h	Φ_h	h
B_1	$*00$	a_2	001		D_3	1
		a_3	011	$\bar{p}_1p_2$	D_2D_3	2
		a_4	101	$\bar{p}_1\bar{p}_2$	D_1D_3	3
B_2	$*01$	a_5	010	p_2	D_2	4
		a_6	110	$\bar{p}_2$	D_1D_2	5
B_3	$*10$	a_7	111		$D_1D_2D_3$	6
		a_1	000	$\bar{p}_1$	–	7

collections are identified by the following codes:$C(Y_1)^1 = 000$, $C(Y_2)^2 = 001$, $C(Y_2)^4 = 101$, $C(Y_2)^6 = 110$, $C(Y_3)^3 = 010$, $C(Y_4)^5 = 010$, and $C(Y_4)^7 = 111$. The transformed ST of MP_0LC_AY Moore FSM S_{22} includes $H_0 = 7$ rows (Table 7.41).

This table is used to derive the system of input memory functions. For example, the following Boolean equation can be derived from Table 7.41: $D_1 = F_3 \vee F_5 \vee F_6 = B_1\bar{p}_1\bar{p}_2 \vee B_2\bar{p}_2 \vee B_3p_1 = \bar{T}_1\bar{T}_3\bar{p}_1\bar{p}_2 \vee T_3\bar{p}_2 \vee T_2\bar{T}_3p_1$.

The table for block CCS includes columns a_m, $K(a_m)$, p_1, p_2, $K(p_1)$, $K(p_2)$, τ_m, m. In this table, the column $K(p_q)$ contains the code $K(x_l)$ of some logical condition $x_l \in X$, replaced by the variable p_q for the state a_m. The column τ_m of the table includes variables $\tau_r \in \tau$ equal to 1 in the row m. For the MP_0LC_AYMoore FSM S_{22}, the block CCS is specified by Table 7.42.

It is enough $Q = 2$ variables from the set $Z = \{z_1, z_2\}$ for encoding of collections $Y_t \subseteq Y$. If $K(Y_1) = 00$, ..., $K(Y_4) = 11$, then the table of microoperations for discussed example is the same as Table 7.15.

Table 7.42 Specification of block CCS for MP_0LC_AY Moore FSMS_{22}

a_m	$K(a_m)$	p_1	p_2	$K(p_1)$	$K(p_2)$	τ_m	m
a_1	000	x_1	x_2	0	0	–	1
a_2	001	–	x_3	–	1	τ_2	2
a_3	011	–	x_3	–	1	τ_2	3
a_4	101	–	x_3	–	1	τ_2	4
a_5	101	x_2	–	1	–	τ_1	5
a_6	110	x_2	–	1	–	τ_1	6
a_7	111	–	x_3	–	1	τ_2	7

Table 7.43 Specification of block TSM for MP_0LC_AY Moore FSM S_{22}

a_m	Y_t	$C(Y_t)^m$	$K(Y_t)$	Z_m	m
a_1	Y_1	000	00	–	1
a_2	Y_2	001	01	z_2	2
a_3	Y_3	011	10	z_1	3
a_4	Y_2	101	01	z_2	4
a_5	Y_4	010	11	z_1z_2	5
a_6	Y_2	110	01	z_1	6
a_7	Y_4	111	11	z_1z_2	7

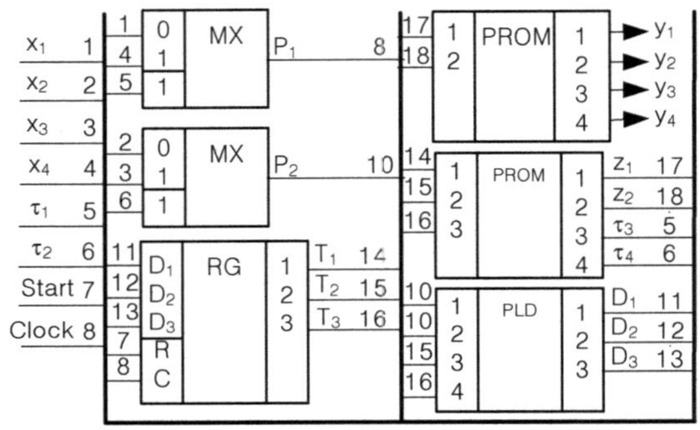

Fig. 7.30 Logic circuit of MP_0LC_AY Moore FSM S_{22}

The table of block TSM includes columns a_m, Y_t, $C(Y_t)^m$, $K(Y_t)$, Z_m, m. The column Z_m includes variables $z_q \in Z$, equal to 1 in the code $K(Y_t)$. For the MP_0LC_AY Moore FSM S_{22}, the block TSM is specified by Table 7.43. Obviously, this table can be viewed as a truth table for functions Z. The best way for its implementation is use of embedded memory blocks. As follows from Table 7.42 and Table 7.43, both code transformers have the same inputs. It means that both systems τ and Z can be implemented using the common block TSM.

The logic circuit of MP_0LC_AY Moore FSM S_{22} is shown in Fig. 7.30. The discussed examples give a key for logic synthesis of any multilevel FSM model represented by Table 7.33 and Table 7.39.

References

1. Baranov, S.I.: Logic Synthesis of Control Automata. Kluwer Academic Publishers, Dordrecht (1994)
2. Barkalov, A., Barkalov Jr., A.: Design of mealy finite-state machines with the transformation of object codes. International Journal of Applied Mathematics and Computer Science 15(1), 151–158 (2005)

3. Barkalov, A., Barkalov, A.: Synthesis of finite state machines with transformation of the object's codes. In: Proc. of the Inter. Conf. TCSET 2004, Lviv, Ukraina, pp. 61–64. Lviv Polytechnic National University, Publishing House of Lviv Polytechnic, Lviv (2004)
4. Barkalov, A., Titarenko, L.: Synthesis of Operational and Control Automata. UNITECH, Donetsk (2009)
5. Barkalov, A., Titarenko, L., Barkalov Jr., A.: Moore fsm synthesis with coding of compatible microoperations fields. In: Proc. of IEEE East-West Design & Test Symposium - EWDTS 2007, Yerevan, Armenia, pp. 644–646. Kharkov National University of Radioelectronics, Kharkov (2007)
6. Barkalov, A., Wêgrzyn, A., Barkalov Jr., A.: Synthesis of control units with transformation of the codes of objects. In: Proc. of the IXth Inter. Conf. CADSM 2007, Lviv - Polyana, Ukraine, pp. 260–261. Lviv Polytechnic National University, Publishing House of Lviv Polytechnic National University, Lviv (2007)
7. Barkalov, A.A.: Principles of optimization of logic circuit of Moore FSM. Cybernetics and System Analysis (1), 65–72 (1998) (in Russian)
8. Minns, P., Elliot, I.: FSM-based digital design using Verilog HDL. John Wiley and Sons, Chichester (2008)

Chapter 8
FSM Synthesis with Elementary Chains

Abstract. The chapter is devoted to original methods oriented on optimization of Moore FSM interpreting graph-schemes of algorithms with long sequences of operator vertices having only one input. These sequences are named elementary operational linear chains (EOLC). These FSM models include the counter keeping, either microinstruction addresses or code of EOLC component. In the beginning the Moore FSM models with code sharing are analysed, where the register keeps EOLC codes. The methods of EOLC encoding and transformation are discussed; these methods permit to decrease the number of macrocells in the block generating input memory functions. The second part of the chapter is devoted to reduction of the number of embedded memory blocks in the FSM block generating microoperations. These methods are based on transformation of microinstruction address represented as concatenation of EOLC code and code of its component into either linear microinstruction address or code of collection of microoperations. The last part of the chapter discusses synthesis methods for multilevel FSM models with EOLC.

8.1 Basic Models of FSM with Elementary Chains

Definitions of the OLC, its input and output can be found in Section 1.4. An elementary operational linear chain (EOLC) is a particular case of OLC. Such a chain has only one input. These control units are known as compositional microprogram control units [2], but they are based on the model of Moore FSM. Let us name such control units as $P_{EC}Y$ Moore FSM. The structural diagram of $P_{EC}Y$ Moore FSM is shown in Fig. 8.1.

In $P_{EC}Y$ Moore FSM, the block BP generates input memory functions to change the content of a counter CT

$$\Phi = \Phi(T,X), \tag{8.1}$$

whereas the block BY keeps collections of microoperations (microinstructions) $Y(b_t) \subseteq Y$, as well as variables y_0 (to control the mode of counter synchronization)

A. Barkalov and L. Titarenko: Logic Synthesis for FSM-Based Control Units, LNEE 53, pp. 193–227.
springerlink.com

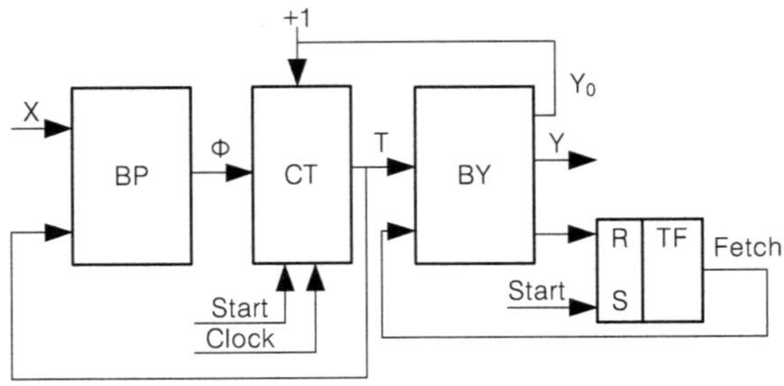

Fig. 8.1 Structural diagram of P_{EC}Y Moore FSM

and y_E (to control the fetching of microinstructions). These functions are represented as

$$Y = Y(T), \tag{8.2}$$

$$y_0 = y_0(T), \tag{8.3}$$

$$y_E = y_E(T). \tag{8.4}$$

Let $C_E = \{\alpha_1, \ldots, \alpha_{G_E}\}$ be a set of EOLC. If a transition is executed between the elements of the same EOLC, then the counter is incremented according with (1.28). If a transition is executed between the output of EOLC $\alpha_i \in C_E$ and the input of EOLC $\alpha_j \in C_E$ (of course, it can be the same EOLC), then an address of transition is generated by the block BP as it is determined by (1.29). This FSM operates in the following manner.

The pulse "Start" initializes the following actions: the zero address of the first microinstruction is loaded into counter CT; the flip-flop TF is set up (Fetch=1). A current microinstruction is read out the block BY. If $y_0 = 1$ for this microinstruction, then 1 is added to the content of counter CT. Otherwise, the output of current EOLC is reached and the block BP generates the next microinstruction address. If $y_E = 1$, then the output of control algorithm (corresponding to the microprogram) is reached. In this case the flip-flop TF is cleared and microinstruction fetching is terminated.

The synthesis method of P_{EC}Y Moore FSM includes the following steps [2]:

1. Transformation of the initial GSA Γ.
2. Construction of the EOLC set for transformed GSA Γ.
3. Natural addressing of microinstructions.
4. Construction of FSM structure table.
5. Specification of block BY.
6. Synthesis of logic circuit with given logic elements.

Let us discuss an example of design for the Moore FSM S_{25} represented by the graph-scheme of algorithm Γ_6 (Fig. 8.2).

1. **Transformation of initial GSA** Γ is executed in the following manner [2]:

   - if there is an arc $\langle b_0, b_q \rangle \in E$, where $b_q \in B_2$, some vertex $b_t \in B_1$ is introduced into GSA Γ, where $Y(b_t) = \emptyset$, and the initial arc is replaced by a pair of new arcs $\langle b_0, b_t \rangle$ and $\langle b_t, b_q \rangle$;

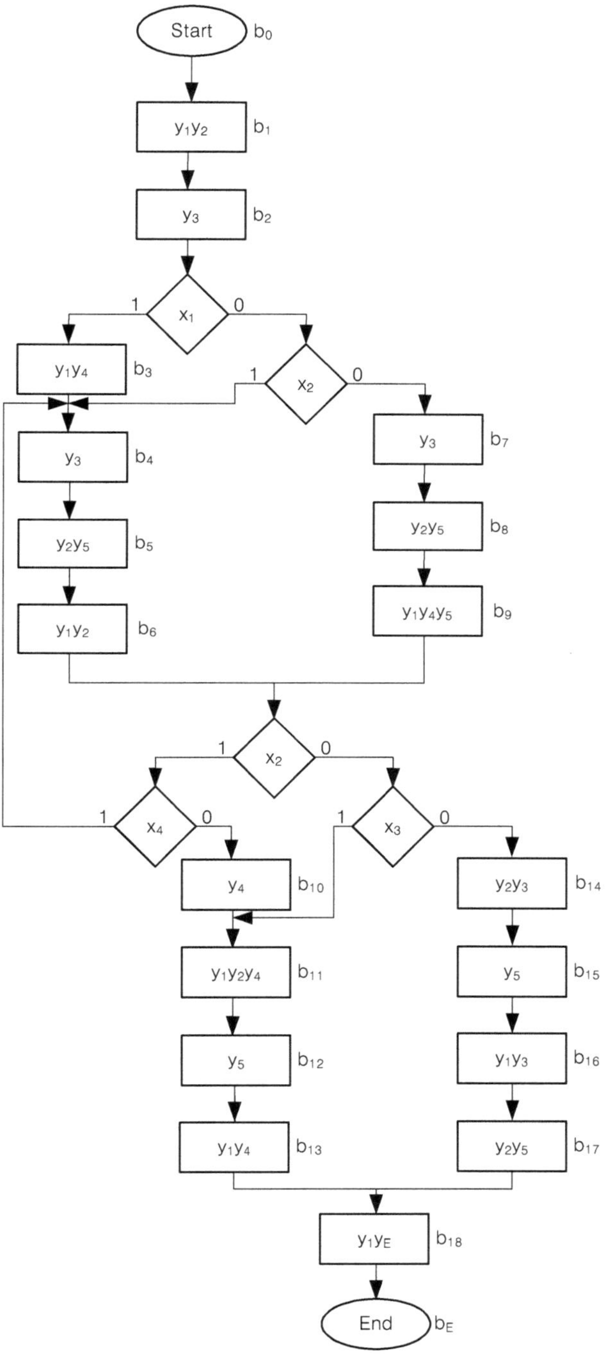

Fig. 8.2 Initial graph-scheme of algorithm Γ_6

- if there is an arc $\langle b_t, b_E \rangle \in E$, where $b_t \in B_2$, then some vertex $b_q \in B_1$ with y_E is introduced into GSA and the initial arc is replaced by two arcs $\langle b_t, b_q \rangle$ and $\langle b_q, b_E \rangle$;
- if there is an arc $\langle b_t, b_E \rangle \in E$, where $b_t \in B_1$, then the variable y_E is inserted into the vertex b_t.

2. **Construction of the set of EOLC.** There are two stages in this step execution. The first stage is reduced to construction of the set of EOLC inputs $I(\Gamma)$. The second stage is connected with construction of EOLC for each element of the set $I(\Gamma)$. In the discussed example, this set includes eight elements $I(\Gamma_6) = \{b_1, b_3, b_4, b_7, b_{10}, b_{11}, b_{14}, b_{18}\}$. The EOLC α_1 is constructed in the following manner. Let us take the vertex $b_1 \in I(\Gamma_6)$, which is treated as the input I_1 of EOLC α_1. Let us analyze transitions from the vertex b_1 . There is the arc $\langle b_1, b_2 \rangle$ in the GSA Γ_6 . The vertex $b_2 \in B_1$ and this vertex does not belong to the set of inputs. Therefore, the vertex b_2 is the second component of EOLC α_1. Let us analyze transitions from the vertex b_2. There is the arc $\langle b_2, b(x_1) \rangle$in the GSA Γ_6, such that $b(x_1) \in B_2$. Thus, the vertex b_2 is the output O_1 of EOLC $\alpha_1 = \langle b_1, b_2 \rangle$. Construction of any EOLC is terminated, if an analyzed vertex belongs to the set $I(\Gamma)$, or if it is the final vertex b_E. Let the symbol L_g stand for the number of components in EOLC α_g. In case of the GSA Γ_6, the following set of EOLC $C_E = \{\alpha_1, \ldots, \alpha_8\}$ can be found, where $\alpha_1 = \langle b_1, b_2 \rangle$, $I_1 = b_1$, $O_1 = b_2$, $L_1 = 2$; $\alpha_2 = \langle b_3 \rangle$, $I_2 = O_2 = b_3$, $L_2 = 1$; $\alpha_3 = \langle b_4, b_5, b_6 \rangle$, $I_3 = b_4$, $O_3 = b_6$, $L_3 = 3$; $\alpha_4 = \langle b_7, b_8, b_9 \rangle$, $I_4 = b_7$, $O_4 = b_9$, $L_4 = 3$; $\alpha_5 = \langle b_{10} \rangle$, $I_5 = O_5 = b_{10}$, $L_5 = 1$; $\alpha_6 = \langle b_{11}, b_{12}, b_{13} \rangle$, $I_6 = b_{11}$, $O_6 = b_{13}$, $L_6 = 3$; $\alpha_7 = \langle b_{14}, \ldots, b_{17} \rangle$, $I_7 = b_{14}$, $O_7 = b_{17}$, $L_7 = 4$; $\alpha_8 = \langle b_{18} \rangle$, $I_8 = O_8 = b_{18}$, $L_8 = 1$.
3. **Natural addressing of microinstructions.** This step is reduced to construction of the table of addressing, similar to Karnaugh map. The number of address variables is determined as

$$R_E = \lceil \log_2 M_E \rceil . \tag{8.5}$$

In (8.5), the symbol M_E denotes the number of operator vertices in the transformed GSA. In the discussed example, we have $R_E = 5$ and $T = \{T_1, \ldots, T_5\}$. The addresses of microinstructions are shown in Fig. 8.3.

T_1T_2 \ $T_1T_2T_3$	000	001	010	011	100	101	110	111
00	b_1	b_2	b_3	b_4	b_5	b_6	b_7	b_8
01	b_9	b_{10}	b_{11}	b_{12}	b_{13}	b_{14}	b_{15}	b_{16}
11	b_{17}	b_{18}	*	*	*	*	*	*
10	*	*	*	*	*	*	*	*

Fig. 8.3 Microinstruction addresses for $P_{EC}Y$ Moore FSM S_{25}

4. **Construction of FSM structure table** assumes finding the system of generalized formulae of transition (GFT) for EOLC outputs $\alpha_g \in C_E$. In the discussed case, this system includes $G_E - 1 = 7$ following formulae:

$$
\begin{array}{ll}
\alpha_1 \rightarrow x_1 I_2 \vee \bar{x}_1 x_2 I_3 \vee \bar{x}_1 \bar{x}_2 I_4; & \alpha_5 \rightarrow I_6; \\
\alpha_2 \rightarrow I_3; & \alpha_6 \rightarrow I_8; \\
\alpha_3 \rightarrow x_2 x_4 I_3 \vee x_2 \bar{x}_4 I_5 \vee \bar{x}_2 x_3 I_6 \vee \bar{x}_2 \bar{x}_3 I_7; & \alpha_7 \rightarrow I_8. \\
\alpha_4 \rightarrow x_2 x_4 I_3 \vee x_2 \bar{x}_4 I_5 \vee \bar{x}_2 x_3 I_6 \vee \bar{x}_2 \bar{x}_3 I_7; &
\end{array}
\quad (8.6)
$$

This system serves to construct the FSM structure table having the following columns: O_g is an output of EOLC $\alpha_g \in C_E$; $A(O_g)$ is an address of the output O_g; I_j is an input of EOLC $\alpha_j \in C_E$; $A(I_j)$ is an address of input I_j; X_h is a set of logical conditions determined the transition from O_g into I_j; Φ_h is a set of input memory functions. For the $P_{EC}Y$ Moore FSM S_{25}, the structure table is shown in Table 8.1.

Table 8.1 Structure table of $P_{EC}Y$ Moore FSM S_{25}

O_g	$A(O_g)$	I_j	$A(I_j)$	X_h	Φ_h	h
O_1	00001	I_2	00010	x_1	D_4	1
		I_3	00011	$\bar{x}_1 x_2$	$D_4 D_5$	2
		I_4	00110	$\bar{x}_1 \bar{x}_2$	$D_3 D_4$	3
O_2	00010	I_3	00011	1	$D_4 D_5$	4
O_3	00101	I_3	00011	$x_2 x_4$	$D_4 D_5$	5
		I_5	01101	$x_2 \bar{x}_4$	$D_2 D_3 D_5$	6
		I_6	01010	$\bar{x}_2 x_3$	$D_4 D_5$	7
		I_7	01101	$\bar{x}_2 \bar{x}_3$	$D_2 D_3 D_5$	8
O_4	01000	I_3	00011	$x_2 x_4$	$D_4 D_5$	9
		I_5	01101	$x_2 \bar{x}_4$	$D_2 D_3 D_5$	10
		I_6	01010	$\bar{x}_2 x_3$	$D_4 D_5$	11
		I_7	01101	$\bar{x}_2 \bar{x}_3$	$D_2 D_3 D_5$	12
O_5	01001	I_6	01010	1	$D_2 D_4$	13
O_6	01100	I_8	10101	1	$D_1 D_3 D_5$	14
O_7	10000	I_8	10101	1	$D_1 D_3 D_5$	15

This structure table is the base to construct functions of system (8.1). For example, the following Boolean expression can be extracted from Table 8.1: $D_1 = F_{14} \vee F_{15} = \bar{T}_1 T_2 T_3 \bar{T}_4 \bar{T}_5 \vee T_1 \bar{T}_2 \bar{T}_3 \bar{T}_4 \bar{T}_5$.

5. **Specification of block BY** is reduced to the replacement of vertices $b_t \in B_1$ by collections of microoperations $Y(b_t)$. If a vertex $b_t \neq O_g$, then the variable y_0 is inserted into the corresponding set $Y(b_t)$. For example, let us discuss execution of this step for the EOLC $\alpha_6 = \langle b_{11}, b_{12}, b_{13} \rangle$. In this case the microoperations y_0, y_1, y_2, and y_4 are written into the cell with address 01010, the microoperations y_0, y_5 are written into the cell with address 01011, and the microoperations y_1, y_4

Table 8.2 Specification of block BY for $P_{EC}Y$ Moore FSM S_{25}

$A(b_t)$	$Y(b_t)$	t	$A(b_t)$	$Y(b_t)$	t	$A(b_t)$	$Y(b_t)$	t
00000	$y_0y_1y_2$	1	00110	y_0y_3	7	01100	y_1y_4	13
00001	y_3	2	00111	$y_0y_2y_5$	8	01101	$y_0y_2y_3$	14
00010	$y_0y_1y_4$	3	01000	$y_1y_4y_5$	9	01110	y_0y_5	15
00011	y_0y_3	4	01001	y_0y_3	10	01111	$y_0y_1y_3$	16
00100	$y_0y_2y_5$	5	01010	$y_0y_1y_2y_4$	11	10000	y_2y_5	17
00101	y_1y_2	6	01011	y_0y_5	12	10001	y_1y_E	18

are written into the cell with address 01100. The outcome of this step is shown in Table 8.2.

Obviously, functions of systems (8.2) – (8.4) belong to the class of regular functions and the best way for their implementation is use of embedded memory blocks BRAM. If there are no BRAMs in FPLDs in use, then some table similar to Table 8.2 is used to construct $N+2$ Karnaugh maps. These maps are used to get minimized forms of functions $y_n \in Y$, y_0 and y_E. For example, from the Karnaugh map for the function $y_1 \in Y$ the following minimized sum-of-products $y_1 = \bar{T}_1\bar{T}_3\bar{T}_4 \vee T_1T_3 \vee T_2T_3T_4T_5 \vee \bar{T}_2T_3\bar{T}_4T_5 \vee T_2\bar{T}_4\bar{T}_5$ can be derived (taking into account the "don't care" input assignments from 10011 to 11111).

6. **Implementation of FSM logic circuit** is reduced to implementation of corresponding circuits for systems (8.1) – (8.4) using given logic elements. We do not discuss this step for the $P_{EC}Y$ Moore FSM S_{25}.

The main drawback of $P_{EC}Y$ Moore FSM is the redundant number of feedback variables, as it follows from (8.5). This number can be decreased using the principle of *code sharing* [2],which can be explained as the following. Let GSA Γ include G_E different EOLC, and let each of them include M_g components. Let $Q_E = \max(M_1,\ldots,M_{G_E})$. Obviously, it is enough R_{EO} variables for EOLC encoding:

$$R_{EO} = \lceil \log_2 G_E \rceil. \tag{8.7}$$

These variables form a set τ. Next, it is enough R_{CO} variables for encoding of EOLC components:

$$R_{CO} = \lceil \log_2 Q_E \rceil. \tag{8.8}$$

These variables form a set T. Let $K(\alpha_g)$ and $K(b_t)$ be respectively a code of EOLC $\alpha_g \in C_E$ and its component $b_t \in B_1$. Then expression (1.33) determines an address $A(b_t)$ of microinstruction corresponded to the vertex $b_t \in B_1$.

Let microinstructions corresponded to components of each EOLC be addressed using the principle of natural addressing. Let first components of each EOLC have zero codes. In this case the GSA Γ can be interpreted by $P_{ES}Y$ Moore FSM having the structural diagram shown in Fig. 8.4. The $P_{ES}Y$ Moore FSM operates in the following manner.

Pulse ”Start” initiates loading of zero codes into both the register RG and counter CT. Simultaneously, the flip-flop TF is set up and it allows generation of

Fig. 8.4 Structural diagram of $P_{ES}Y$ Moore FSM

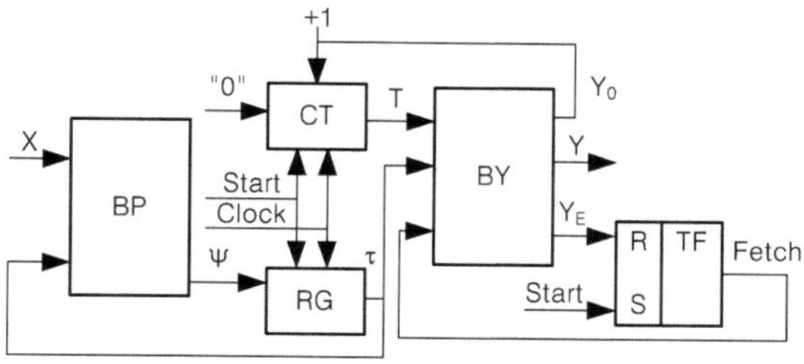

microinstructions by the block BY. If contents of RG and CT form an address of microinstruction, which does not correspond to the output of some EOLC, then pulse "Clock" causes increment of CT. It corresponds to unconditional transitions between adjacent components of the same EOLC. Otherwise, pulse "Clock" causes reset of CT, whereas the block BP loads into RG an input address of some other EOLC. The following functions are used to form a microinstruction address:

$$\psi = \psi(\tau, X). \tag{8.9}$$

All other operation principles are the same for both $P_{ES}Y$ and $P_{EC}Y$ Moore FSM. Procedures of their synthesis include the same steps, but sometimes there is some difference in the execution of these steps. Let us discuss an example of logic synthesis for the $P_{ES}Y$ Moore FSM S_{25} represented by the GSA Γ_6 (Fig. 8.2).

Obviously, there are the same outcomes for such steps as GSA transformation and EOLC construction for equivalent $P_{ES}Y$ and $P_{EC}Y$ Moore FSMs.

Natural microinstruction addressing is reduced to the natural addressing of components of EOLC $\alpha_g \in C_E$. In the discussed example, the following values and sets can be found: $R_{EO} = 3$, $\tau = \{\tau_1, \tau_2, \tau_3\}$, $R_{CO} = 2$, and $T = \{T_1, T_2\}$. Let us encode EOLC $\alpha_g \in C_E$ in a trivial way, namely: $K(\alpha_1) = 000, \ldots, K(\alpha_8) = 111$. First components of all EOLC have the code 00, second components of all EOLC have the code 01, third components of all EOLC have the code 10, and fourth components of all EOLC have the code 11. This procedure allows getting the microinstruction addresses shown in Fig. 8.5.

T_1T_2 \ $\tau_1\tau_2\tau_3$	000	001	010	011	100	101	110	111
00	b_1	b_3	b_4	b_7	b_{10}	b_{11}	b_{14}	b_{18}
01	b_2	*	b_5	b_8	*	b_{12}	b_{15}	*
11	*	*	b_6	b_9	*	b_{13}	b_{16}	*
10	*	*	*	*	*	*	b_{17}	*

Fig. 8.5 Microinstruction addresses for $P_{ES}Y$ Moore FSM S_{25}

Construction of FSM structure table. This step is executed in two stages. The first of them is construction of the system of generalized formulae of transitions. The GFT for our example is represented by system (8.6). The second stage is construction of FSM structure table corresponding to this GFT. For the $P_{ES}Y$ Moore FSM S_{25}, it is Table 8.3. This table is the base to derive system (8.8). For example, the following sum-of-products can be derived from Table 8.3: $D_1 = F_6 \vee F_7 \vee F_8 \vee F_{10} \vee \ldots \vee F_{15} = \bar{\tau}_1 \tau_2 \bar{\tau}_3 x_2 \bar{x}_4 \vee \ldots \vee \tau_1 \tau_2 \bar{\tau}_3$.

Specification of block BY. Obviously, this step is executed in the same manner for both models of $P_{ES}Y$ and $P_{EC}Y$ Moore FSMs. Of course, the same microinstructions have different addresses for equivalent $P_{ES}Y$ and $P_{EC}Y$ Moore FSMs. If the logic circuit of block BY is implemented using embedded memory blocks, then corresponding memory cells for both models have the same content. For example, the cell with address $A(b_5) = 01001$ includes the code corresponding to microoperations $y_0y_2y_5$(for the $P_{ES}Y$ Moore FSM S_{25}), the same code is contained by the cell with address $A(b_5) = 00100$ (for the $P_{EC}Y$ Moore FSM S_{25}).

Synthesis of FSM logic circuit is reduced to implementation of systems (8.2) – (8.4) and (8.9) using some logic elements. The logic circuit of the $P_{ES}Y$ Moore FSM S_{25} is shown in Fig 8.6. This circuit includes a block generated pulses of synchronization for the register and counter. The following functions are used to control the synchronization:

$$C_1 = y_0 \cdot \text{Clock}; \tag{8.10}$$

$$C_2 = \bar{y}_0 \cdot \text{Clock}. \tag{8.11}$$

The pulse C_1 is used to increment the counter for execution of unconditional jumps. The pulse C_2 is used to clear the counter and to load a parallel code determined by (8.9) into the register.

Analysis of $P_{ES}Y$ Moore FSM shows that there is no dependence between codes of EOLC and codes of their components. It allows application for $P_{ES}Y$ Moore FSM all known methods used for optimization of PY Moore FSM. Of course, these methods should be adapted to the peculiarities of $P_{ES}Y$ Moore FSM. Besides, application of these models has sense only if the following condition takes place:

$$R_{EO} + R_{CO} = R_E. \tag{8.12}$$

If condition (8.12) is violated, then the total size of used blocks BRAM increases drastically in comparison with its minimal value, determined as

$$V_{\min} = 2^{R_E}(N+2). \tag{8.13}$$

This formula assumes application of the hot-one encoding of microoperations. It should be modified if the block BY is implemented using either maximal encoding of collections of microoperations or encoding of the classes of compatible microoperations. The hardware amount in logic circuit of $P_{ES}Y$ Moore FSM can be reduced using some optimization methods [3–7, 9]. Let us discuss these methods in details.

Table 8.3 Structure table of $P_{ES}Y$ Moore FSM S_{25}

α_g	$K(\alpha_g)$	I_m^j	$A(I_m^j)$	X_h	Ψ_h	h
α_1	000	I_2	00100	x_1	D_3	1
		I_3	01000	$\bar{x}_1x_2$	D_2	2
		I_4	01100	$\bar{x}_1\bar{x}_2$	D_2D_3	3
α_2	001	I_3	01000	1	D_2	4
α_3	010	I_3	01000	x_2x_4	D_2	5
		I_5	10000	$x_2\bar{x}_4$	D_1	6
		I_6	10100	$\bar{x}_2x_3$	D_1D_3	7
		I_7	11000	$\bar{x}_2\bar{x}_3$	D_1D_2	8
α_4	011	I_3	01000	x_2x_4	D_2	9
		I_5	10000	$x_2\bar{x}_4$	D_1	10
		I_6	10100	$\bar{x}_2x_3$	D_1D_3	11
		I_7	11000	$\bar{x}_2\bar{x}_3$	D_1D_2	12
α_5	100	I_6	10100	1	D_1D_3	13
α_6	101	I_8	11100	1	$D_1D_2D_3$	14
α_7	110	I_8	11100	1	$D_1D_2D_3$	15

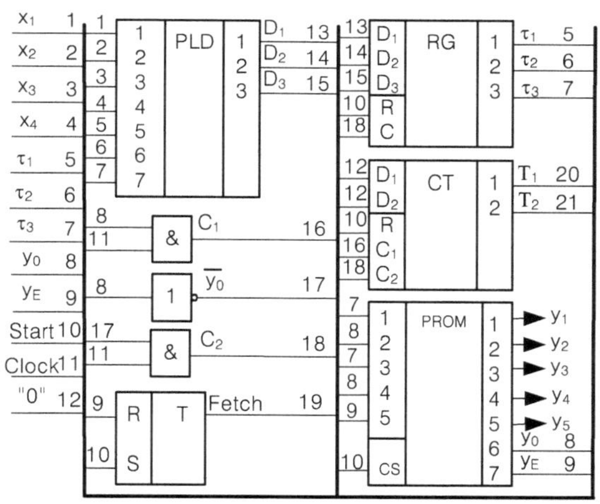

Fig. 8.6 Logic circuit of $P_{ES}Y$ Moore FSM S_{25}

8.2 Optimization of Block of Input Memory Functions

All discussed methods are based on existence of pseudoequivalent elementary operational linear chains in a GSA Γ to be interpreted [2]. Elementary OLCs $\alpha_i, \alpha_j \in C_E$ are pseudoequivalent EOLCs, if their outputs are connected with the input of the same vertex of GSA Γ. There are two optimization methods for the logic circuit of block BP, namely:

1. The optimal encoding of EOLC leading to $P_{ES1}Y$ Moore FSM.
2. The code transformation of pseudoequivalent EOLC into the codes of their classes leading to $P_{ES2}Y$ Moore FSM.

For the GSA Γ_6 (Fig. 8.2), the partition Π_E includes the following classes of pseudoequivalent EOLC: $B_1 = \{\alpha_1\}$, $B_2 = \{\alpha_2\}$, $B_3 = \{\alpha_3, \alpha_4\}$, $B_4 = \{\alpha_5\}$, and $B_5 = \{\alpha_6, \alpha_7\}$.

Optimal EOLC encoding is executed in the way minimizing the number of terms in system (8.14), represented in the following form:

$$B_i = \bigvee_{g=1}^{G_E} C_{gi} A_g \quad (i = 1, \ldots, I_E). \tag{8.14}$$

In this system, the symbol C_{gi} stands for a Boolean variable equal to 1 iff $\alpha_g \in \Pi_E$, and $I_E = |\Pi_E|$. Let us point out that codes for EOLC whose outputs are connected with the input of final GSA vertex are treated as "don't care" input assignments. It is possible because the FSM structure table does not include transitions for these outputs.

Structural models are the same for $P_{ES}Y$ and $P_{ES1}Y$ Moore FSMs. The only difference in their design methods is an approach used for EOLC encoding. Let us discuss an example of logic synthesis for the $P_{ES1}Y$ Moore FSM S_{25} specified by the GSA Γ_6 (Fig. 8.2). In comparison with previous example, some difference in the outcomes of synthesis procedure steps appears only in generated addresses of microinstructions.

Natural addressing of microinstructions. Component codes for EOLC $\alpha_g \in C_E$ are the same for both $P_{ES}Y$ and $P_{ES1}Y$ Moore FSMs S_{25}. For the $P_{ES1}Y$ Moore FSM S_{25}, the system (8.14) is represented as the following one:

$$B_1 = A_1; B_2 = A_2; B_3 = A_3 \vee A_4; B_4 = A_5; B_5 = A_6 \vee A_7. \tag{8.15}$$

The algorithm ESPRESSO [8], can be used for optimal EOLC encoding. In the discussed example, it generates the codes shown in Fig. 8.7.

Fig. 8.7 Optimal EOLC codes of $P_{ES1}Y$ Moore FSM S_{25}

$\tau_1 \backslash \tau_2\tau_3$	00	01	11	10
0	α_1	α_2	α_3	α_6
1	α_8	α_5	α_4	α_7

Because the EOLC α_8 does not belong to set $B_i \in \Pi_E$, then its code 100 is treated as a "don't care" input assignment. Taking it into account, the following conjunctions can be found for classes $B_i \in \Pi_E$: $B_1 = \bar{\tau}_2\bar{\tau}_3$; $B_2 = \bar{\tau}_1\bar{\tau}_2\tau_3$; $B_3 = \tau_2\tau_3$; $B_4 = \tau_1\bar{\tau}_2$; $B_5 = \tau_2\bar{\tau}_3$. Thus, the following codes correspond to the classes $B_i \in \Pi_E$: $K(B_1) = {*}00$, $K(B_2) = 001$, $K(B_3) = {*}11$, $K(B_4) = 10{*}$, and $K(B_5) = {*}10$.

The addresses of microinstructions shown in Fig. 8.8 are determined by concatenations of optimal EOLC codes from Fig. 8.7 and their components codes.

Construction of FSM transition table is executed in three stages. The first of them is reduced to construction of the system of GTFs. In the discussed case it is system (8.6). During the second stage, EOLC $\alpha_g \in B_i$ from the left part of each

T_1T_2 \ $\tau_1\tau_2\tau_3$	000	001	010	011	100	101	110	111
00	b_1	b_3	b_{11}	b_4	b_{18}	b_{110}	b_{14}	b_7
01	b_2	*	b_{12}	b_5	*	*	b_{15}	b_8
11	*	*	b_{13}	b_6	*	*	b_{16}	b_9
10	*	*	*	*	*	*	b_{17}	*

Fig. 8.8 Microinstruction addresses for $P_{ES1}Y$ Moore FSM S_{25}

GTF are replaced by corresponding classes $B_i \in \Pi_E$. If after such a replacement the system includes equal formulae, then only one of them remains. For the $P_{ES1}Y$ Moore FSM S_{25}, the system of transformed GTF is the following one:

$$\begin{array}{ll} B_1 \to x_1I_2 \vee \bar{x}_1x_2I_3 \vee \bar{x}_1\bar{x}_2I_4; & B_4 \to I_6; \\ B_2 \to I_3; & B_5 \to I_8. \\ B_3 \to x_2x_4I_3 \vee x_2\bar{x}_4I_5 \vee \bar{x}_2x_3I_6 \vee \bar{x}_2\bar{x}_3I_7; & \end{array} \tag{8.16}$$

The third stage is connected with construction of the table having columns B_i, $K(B_i)$, I_g, $A(I_g)$, X_h, Ψ_h, h (Table 8.4). Relationship of this table with system (8.16) is evident.

This table is used to derive system (8.8). For example, the following sum-of-products $D_1 = F_3 \vee F_6 \vee F_8 \vee F_9 = \bar{\tau}_2\bar{\tau}_3\bar{x}_1\bar{x}_2 \vee \tau_1\tau_2x_2\bar{x}_4 \vee \tau_2\tau_3\bar{x}_2\bar{x}_3 \vee \tau_2\bar{x}_3$ can be derived from Table 8.4.

Specification of block BY is reduced to replacement of vertices b_t by corresponding collections $Y(b_t)$ and variables y_0, y_E.

Logic synthesis of FSM circuit is reduced to implementation of obtained functions using some macrocells, whereas the block BY is implemented using embedded memory blocks. Logic circuits for $P_{ES}Y$ and $P_{ES1}Y$ Moore FSMs are practically identical, but the block BY of $P_{ES1}Y$ Moore FSM S_{25} includes only 10 terms (Fig. 8.6), but it includes 15 terms in the case of $P_{ES}Y$ Moore FSM S_{25}.

The number of inputs and terms of block BP can be decreased using the *method of transformation of EOLC codes into codes of the classes of pseudoequivalent EOLC*. In this case each class $B_i \in \Pi_E$ is identified by its binary code $K(B_i)$ having R_B bits:

$$R_B = \lceil \log_2 I_E \rceil . \tag{8.17}$$

Additional variables $z_r \in Z$, where $|Z| = R_B$, are used for EOLC encoding. Special block of code transformer BTC is used for encoding of the classes $B_i \in \Pi_E$. It turns $P_{ES1}Y$ Moore FSM into $P_{ES2}Y$ Moore FSM (Fig. 8.9).

The principles of this FSM operation are clear. In this model, the blocks BTC and BP generate functions

$$Z = Z(\tau), \tag{8.18}$$

Table 8.4 Structure table of $P_{ES1}Y$ Moore FSM S_{25}

B_i	$K(B_i)$	I_g	$A(I_g)$	X_h	Ψ_h	h
B_1	$*00$	I_2	00100	x_1	D_3	1
		I_3	01100	$\bar{x}_1x_2$	D_2D_3	2
		I_4	11100	$\bar{x}_1\bar{x}_2$	$D_1D_2D_3$	3
B_2	001	I_3	01100	1	D_2D_3	4
B_3	$*11$	I_3	01100	x_2x_4	D_2D_3	5
		I_5	10100	$x_2\bar{x}_4$	D_1D_3	6
		I_6	01000	$\bar{x}_2x_3$	D_2	7
		I_7	11000	$\bar{x}_2\bar{x}_3$	D_1D_2	8
B_4	$10*$	I_6	01000	1	D_2	9
B_5	$*10$	I_8	10000	1	D_1	10

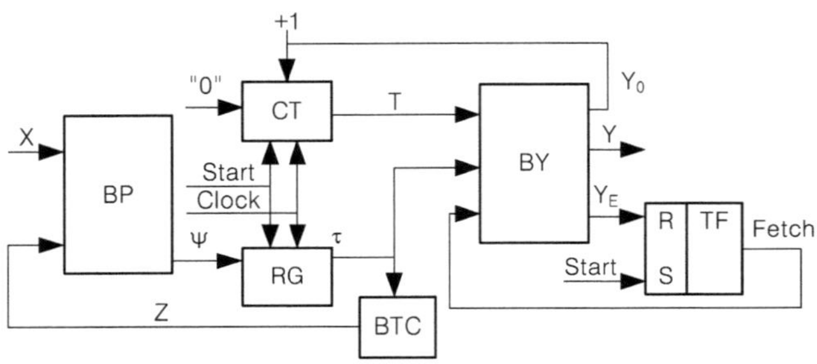

Fig. 8.9 Structural diagram of $P_{ES2}Y$ Moore FSM

$$\Psi = \Psi(Z,X). \tag{8.19}$$

The synthesis method for $P_{ES2}Y$ Moore FSM includes two additional steps, namely encoding of the classes $B_i \in \Pi_E$ and construction of a table specified the block BTC. Let us discuss an example of logic design for the $P_{ES2}Y$ Moore FSM S_{26} specified by the GSA Γ_7 (Fig. 8.10).

1. **Construction of EOLC set** results in the set $C_E = \{\alpha_1,\ldots,\alpha_7\}$, where $\alpha_1 = \langle b_1,b_2\rangle$, $I_1 = b_1$, $O_1 = b_2$, $L_1 = 2$; $\alpha_2 = \langle b_3,\ldots,b_6\rangle$, $I_2 = b_3$, $O_2 = b_6$, $L_2 = 4$; $\alpha_3 = \langle b_7,b_8\rangle$, $I_3 = b_7$, $O_3 = b_8$, $L_3 = 2$; $\alpha_4 = \langle b_9,\ldots,b_{12}\rangle$, $I_4 = b_9$, $O_4 = b_{12}$, $L_4 = 4$; $\alpha_5 = \langle b_{13},b_{14},b_{15}\rangle$, $I_5 = b_{13}$, $O_5 = b_{15}$, $L_5 = 3$; $\alpha_6 = \langle b_{16},b_{17}\rangle$, $I_6 = b_{16}$, $O_6 = b_{17}$, $L_6 = 2$; $\alpha_7 = \langle b_{18},b_{19}\rangle$, $I_7 = b_{18}$, $O_7 = b_{19}$, $L_7 = 2$. As it follows from analysis of the GSA Γ_7, there are two classes of pseudoequivalent EOLC, namely $\Pi_E = \{B_1,B_2\}$, where $B_1 = \{\alpha_1\}$, $B_2 = \{\alpha_2,\ldots,\alpha_6\}$.
2. **Natural addressing of microinstructions.** There are $G_E = 7$ EOLC in the GSA Γ_7, they could be encoded using $R_{EO} = 3$ elements from the set $\tau = \{\tau_1,\tau_2,\tau_3\}$. The maximal number of components per EOLC is equal to four ($Q_E = 4$), they could be encoded using $R_{CO} = 2$ elements from the set $T = \{T_1,T_2\}$. If the logic circuit of block BCT is implemented using some macrocells, then EOLC should

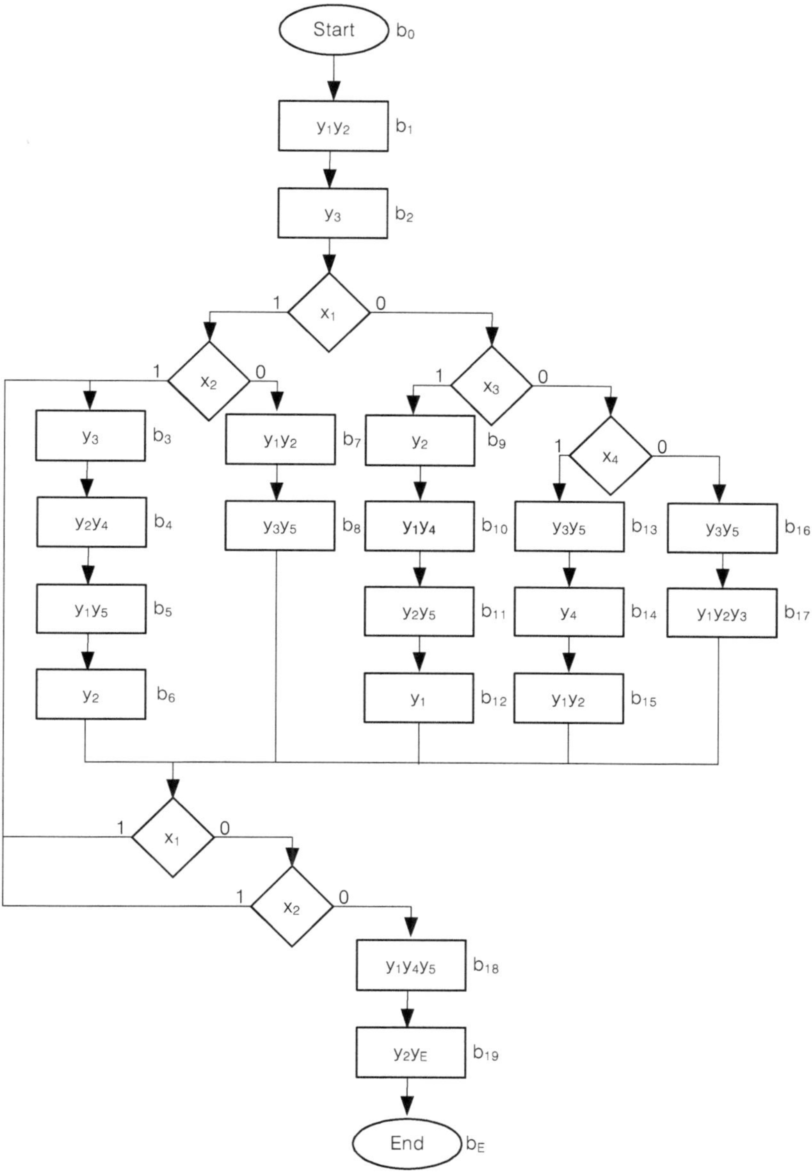

Fig. 8.10 Transformed graph-scheme of algorithm Γ_7

Fig. 8.11 Optimal EOLC codes for Moore FSM S_{26}

τ_1 \ $\tau_2\tau_3$	00	01	11	10
0	a_1	a_2	a_3	a_4
1	*	a_5	a_6	a_7

T_1T_2 \ $\tau_1\tau_2\tau_3$	000	001	010	011	100	101	110	111
00	b_1	b_3	b_9	b_7	*	b_{13}	b_{18}	b_{16}
01	b_2	b_4	b_{10}	b_8	*	b_{14}	b_{19}	b_{17}
11	*	b_5	b_{11}	*	*	b_{15}	*	*
10	*	b_6	b_{12}	*	*	*	*	*

Fig. 8.12 Microinstruction addresses for Moore FSM S_{26}

be encoded using the approach of optimal encoding. It minimizes the number of terms in system (8.14). For the $P_{ES2}Y$ Moore FSM S_{26}, the optimal EOLC codes are shown in Fig. 8.11.

Encoding of EOLC components is executed in a trivial way. For the $P_{ES2}Y$ Moore FSM S_{26}, these codes are shown in Fig. 8.12.

3. **Construction of FSM structure table** is executed in the same manner as for $P_{ES1}Y$ Moore FSM. But there is some additional stage connected with encoding of the classes $B_i \in \Pi_E$. In the discussed case, there are two classes ($I_E = 2$) and they can be encoded using the set $Z = \{z_1\}$. Let $K(B_1) = 0$, $K(B_2) = 1$. The transformed system of GFT is the following one:

$$B_1 \rightarrow x_1x_2I_2 \vee x_1\bar{x}_2I_3 \vee \bar{x}_1x_3I_4 \vee \bar{x}_1\bar{x}_3x_4I_5 \vee \bar{x}_1\bar{x}_3\bar{x}_4I_6;$$
$$B_3 \rightarrow x_1I_2 \vee \bar{x}_1x_2I_2 \vee \bar{x}_1\bar{x}_2I_7.$$

Table 8.5 Structure table of $P_{ES2}Y$ Moore FSM S_{26}

B_i	$K(B_i)$	I_g	$A(I_g)$	X_h	Ψ_h	h
B_1	0	I_2	00100	x_1x_2	D_3	1
		I_3	01100	$x_1\bar{x}_2$	D_2D_3	2
		I_4	01000	$\bar{x}_1x_3$	D_2	3
		I_5	10100	$\bar{x}_1\bar{x}_3x_4$	D_1D_3	4
		I_6	11100	$\bar{x}_1\bar{x}_3\bar{x}_4$	$D_1D_2D_3$	5
B_2	1	I_2	00100	x_1	D_3	6
		I_2	00100	$\bar{x}_1x_2$	D_3	7
		I_7	11000	$\bar{x}_1\bar{x}_2$	D_1D_2	8

This system is used to construct the structure table of $P_{ES2}Y$ Moore FSM S_{26} (Table 8.5).

This table is used to derive system (8.19). For example, the following SOP $D_1 = F_4 \vee F_5 \vee F_8 = \bar{z}_1\bar{x}_1\bar{x}_3 \vee z_1\bar{x}_1\bar{x}_2$ can be derived from Table 8.5 (after minimization).

4. **Specification of block BY** is executed according with approaches discussed before. For example, the address 00101 corresponds to the vertex b_4 (Fig. 8.12), then some memory cell having this address should contain the microoperations y_2 and y_4, as well as the additional variable y_0.
5. **Specification of block BTC** is reduced to construction of the table with columns α_g, $K(\alpha_g)$, B_i, $K(B_i)$, Z_g, g. In this table, the column Z_g contains variables $z_r \in Z$ equal to 1 in the code $K(B_i)$, where $\alpha_g \in B_i$. For the $P_{ES2}Y$ Moore FSM S_{26}, this table has $G_E = 6$ rows (Table 8.6).

   If the logic circuit of block BTC is implemented using embedded memory blocks BRAM, then the code $K(\alpha_g)$ is treated as a memory address, whereas the corresponding memory cell contains the code $K(B_i)$. If the logic circuit of block BTC is implemented using some macrocells, then the table of block BTC is used to derive system (8.18). The following minimized Boolean equation $z_1 = \tau_3 \vee \bar{\tau}_1\tau_2$ can be derived from Table 8.6. Obviously, in the second case classes $B_i \in \Pi_E$ should be encoded in the way minimizing the number of terms in system (8.18). If in our example classes $B_i \in \Pi_E$ have the codes $K(B_1) = 1$ and $K(B_2) = 0$, then the previous equation is simplified and represented as $z_1 = \bar{\tau}_2\bar{\tau}_3$.

Table 8.6 Specification of block BTC for $P_{ES2}Y$ Moore FSM S_{26}

α_g	$K(\alpha_g)$	B_i	$K(B_i)$	Z_g	g
α_1	000	B_1	0	–	1
α_2	001	B_2	1	Z_1	2
α_3	011	B_2	1	Z_1	3
α_4	010	B_2	1	Z_1	4
α_5	101	B_2	1	Z_1	5
α_6	111	B_2	1	Z_1	6

6. **Implementation of FSM logic circuit** is reduced to implementation of obtained tables and functions using some macrocells and embedded memory blocks.

   Let us point out that application of the block BTC guarantees reduction for both numbers of ST rows and inputs of the block BP up to the corresponding values of equivalent Mealy FSM. From the other hand, this block consumes some additional chip recourses. Therefore, the final choice between these two models is determined by their hardware amounts. For the $P_{ES1}Y$ Moore FSM S_{26}, the system $B(\tau)$ is the following one: $B_1 = \bar{\tau}_2\bar{\tau}_3$, $B_2 = \tau_3 \vee \bar{\tau}_1\tau_2$. Thus, the structure table of $P_{ES1}Y$ Moore FSM S_{26} includes 11 rows, whereas its block BP has three inputs. Let us point out that the maximal number of rows in the structure table of the Moore FSM S_{26} is equal to 17.

8.3 Optimization of Block of Microoperations

If condition (8.12) is violated, then required size of the block BY increases in m_K times, where

$$m_K = 2^R_{EO} + R_{CO} - R_E. \tag{8.20}$$

In this case, there is no sense in application of the code sharing method. From the other hand, this method gives a potential possibility for the block BP hardware amount decrease. To keep using this positive feature if condition (8.12) is violated, it can be used transformation of the microinstruction address $A(b_t)$, represented as (1.33), in an address having R_E bits [2]. To execute such a transformation, a block of address transformer BAT is introduced in $P_{ES}Y$ Moore FSM model. It results in a model of $P_{ES}Y_{T1}$ Moore FSM shown in Fig. 8.13.

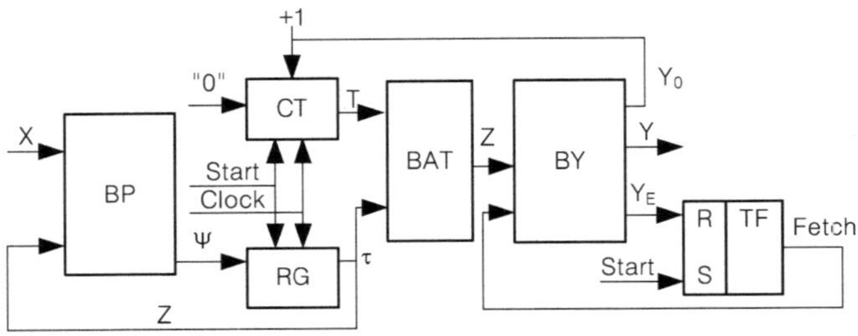

Fig. 8.13 Structural diagram of $P_{ES}Y_{T1}$ Moore FSM

Let us discuss an example of logic synthesis for the $P_{ES}Y_{T1}$ Moore FSM S_{27} specified by an initial GSA Γ_8 (Fig. 8.14).

1. **Construction of EOLC set.** Applying the already known approaches to the GSA Γ_8, the following EOLC set $C_E = \{\alpha_1, \ldots, \alpha_8\}$ can be found, where $\alpha_1 = \langle b_1, b_2 \rangle$, $L_1 = 2$; $\alpha_2 = \langle b_3 \rangle$, $L_2 = 1$; $\alpha_3 = \langle b_4, b_5 \rangle$, $L_3 = 2$; $\alpha_4 = \langle b_6, b_7 \rangle$, $L_4 = 2$; $\alpha_5 = \langle b_8, b_9 \rangle$, $L_5 = 2$; $\alpha_6 = \langle b_{15}, \ldots, b_{19} \rangle$, $L_6 = 5$; $\alpha_7 = \langle b_{12}, b_{13}, b_{14} \rangle$, $L_7 = 3$; $\alpha_8 = \langle b_{15}, \ldots, b_{19} \rangle$, $L_8 = 5$. Therefore, the GSA Γ_8 includes $G_E = 8$ elementary OLCs, encoded using $R_{EO} = 3$ elements. The maximal number of components for these EOLC is equal to $Q_E = 5$, thus it is enough $R_{CO} = 3$ variables for component codes. Besides, there are $M_E = 19$ operator vertices in the GSA Γ_8, thus there are 19 microinstructions and it is enough $R_E = 5$ bits for their addressing. It can be found that condition (8.12) is violated, because $R_{EO} + R_{CO} = 6 > R_E$. Therefore, there is sense in applying of the address transformer BAT.
2. **Natural microinstruction addressing.** Let us encode the EOLC $\alpha_g \in C_E$ in the trivial way: $K(\alpha_1) = 000, \ldots, K(\alpha_8) = 111$. The first component of any EOLC has the address 000, the second has the address 001 and so on (Fig. 8.15).
3. **Linear microinstruction addressing.** It is enough $R_E = 5$ variables to address the microinstructions $Y(b_t)$. It determines the following set of address variables

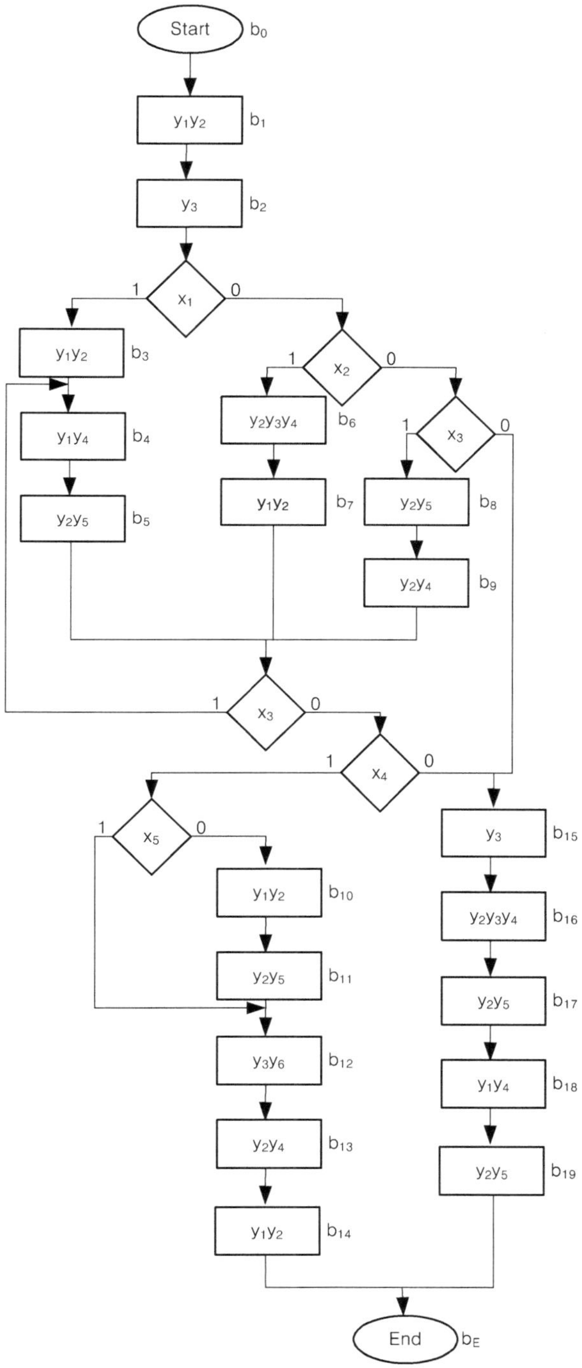

Fig. 8.14 Initial graph-scheme of algorithm Γ_8

$T_1T_2T_3$ \ $\tau_1\tau_2\tau_3$	000	001	010	011	111	101	110	100
000	y_0	y_0	y_0	y_0	*	*	y_0	y_0
001	0	y_0	0	0	*	*	y_0	y_0
010	*	0	*	*	*	*	y_0	y_0
011	*	*	*	*	*	*	y_0	y_0
111	*	*	*	*	*	*	*	*
101	*	*	*	*	*	*	*	*
110	*	*	*	*	*	*	*	*
100	*	*	*	*	*	*	y_E	y_E

Fig. 8.15 Microinstruction addresses for $P_{ES}Y_{T1}$ Moore FSM S_{27}

z_4z_5 \ $z_1z_2z_3$	000	001	010	011	100	101	110	111
00	b_1	b_5	b_9	b_{13}	b_{17}	*	*	*
01	b_2	b_6	b_{10}	b_{14}	b_{18}	*	*	*
11	b_3	b_7	b_{11}	b_{15}	b_{19}	*	*	*
10	b_4	b_8	b_{12}	b_{16}	*	*	*	*

Fig. 8.16 Linear microinstruction addresses of $P_{ES}Y_{T1}$ Moore FSM S_{27}

$Z = \{z_1, \ldots, z_5\}$. The microinstruction addresses of the $P_{ES}Y_{T1}$ Moore FSM S_{27} are shown in Fig. 8.16.

4. **Construction of FSM structure table.** This step is executed in the traditional way. In the first place, the system of GFT is constructed for EOLC, whose outputs are not connected with the input of the final vertex b_E. For the $P_{ES}Y_{T1}$ Moore FSM S_{27}, the following system is constructed:

$$\begin{aligned}
\alpha_1 &\rightarrow x_1I_2 \vee \bar{x}_1x_2I_4 \vee \bar{x}_1\bar{x}_2x_3I_5 \vee \bar{x}_1\bar{x}_2\bar{x}_3I_8;\\
\alpha_2 &\rightarrow I_3;\\
\alpha_3, \alpha_4, \alpha_5 &\rightarrow x_3I_3 \vee \bar{x}_3x_4x_5I_7 \vee \bar{x}_3x_4\bar{x}_5I_6 \vee \bar{x}_3x_4\bar{x}_5I_5^1 \vee \bar{x}_3\bar{x}_4I_8;\\
\alpha_6 &\rightarrow I_7.
\end{aligned} \tag{8.21}$$

Next, this system is used to construct a table with columns α_g, $K(\alpha_g)$, α_m, $K(\alpha_m)$, X_h, Ψ_h, h. For the $P_{ES}Y_{T1}$ Moore FSM S_{27}, the structure table includes $H = 18$ rows (Table 8.7).

Table 8.7 Structure table of $P_{ES}Y_{T1}$ Moore FSM S_{27}

α_g	$K(\alpha_g)$	α_m	$K(\alpha_m)$	X_h	Ψ_h	h
α_1	000	α_2	001	x_1	D_3	1
		α_4	011	$\bar{x}_1x_2$	D_2D_3	2
		α_5	100	$\bar{x}_1\bar{x}_2x_3$	D_1	3
		α_8	111	$\bar{x}_1\bar{x}_2\bar{x}_3$	$D_1D_2D_3$	4
α_2	001	α_3	010	1	D_2	5
α_3	010	α_3	010	x_3	D_2	6
		α_7	110	$\bar{x}_3x_4x_5$	D_1D_2	7
		α_6	101	$\bar{x}_3x_4\bar{x}_5$	D_1D_3	8
		α_8	111	$\bar{x}_3\bar{x}_4$	$D_1D_2D_3$	9
α_4	011	α_3	010	x_3	D_2	10
		α_7	110	$\bar{x}_3x_4x_5$	D_1D_2	11
		α_6	101	$\bar{x}_3x_4\bar{x}_5$	D_1D_3	12
		α_8	111	$\bar{x}_3\bar{x}_4$	$D_1D_2D_3$	13
α_5	100	α_3	010	x_3	D_2	14
		α_7	110	$\bar{x}_3x_4x_5$	D_1D_2	15
		α_6	101	$\bar{x}_3x_4\bar{x}_5$	D_1D_3	16
		α_8	111	$\bar{x}_3\bar{x}_4$	$D_1D_2D_3$	17
α_6	101	α_8	110	1	D_1D_2	18

The structure table is used to derive system Ψ. For example, the following SOP can be derived in our case: $D_1 = F_3 \vee F_4 \vee F_7 \vee F_8 \vee F_9 \vee F_{11} \vee F_{12} \vee F_{13} \vee$ $\vee F_{15} \vee \ldots \vee F_{18} = \bar{\tau}_1\bar{\tau}_2\bar{\tau}_3\bar{x}_1\bar{x}_2x_3 \vee \ldots \vee \tau_1\bar{\tau}_2\tau_3$.

5. **Specification of address transformer** is reduced to construction of the table with columns b_q, α_g, $K(\alpha_g)$, $K(b_q)$, $A(b_q)$, Z_q, q, having M_E rows. The column Z_q includes variables $z_r \in Z$ equal to 1 in the linear microinstruction address from the row q of the table($q = 1,\ldots,M_E$). For the $P_{ES}Y_{T1}$ Moore FSM S_{27}, this table includes $M_E = 19$ rows (Table 8.8).

   Some of the bits for component codes are marked by the symbol "*" in Table 8.8. This symbol points on the "don't care" variables $z_r \in Z$. The block BAT generates functions

$$Z = Z(\tau, T). \qquad (8.22)$$

   For example, the following SOP $z_1 = \tau_1\tau_2\tau_3T_2 \vee \tau_1\tau_2\tau_3T_1$ can be derived from Table 8.8.

6. **Specification of block BY.** This step is reduced to replacement of vertices $b_t \in B_1$ by corresponding contents. If some vertex is not an output of some EOLC, then the variable y_0 is written in the corresponding cell. If the output of EOLC is connected with the final vertex b_E, then the variable y_E is written in the corresponding cell. In the discussed case, for example, the vertex b_5 corresponds to address 00001, the cell with this address should contain y_2, y_5; the vertex b_{19} corresponds to address 10001, the cell with this address should contain y_2, y_5, y_E, and so on.

Table 8.8 Specification of block BAT for $P_{ES}Y_{T1}$ Moore FSM S_{27}

b_q	α_g	$K(\alpha_g)$	$K(b_q)$	$A(b_q)$	Z_q	q
b_1	α_1	000	$**0$	00000	–	1
b_2	α_1	000	$**1$	00001	z_5	2
b_3	α_2	001	$***$	00010	z_4	3
b_4	α_3	010	$**0$	00011	z_4z_5	4
b_5	α_3	010	$**1$	00100	z_3	5
b_6	α_4	011	$**0$	00101	z_3z_5	6
b_7	α_4	011	$**1$	00110	z_3z_4	7
b_8	α_5	100	$**0$	00111	$z_3z_4z_5$	8
b_9	α_5	100	$**1$	01000	z_2	9
b_{10}	α_6	101	$**0$	01001	z_2z_5	10
b_{11}	α_6	101	$**1$	01010	z_2z_4	11
b_{12}	α_7	110	$*00$	01011	$z_2z_4z_5$	12
b_{13}	α_7	110	$**1$	01100	z_2z_3	13
b_{14}	α_7	110	$*1*$	01101	$z_2z_3z_5$	14
b_{15}	α_8	111	000	01110	$z_2z_3z_4$	15
b_{16}	α_8	111	$*01$	01111	$z_2z_3z_4z_5$	16
b_{17}	α_8	111	$*10$	10000	z_1	17
b_{18}	α_8	111	$*11$	10001	z_1z_5	18
b_{19}	α_8	111	$1**$	10010	z_1z_4	19

7. **Logic synthesis of FSM circuit** is reduced to implementation of obtained systems and tables using some macrocells and embedded memory blocks. The logic circuit of $P_{ES}Y_{T1}$ Moore FSM S_{27} is shown in Fig. 8.17.

As can be seen from Fig. 8.17, the data inputs of the counter are connected with the wire 13 with logic "0". If $y_0 = 0$, then the pulse "Clock" causes the pulse C_2 and the counter is reset.

There are two approaches for reduction of the number of macrocells in the logic circuit of block BP. The first approach is the optimal EOLC encoding leading to $P_{ES1}Y_{T1}$ Moore FSM. The second method assumes simultaneous application of the blocks BAT and CCS resulted in $P_{ES2}Y_{T1}$ Moore FSM. There are two approaches for reduction of the number of embedded memory blocks in the logic circuit of block BY. The addresses (1.33) can be transformed into the addresses of expanded microinstructions (the first approach) or the addresses of collections of microoperations (the second approach) [2].

An expanded microinstruction $Y_E(b_q)$ is the set of microoperations $y_n \in Y$ and additional variables y_0, y_E, written into the vertex $b_q \in B_1$. Let $Y_E(\Gamma)$ be a set of expanded microinstructions of GSA Γ and $Q_{E1} = |Y_E(\Gamma)|$. It is enough R_{E1} variables to encode the expanded microinstructions, where

$$R_{E1} = \lceil \log_2 Q_{E1} \rceil . \tag{8.23}$$

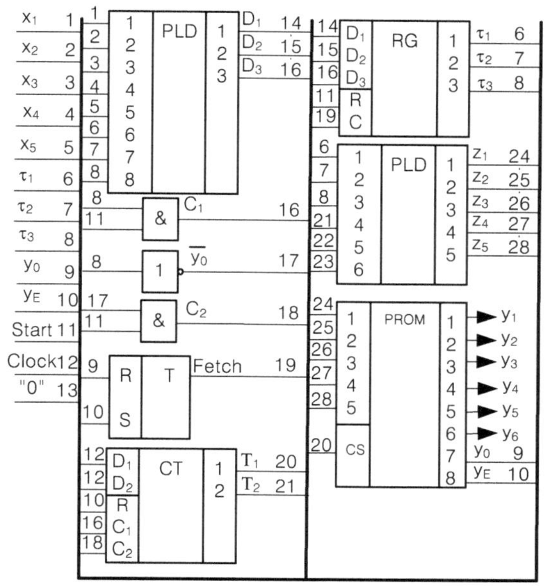

Fig. 8.17 Logic circuit of $P_{ES}Y_{T1}$ Moore FSM S_{27}

These variables form a set Z. Let the following condition take place

$$R_{E1} < R_E. \tag{8.24}$$

In this case the size of block BY can be decreased in comparison with $V_{\min}$ (8.13) due to transformation of addresses $A(b_q)$ into addresses of expanded microinstructions. The block BAT is used for such a transformation. Its application results in $P_{ES}Y_{T2}$, $P_{ES1}Y_{T2}$, and $P_{ES2}Y_{T2}$ models of Moore FSM.

There are $Q_{E1} = 13$ expanded microinstructions in the GSA Γ_8, namely: $Y_1 = \{y_0, y_1, y_2\}$, $Y_2 = \{y_1, y_2\}$, $Y_3 = \{y_1, y_2, y_E\}$, $Y_4 = \{y_0, y_3\}$, $Y_5 = \{y_3\}$, $Y_6 = \{y_0, y_1, y_4\}$, $Y_7 = \{y_0, y_2, y_5\}$, $Y_8 = \{y_2, y_5\}$, $Y_9 = \{y_2, y_5, y_E\}$, $Y_{10} = \{y_0, y_2, y_3, y_4\}$, $Y_{11} = \{y_2, y_4\}$, $Y_{12} = \{y_0, y_2, y_4\}$, $Y_{13} = \{y_0, y_3, y_6\}$. It is enough $R_{E1} = 4$ variables for encoding microinstructions from the set $Y_E(\Gamma_8) = \{Y_1, \ldots, Y_{13}\}$. It means that condition (8.24) takes place and the address transformation can be applied.

Let us discuss an example of logic synthesis for the $P_{ES}Y_{T2}$ Moore FSM S_{27}. Its structural diagram is the same as the one for $P_{ES}Y_{T1}$ Moore FSM shown in Fig. 8.13. It is necessary to execute the encoding of expanded microinstructions instead of linear addressing of microinstructions. Let us encode expanded microinstructions $Y_t \subseteq Y_E(\Gamma_8)$ using the method of frequency encoding. In the discussed example, this approach leads to the codes shown in Fig. 8.18.

The block BAT is specified by corresponding table, which is constructed in the same way as for the $P_{ES}Y_{T1}$ Moore FSM S_{27}. But now its column $A(b_q)$ is replaced by the column $K(Y_q)$, where Y_q is an expanded microinstruction written in some vertex $b_q \in B_1$. This table is represented by Table 8.9. The table is used to derive

Fig. 8.18 Expanded microinstruction codes for $P_{ES}Y_{T2}$ Moore FSM S_{27}

z_3z_4 \ z_1z_2	00	01	11	10
00	Y_1	Y_7	Y_9	Y_6
01	Y_{10}	Y_3	Y_{11}	Y_4
11	Y_5	Y_{12}	*	*
10	Y_2	Y_{13}	*	Y_8

system (8.22). If there are some insignificant input assignments for codes of EOLC or their components, they are used for optimization of functions $z_r \in Z$.

The block BY can be specified by the Karnaugh map shown in Fig. 8.19.

Fig. 8.19 Specification of block BY of $P_{ES}Y_{T2}$ Moore FSM S_{27}

z_3z_4 \ z_1z_2	00	01	11	10
00	$y_0y_1y_2$	$y_0y_2y_5$	$y_0y_1y_4$	$y_2y_5y_E$
01	$y_0y_2y_3$	$y_1y_2y_E$	y_0y_3	y_2y_4
11	y_1y_2	$y_0y_3y_6$	y_2y_5	*
10	y_3	$y_0y_2y_4$	*	*

This Karnaugh map can be used either to minimize functions y_0, y_E, and $y_n \in Y$, or to program embedded memory blocks BRAM. Let us point out that the structure table of $P_{ES}Y_{T2}$ Moore FSM S_{27} is the same as Table 8.7.

The logic circuit of $P_{ES}Y_{T2}$ Moore FSM S_{27} is shown in Fig. 8.20. Comparison of Fig. 8.17 and Fig. 8.20 shows that they are practically the same, but for the $P_{ES}Y_{T2}$ Moore FSM S_{27} the block BAT has fewer outputs, whereas the block BY has fewer inputs than their counterparts of the $P_{ES}Y_{T1}$ Moore FSM S_{27}.

Let an initial GSA Γ include Q_{E2} different collections of microoperations creating the set $Y_N(\Gamma)$. These collections can be encoded using

$$R_{E2} = \lceil \log_2 Q_{E2} \rceil \tag{8.25}$$

variables, forming the set Z. If both condition (8.24) and condition

$$R_{E2} < R_{E1} \tag{8.26}$$

take place, then the number of blocks BRAM in the FSM logic circuit can be decreased in comparison with equivalent $P_{ES}Y_{T2}$ Moore FSM. But it is necessary to use an additional block CCS to generate variables y_0 and y_E. Such an approach results in the model of $P_{ES}Y_{T3}$ Moore FSM shown in Fig. 8.21.

Table 8.9 Specification of block BAT for $P_{ES}Y_{T2}$ Moore FSM S_{27}

b_q	α_g	$K(\alpha_g)$	$K(b_q)$	$K(Y_q)$	Z_q	q
b_1	α_1	000	$**0$	0000	–	1
b_2	α_1	000	$**1$	0010	z_3	2
b_3	α_2	001	$***$	0000	–	3
b_4	α_3	010	$**0$	1100	z_1z_2	4
b_5	α_3	010	$**1$	1111	$z_1z_2z_3z_4$	5
b_6	α_4	011	$**0$	0001	z_4	6
b_7	α_4	011	$**1$	0011	z_3z_4	7
b_8	α_5	100	$**0$	0100	z_2	8
b_9	α_5	100	$**1$	1001	z_1z_4	9
b_{10}	α_6	101	$**0$	0000	–	10
b_{11}	α_6	101	$**1$	0100	z_2	11
b_{12}	α_7	110	$*00$	0111	$z_2z_3z_4$	12
b_{13}	α_7	110	$**1$	0110	z_2z_3	13
b_{14}	α_7	110	$*1*$	0101	z_2z_4	14
b_{15}	α_8	111	000	1101	$z_1z_2z_4$	15
b_{16}	α_8	111	$*01$	0001	z_4	16
b_{17}	α_8	111	$*10$	0100	z_2	17
b_{18}	α_8	111	$*11$	1100	z_1z_2	18
b_{19}	α_8	111	$1**$	1000	z_1	19

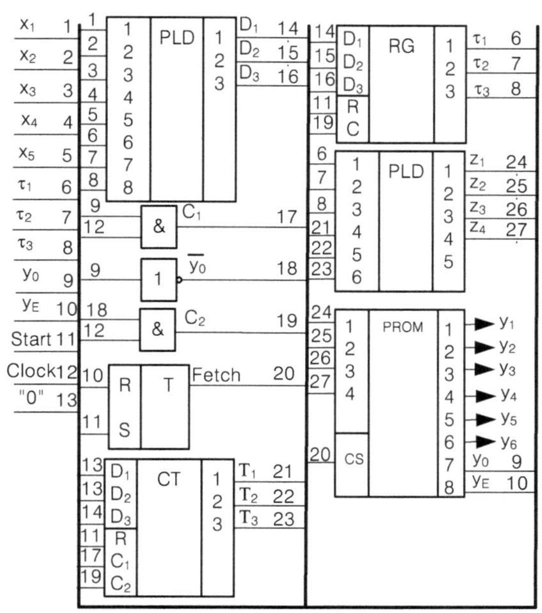

Fig. 8.20 Logic circuit of $P_{ES}Y_{T2}$ Moore FSM S_{27}

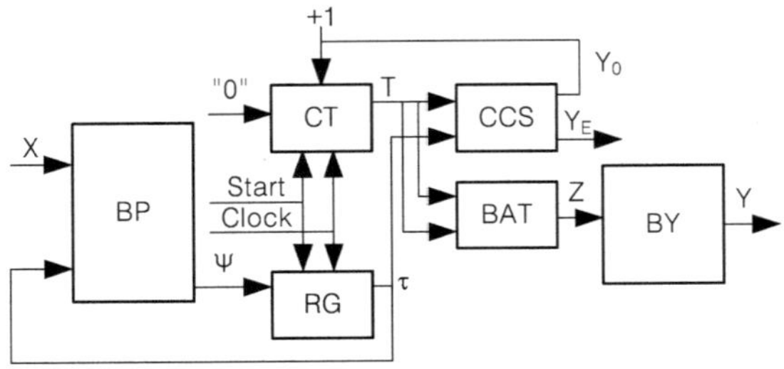

Fig. 8.21 Structural diagram of $P_{ES}Y_{T3}$ Moore FSM

In the discussed case, the set $Y_N(\Gamma)$ includes $Q_{E2} = 7$ collections of microoperations, namely: $Y_1 = \{y_1, y_2\}$, $Y_2 = \{y_3\}$, $Y_3 = \{y_1, y_4\}$, $Y_4 = \{y_2, y_5\}$, $Y_5 = \{y_2, y_3, y_4\}$, and $Y_6 = \{y_2, y_4\}$, $Y_7 = \{y_3, y_6\}$. It means that $R_{E2} = 3$ and conditions (8.24) and (8.26) take place. Thus, the application of address transformation allows quadruple decrease for the block BY size in comparison with $V_{\min}$ determined by formula (8.13). To optimize the logic circuit of BAT, it is necessary to apply the method of frequency encoding for collections of microoperations. Remind that such an encoding is executed in the following manner: the more times some collection appears in operator vertices of GSA, the more zeros its code contains. In the discussed case, there is the set of coding variables $Z = \{z_1, z_2, z_3\}$ and optimal codes are shown in Fig. 8.22.

Fig. 8.22 Optimal codes for collections of microoperations

z_1 \ z_2z_3	00	01	11	10
0	Y_1	Y_4	Y_5	Y_2
1	Y_3	Y_6	*	Y_7

The table of block BAT is built in the same manner as for previous examples. For the $P_{ES}Y_{T3}$ Moore FSM S_{27} this table has $M_E = 19$ rows (Table 8.10).

The logic circuit of block CCS is implemented using the Karnaugh map (Fig. 8.23).

This map is filled in the following manner. For example, the cell 001001 of the map corresponds to the vertex b_4 of GSA Γ_8. This vertex should include the variable y_0. Therefore, the variable y_0 is written in the cell 001001. Next, the cell 000100 does not correspond to any vertex of the GSA Γ_8. Therefore, this cell is marked by the symbol "*" (don't care) and so on. The following equations can be derived from this Karnaugh map:

$$\begin{aligned} y_0 &= \bar{T}_1\bar{T}_3 \vee \tau_1\bar{T}_1 \vee \bar{\tau}_1\bar{\tau}_2\tau_3\bar{T}_2; \\ y_E &= T_1. \end{aligned} \qquad (8.27)$$

Table 8.10 Specification of block BAT for $P_{ES}Y_{T3}$ Moore FSM S_{27}

b_q	α_g	$K(\alpha_g)$	$K(b_q)$	$K(Y_t)$	z_q	q
b_1	α_1	000	$**0$	000	–	1
b_2	α_1	000	$**1$	010	z_2	2
b_3	α_2	001	$*00$	000	–	3
b_4	α_2	001	$**1$	100	z_1	4
b_5	α_2	001	$*1*$	001	z_3	5
b_6	α_3	010	$**0$	011	z_2z_3	6
b_7	α_3	010	$**1$	000	–	7
b_8	α_4	011	$**0$	001	z_3	8
b_9	α_4	011	$**1$	101	z_1z_3	9
b_{10}	α_5	100	000	000	–	10
b_{11}	α_5	100	$*01$	001	z_3	11
b_{12}	α_5	100	$*10$	110	z_1z_2	12
b_{13}	α_5	100	$*11$	101	z_1z_3	13
b_{14}	α_5	100	$1**$	000	–	14
b_{15}	α_6	101	000	010	z_2	15
b_{16}	α_6	101	$*01$	011	z_2z_3	16
b_{17}	α_6	101	$*10$	001	z_3	17
b_{18}	α_6	101	$*11$	100	z_1	18
b_{19}	α_6	101	$1**$	001	z_3	19

$T_1T_2T_3$ \ $\tau_1\tau_2\tau_3$	000	001	011	010	110	111	101	100
000	y_0	y_0	y_0	y_0	*	*	y_0	y_0
001	0	y_0	0	0	*	*	y_0	y_0
011	*	0	*	*	*	*	y_0	y_0
010	*	*	*	*	*	*	y_0	y_0
110	*	*	*	*	*	*	*	*
111	*	*	*	*	*	*	*	*
101	*	*	*	*	*	*	*	*
100	*	*	*	*	*	*	y_E	y_E

Fig. 8.23 Specification of block CCS for $P_{ES}Y_{T3}$ Moore FSM S_{27}

To find the content of embedded memory blocks BRAM, it is enough to replace the symbols of collections of microoperations in a map similar to the one shown in Fig. 8.22 by corresponding microoperations (Fig. 8.24 in our case).

Implementation of FSM logic circuit is reduced to implementation of obtained tables and systems of functions using some logic elements. For the $P_{ES}Y_{T3}$ Moore

Fig. 8.24 Specification of block BY for $P_{ES}Y_{T3}$ Moore FSM S_{27}

z_1 \ z_2z_3	00	01	11	10
0	y_1y_2	y_2y_5	$y_2y_3y_4$	y_3
1	y_1y_4	y_2y_4	*	y_3y_6

FSM S_{27}, the logic circuit is similar to the one shown in Fig. 8.20, but there are two alterations. Firstly, the block BAT generates only three functions (z_1, z_2, z_3) served as the block BRAM inputs. Secondly, the outputs y_0, y_E are generated by the block CCS having only three inputs (T_1, T_2, T_3), as follows from (8.27).

Logic circuits for models of $P_{ES1}Y_{T2}$ and $P_{ES2}Y_{T2}$ Moore FSM are implemented in the same manner. It could be a good exercise for a reader.

8.4 Synthesis for Multilevel Models of FSM with Elementary Chains

As it is for FSM with the register keeping state codes, the hardware amount for FSM with the counter keeping microinstruction (or component) addresses can be decreased using the methods of logical condition replacement and encoding of the collections of microoperations. It results in multilevel models of $P_{EC}Y$ and $P_{ES}Y$ Moore FSM (Table 8.11).

In this table, the symbol F (the level LC) stands for the block BAT, which can generate either codes of collections of microoperations or codes of the classes of compatible microoperations. In the last case, it is possible to use the verticalization for initial GSA. The symbol $P_{EC}Y$ [1] stands for $P_{EC}Y$ Moore FSM with optimal EOLC encoding, whereas the symbol $P_{EC}Y$ [2] stands for FSM with transformation of EOLC addresses into codes of the classes of pseudoequivalent EOLC. The following numbers of different Moore FSM models can be found from Table 8.11:

1. $NL_2 = 6$ (It determines the basic models of the PY type).
2. $NL_3 = 18G + 6K + 18$ (The first member of the formula determines the number of models with the logical condition replacement, whereas the rest determines the number of models with the block BAT).

Table 8.11 Multilevel models of FSM with EOLC

LA			LB			LC	LD		
M_1	M_1C	M_{1L}	P_{EC}	P_{EC1}	P_{EC3}	Y	Y_{T1}	Y_{T2}	Y_{T3}
	⋮		P_{ES}	P_{ES1}	P_{ES2}	F		D_1	
M_G	M_GC	M_GL						⋮	
								D_K	

3. $NL_4 = 54G + 18GK$ (It determines models using both logical condition replacement and the block BAT).

As it can be found, Table 8.11 specifies $72G + 6K + 18GK + 24$ different models of FSM with elementary operational linear chains. In the case of FSM with average complexity [1], where $G = K = 6$, there are 1150 different models for interpretation of the same GSA Γ. If values of both G and K increase up to 8, then the number of models increases up to 1740.

As it has been already mentioned, logic synthesis method for a multilevel model consists from some collection of synthesis methods for models with less number of levels. Let us discuss an example of logic synthesis for the $M_2P_{ES1}LFY_{T1}$ Moore FSM S_{28} specified by the GSA Γ_9 (Fig. 8.25).

1. **Construction of FSM structural diagram.** According to the formula $M_2P_{ES1}LFY_{T1}$, its structural diagram includes the block BM with outputs p_1 and p_2; the block BP, synthesized on the base of optimal EOLC encoding; the block BAT used for transformation of microinstruction addresses form the code sharing format into their linear format; the block CCS generating codes of logical conditions for the block BM; and at last the block BY generating microoperations $y_n \in Y$ and additional variables y_0, y_E (Fig. 8.26).
2. **Transformation of initial GSA.** Because of the member M_2 in the FSM formula $M_2P_{ES1}LFY_{T1}$, the initial GSA Γ_9 should be transformed in such a way, that transitions from each FSM state will depend on not more than two logical conditions. In the discussed example, this transformation is reduced to introduction of vertices b_{19} (to eliminate the ambiguity for initial state code after pulse "Start"), b_{20} and b_{21} (to satisfy the condition $G = 2$). Besides, the variable y_E is inserted into the vertex b_{13}. The transformed GSA Γ_9 is shown in Fig. 8.27.
3. **Construction of EOLC set.** The following set of EOLC $C_E = \{\alpha_1, \ldots, \alpha_{11}\}$ can be found in our example. The characteristics of EOLC are shown in Table 8.12. Its first column contains the number of EOLC component, whereas its first row includes inputs I_g of corresponding EOLC. The following values can be found from the table: $G_E = 11$, $R_{EO} = 4$, $Q_E = 4$, $R_{CO} = 2$, $M_E = 21$, $R_E = 5$, and $R_{EO} + R_{CO} = 6$. It is more than R_E, and therefore, there is necessity in the block BAT.
4. **Optimal EOLC encoding.** Let us find the partition Π_E for the EOLC set $\{\alpha_1, \ldots, \alpha_{10}\}$. This set does not include EOLCs whose outputs are connected with the final vertex b_E.

   In the discussed example, there is the partition $\Pi_E = \{B_1, \ldots, B_7\}$, where $B_1 = \{\alpha_1\}$, $B_2 = \{\alpha_2\}$, $B_3 = \{\alpha_3\}$, $B_4 = \{\alpha_4, \alpha_5, \alpha_7, \alpha_{10}\}$, $B_5 = \{\alpha_6\}$, $B_6 = \{\alpha_8\}$, and $B_7 = \{\alpha_9\}$. Thus, only EOLC $\alpha_g \in B_4$ should be included in one generalized interval of Boolean four-dimensional space, other EOLC can be encoded in the arbitrary manner. The outcome for EOLC encoding for the $M_2P_{ES1}LFY_{T1}$ Moore FSM S_{28} is shown in Fig. 8.28.

   The following codes $K(B_i)$ for classes $B_i \in \Pi_E$ can be found from the Karnaugh map (Fig. 8.28): $K(B_1) = 0000$, $K(B_2) = 001*$, $K(B_3) = *100$,

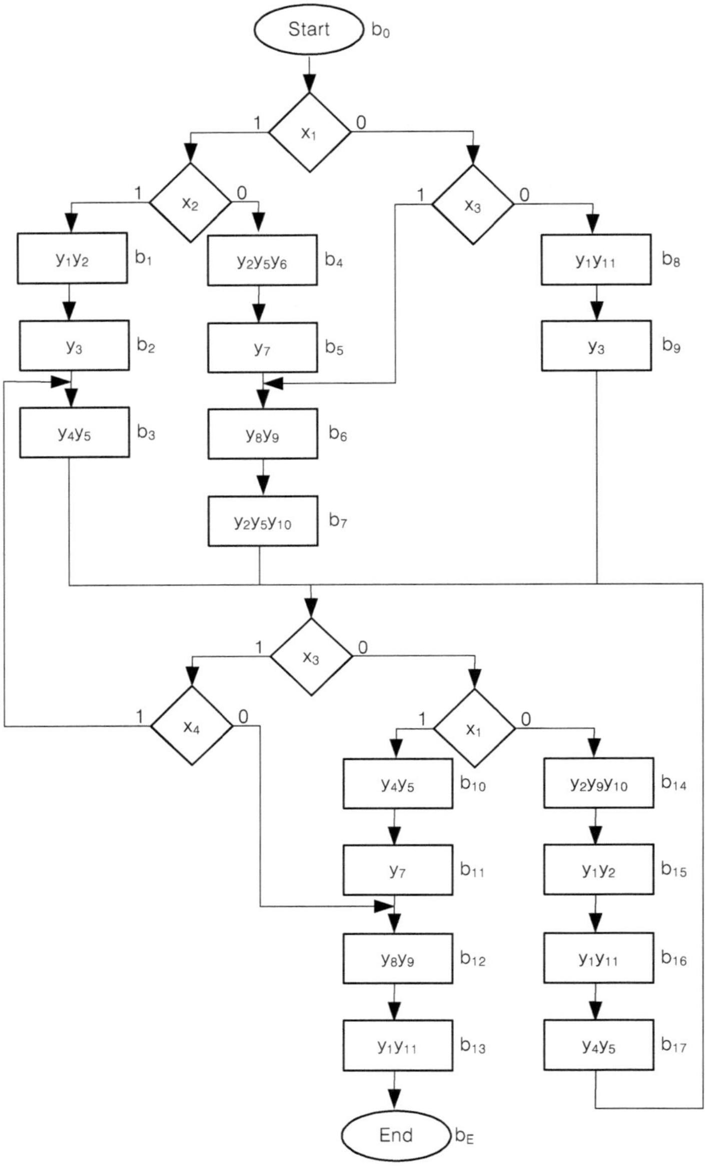

Fig. 8.25 Initial graph-scheme of algorithm Γ_9

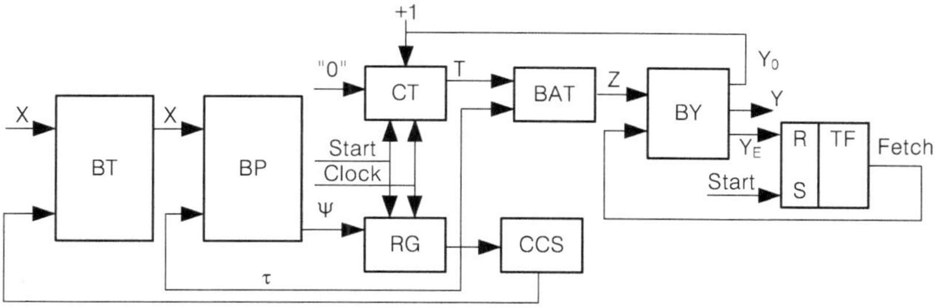

Fig. 8.26 Structural diagram of $M_2P_{ES1}LFY_{T1}$ Moore FSM

Table 8.12 Characteristics of EOLC for transformed GSA Γ_9

	α_1	α_2	α_3	α_4	α_5	α_6	α_7	α_8	α_9	α_{10}	α_{11}
1	b_{19}	b_1	b_4	b_6	b_8	b_{10}	b_{14}	b_{20}	b_{21}	b_3	b_{12}
2		b_2	b_5	b_7	b_9	b_{11}	b_{15}				
3							b_{16}				
4							b_{17}				

$K(B_4) = ***1$, $K(B_5) = *11*$, $K(B_6) = 1*00$, and $K(B_7) = 1*1*$. Let us point out that the code $K(\alpha_{11})$ is treated as insignificant input assignment.

5. **Addressing of microinstructions based on code sharing.** As usually, the first EOLC components have code 00, the second 01, the third 10, and the fourth 11. These codes together with the codes $K(\alpha_g)$ shown in Fig. 8.28 produce the microinstruction addresses shown in Fig. 8.29.
6. **Logical condition replacement.** The following sets $X(B_1) = \{x_1, x_2\}$, $X(B_2) = \emptyset$, $X(B_3) = \emptyset$, $X(B_4) = \{x_3, x_4\}$, $X(B_5) = \emptyset$, $X(B_6) = \{x_3\}$, $X(B_7) = \{x_1\}$ can be found in our example. The table of logical condition distribution is shown in Table 8.13. There are no difficulties in this table construction.

Table 8.13 Distribution of logical conditions for $M_2P_{ES1}LFY_{T1}$ Moore FSM S_{28}

	B_1	B_2	B_3	B_4	B_5	B_6	B_7
p_1	x_1	–	–	x_3	–	x_3	x_1
p_2	x_2	–	–	x_4	–	–	–

The following sets $X(p_1) = \{x_1, x_3\}$, $X(p_2) = \{x_2, x_4\}$ can be extracted from Table 8.13. Therefore, the logical conditions $x_l \in X(p_1)$ are encoded using the variable z_1, and the logical conditions $x_l \in X(p_2)$ are encoded using the variable z_2. It determines the set $Z_0 = \{z_1, z_2\}$. Let these logical conditions have the following codes: $K(x_1) = K(x_2) = 0$, $K(x_3) = K(x_4) = 1$.

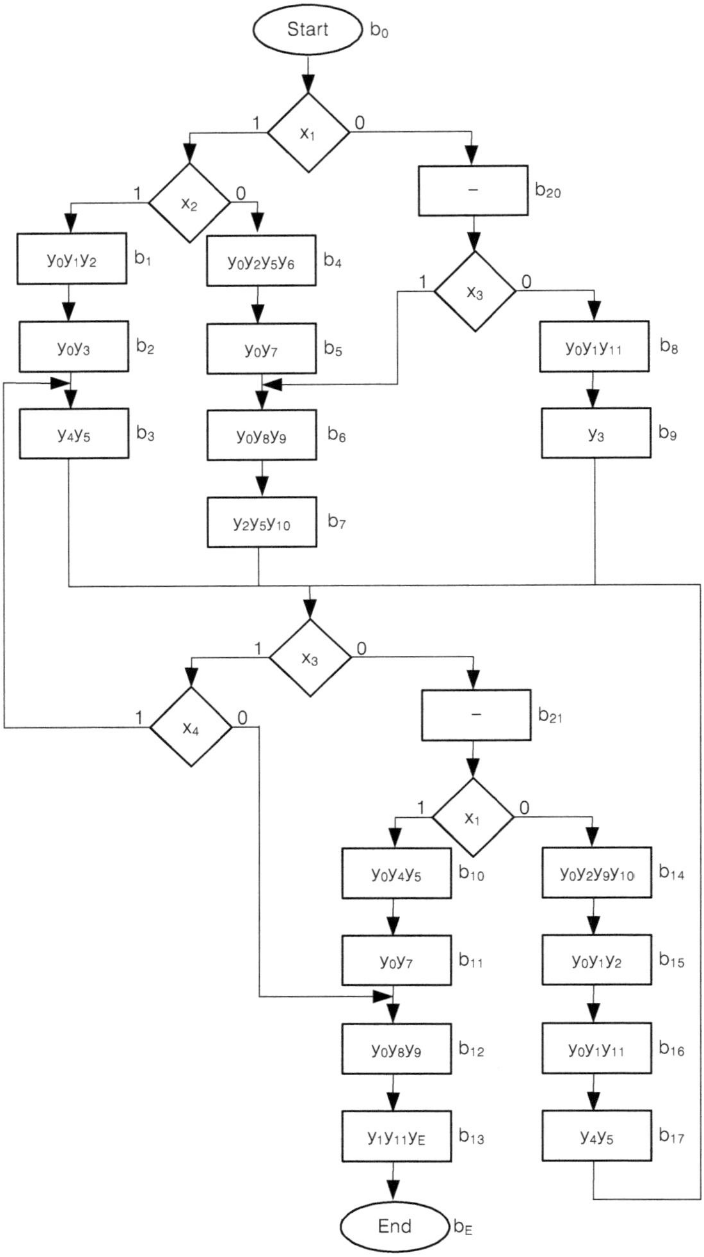

Fig. 8.27 Transformed GSA Γ_9

Fig. 8.28 Optimal codes of EOLC for $M_2P_{ES1}LFY_{T1}$ Moore FSM S_{28}

$\tau_1\tau_2$ \ $\tau_3\tau_4$	00	01	11	10
00	a_1	a_4	*	a_2
01	a_3	a_5	*	a_6
11	*	a_7	*	a_{11}
10	a_8	a_{10}	*	a_9

T_4T_5 \ $\tau_1\tau_2\tau_3\tau_4$	0000	0001	0010	0101	0110	1000	1001	1010	1101	1110	0000
00	b_{19}	b_6	b_1	b_4	b_8	b_{10}	b_{20}	b_3	b_{21}	b_{14}	b_{12}
01		b_7	b_2	b_5	b_9	b_{11}				b_{15}	b_{13}
11										b_{16}	
10										b_{17}	

Fig. 8.29 Microinstruction addresses of $M_2P_{ES1}LFY_{T1}$ Moore FSM S_{28}

$$\begin{array}{ll} B_1 \to x_1x_2I_2 \vee x_1\bar{x}_2I_3 \vee \bar{x}_1I_8; & B_5 \to I_{11}; \\ B_2 \to I_{10}; & B_6 \to x_3I_4 \vee \bar{x}_3I_5; \\ B_3 \to I_4; & B_7 \to x_1I_6 \vee \bar{x}_1I_7. \\ B_4 \to x_3x_4I_{10} \vee x_3\bar{x}_4I_{11} \vee \bar{x}_3I_9; & \end{array} \tag{8.28}$$

Next, the logical conditions in system of GFT are replaced by variables $p_g \in P$, using the logical condition distribution shown in our example in Table 8.13:

$$\begin{array}{ll} B_1 \to p_1p_2I_2 \vee p_1\bar{p}_2I_3 \vee \bar{p}_1I_8; & B_5 \to I_{11}; \\ B_2 \to I_{10}; & B_6 \to p_1I_4 \vee \bar{p}_1I_5; \\ B_3 \to I_4; & B_7 \to p_1I_6 \vee \bar{p}_1I_7. \\ B_4 \to p_1p_2I_{10} \vee p_1\bar{p}_2I_{11} \vee \bar{p}_1I_9; & \end{array} \tag{8.29}$$

This transformed system is used to construct the final transformed ST. For the $M_2P_{ES1}LFY_{T1}$ Moore FSM S_{28}, this table (Table 8.14) is constructed using system (8.29) and EOLC codes from Fig. 8.28.

This table is used to derive the system of input memory functions

$$\Psi = \Psi(p_1, p_1, \tau). \tag{8.30}$$

For example, the following sum-of-products $D_1 = F_3 \vee F_4 \vee F_5 \vee \ldots \vee F_9 \vee F_{13} = \bar{\tau}_1\bar{\tau}_2\bar{\tau}_3\bar{\tau}_4\bar{p}_1 \vee \bar{\tau}_1\bar{\tau}_2\tau_3 \vee \tau_2\bar{\tau}_3\bar{\tau}_4 \vee \tau_4 \vee \tau_2\tau_3 \vee \tau_1\tau_3p_2$ can be derived from Table 8.14. This formula is obtained after minimizing the initial expression derived from the transformed ST.

Table 8.14 Transformed structure table of $M_2P_{ES1}LFY_{T1}$ Moore FSM S_{28}

B_i	$K(B_i)$	α_g	$K(\alpha_g)$	P_h	Ψ_h	h
B_1	0000	α_2	0010	p_2	D_3	1
		α_3	0100	$\bar{p}_2$	D_2	2
		α_8	10100	$\bar{p}_1$	D_1	3
B_2	$001*$	α_{10}	1001	1	D_1D_4	4
B_3	$*100$	α_4	0001	1	D_4	5
B_4	$***1$	α_{10}	1001	p_2	D_1D_4	6
		α_{11}	1110	$\bar{p}_2$	$D_1D_2D_3$	7
		α_9	1010	$\bar{p}_1$	D_1D_3	8
B_5	$*11*$	α_{11}	1110	1	$D_1D_2D_3$	9
$1*00$	$1*00$	α_4	0001		D_4	10
		α_5	0101	$\bar{p}_1$	D_2D_4	11
B_7	$1*1*$	α_6	0110		D_2D_3	12
		α_7	1101	p_2	$D_1D_2D_4$	13

7. **Specification of blocks BM and CCS.** The following equations can be found from the logical condition codes:

$$\begin{aligned} p_1 &= \bar{z}_1x_1 \vee z_1x_3; \\ p_2 &= \bar{z}_2x_2 \vee z_2x_4. \end{aligned} \tag{8.31}$$

System (8.31) specifies the block BM, its analysis shows that $z_1 = 1$ for x_3 and $z_2 = 1$ for x_4. It means that the following system can be derived from Table 8.13:

$$\begin{aligned} z_1 &= B_4 \vee B_6 = \tau_4 \vee \tau_1\bar{\tau}_2\bar{\tau}_3; \\ z_2 &= B_4 = \tau_4. \end{aligned} \tag{8.32}$$

Analysis of system (8.32) shows that the equation z_2 is a part of the equation z_1. It means that it is enough only the variable z_1 to encode the logical conditions. Thus, the block BM is represented by the following system.

$$\begin{aligned} p_1 &= \bar{z}_1x_1 \vee z_1x_3; \\ p_2 &= \bar{z}_1x_2 \vee z_1x_4, \end{aligned} \tag{8.33}$$

In the same time, the block CCS is specified by the equation:

$$z_1 = \tau_4 \vee \tau_1\bar{\tau}_2\bar{\tau}_3. \tag{8.34}$$

8. **Linear microinstruction addressing** is executed in a trivial way. In the discussed example, there are $R_E = 5$, $Z = \{z_2, \ldots, z_6\}$. Let us address microinstructions using maximal possible number of zeros in the addresses. One of the addressing variants is shown in Fig. 8.30.
9. **Specification of block BY.** To specify the block BY, it is enough to replace vertices $b_q \in B_1$ by their contents, taking into account variables y_0 and y_E. In the

z_2z_3 \ $z_4z_5z_6$	000	001	010	011	100	101	110	111
00	b_1	b_2	b_3	b_6	b_4	b_7	b_8	b_{14}
01	b_9	b_{10}	b_{11}	b_{15}	b_{12}	b_{16}	b_{17}	*
11	b_5	b_{18}	b_{19}	b_{12}	b_{20}	*	*	*
10	b_{13}	*	*	*	*	*	*	*

Fig. 8.30 Linear microinstruction addresses of $M_2P_{ES1}LFY_{T1}$ Moore FSM S_{28}

discussed case, the initial specification is represented by Fig. 8.30. For example, the cell with address $A(b_2)$ should contain the collection $Y(b_2) = \{y_3\}$; the cell with address $A(b_8)$ should contain the collection $Y(b_8) = \{y_0, y_1, y_{11}\}$; the cell with address $A(b_{13})$ should contain the collection $Y(b_{13}) = \{y_1, y_{11}, y_E\}$ and so on.

10. **Specification of block BAT.** This block is specified by the table with columns b_q, $K(\alpha_g)$, $K(b_q)$, $A(b_q)$, Z_q, q. In the discussed case, the address $A(b_q)$ is taken

Table 8.15 Specification of block BAT for $M_2P_{ES1}LFY_{T1}$ Moore FSM S_{28}

b_q	$K(\alpha_g)$	$K(b_q)$	$A(b_q)$	Z_q	q
b_1	0010	$*0$	00000	–	1
b_2	0010	$*1$	00001	z_6	2
b_3	$10*1$	$**$	00010	z_5	3
b_4	$*100$	$*0$	00100		4
b_5	$*100$	$*1$	10000	z_2	5
b_6	$00*1$	$*0$	00011	z_5z_6	6
b_7	$00*1$	$*1$	00101	z_6	7
b_8	$01*1$	$*0$	00110	z_5	8
b_9	$01*1$	$*1$	01000	z_3	9
b_{10}	$011*$	$*0$	01001	z_3z_6	10
b_{11}	$011*$	$*1$	01010	z_3z_5	11
b_{12}	$111*$	$*0$	01100	z_3	12
b_{13}	$111*$	$*1$	11000	z_2z_3	13
b_{14}	$110*$	00	00111	z_5z_6	14
b_{15}	$110*$	01	01011	$z_3z_5z_6$	15
b_{16}	$110*$	10	01101	$z_3\ z_6$	16
b_{17}	$110*$	11	01110	$z_3\ z_5$	17
b_{18}	–	–	–	–	18
b_{19}	0000		10010	z_1z_5	19
b_{20}	$1*00$	$**$	10100	z_1	20
b_{21}	$101*$	$**$	10011	$**z_5z_6$	21

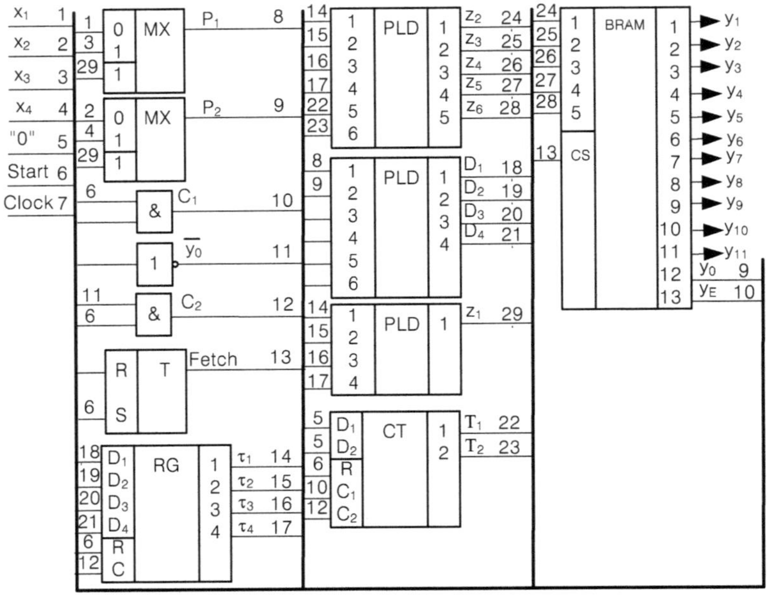

Fig. 8.31 Logic circuit of $M_2P_{ES1}LFY_{T1}$ Moore FSM S_{28}

from Fig. 8.30; the column Z_q contains variables $z_r \in Z$, equal to 1 in address $A(b_q)$. Codes of components and EOLCs should be written taking into account insignificant input assignments. The block BAT of $M_2P_{ES1}LFY_{T1}$ Moore FSM S_{28} is represented by Table 8.15.

This table is used to derive system

$$Z = Z(\tau, T). \tag{8.35}$$

For example, the following SOP $z_1 = E_{19} \vee E_{20} \vee E_{21}$ can be derived from Table 8.15, where the symbol E_q stands for a conjunction of variables $\tau_r \in \tau$ and $T_r \in T$ from the row q of the table ($q = 1, \ldots, M_E$). Taking into account the insignificant input assignment, the final form $z_1 = \bar{\tau}_2\bar{\tau}_3\bar{\tau}_4 \vee \tau_1\bar{\tau}_4 \vee \tau_3\tau_4$ can be obtained.

Acting in the same manner, it is possible to design a logic circuit for any Moore FSM represented by Table 8.11. The huge amount of possible solutions for the same GSA shows necessity of an expert system using for a-priory choice of the best FSM model on the base of preliminary analysis of characteristics for both a GSA to be interpreted and logic elements to be used.

References

1. Baranov, S.I.: Logic Synthesis of Control Automata. Kluwer Academic Publishers, Dordrecht (1994)
2. Barkalov, A., Titarenko, L.: Logic Synthesis for Compositional Microprogram Control Units. Springer, Berlin (2008)
3. Barkalov, A., Titarenko, L., Kołopieńczyk, M.: Optimization of circuit of control unit with code sharing. In: Proc. of IEEE East-West Design & Test Workshop - EWDTW 2006, Sochi, Rosja, pp. 171–174. Kharkov National University of Radioelectronics, Kharkov (2006)
4. Barkalov, A., Titarenko, L., Kołopieńczyk, M.: Optimization of control unit with code sharing. In: Proc. of the 3rd IFAC Workshop: DESDES 2006, Rydzyna, Polska, pp. 195–200. University of Zielona Góra Press, Zielona Góra (2006)
5. Barkalov, A., Wiśniewski, R.: Optimization of compositional microprogram control unit with elementary operational linear chains. Upravlauscie Sistemy i Masiny (5), 25–29 (2004)
6. Barkalov, A., Wiśniewski, R.: Optimization of compositional microprogram control units with sharing of codes. In: Proc. of the Fifth Inter. Conf. CADD'DD 2004, Minsk, Belorus, vol. 1, pp. 16–22. United Institute of the Problems of Informatics, Minsk (2004)
7. Barkalov, A., Wiśniewski, R.: Design of compositional microprogram control units with maximal encoding of inputs. Radioelektronika i Informatika (3), 79–81 (2004)
8. De Micheli, G.: Synthesis and Optimization of Digital Circuits. McGraw-Hill, New York (1994)
9. Wiśniewski, R.: Synthesis of Compositional Microprogram Control Units for Programmable Devices. PhD thesis, University of Zielona Góra (2008)

Chapter 9
Conclusion

Now we are witnesses of the intensive development of design methods oriented on field-programmable logic devices and ASICs. The complexity of digital system to be designed increases drastically, as well as the complexity of FPLD chips used for their design. These devices include billions of transistors and it is not a limit.

Development of digital systems with these complex logic elements is impossible without application of hardware description languages, computer-aided design tools and design libraries. But even application of all these tools does not guarantee that some competitive product will be designed for appropriate time-to-market. To solve this problem, a designer should know not only CAD tools, but the design and optimization methods too. It is especially important in case of such irregular devices as control units. Because of irregularity, their logic circuits are implemented without using of the standard library cells; only macrocells of a particular FPLD chip can be used in FSM logic circuit design. In this case, the knowledge and experience of a designer become a crucial factor of the success. Many experiments conducted with use of standard industrial packages show that outcomes of their operation are, especially in case of complex control units design, far from optimal. Thus, it is necessary to develop own program tools oriented on FSM optimization and use them together with industrial packages. This problem cannot be solved without fundamental knowledge in the area of logic synthesis. Besides, to be able to develop his(her) own new design and optimization methods, a designer should know the existed methods.

We think that FSM models and design methods proposed in our book will help in solution of this very important problem. We hope that our book will be useful for the designers of digital systems and scholars developing synthesis and optimization methods oriented towards the control units and ever-changing field-programmable logic devices used for FSM logic circuit implementation.

A. Barkalov and L. Titarenko: Logic Synthesis for FSM-Based Control Units, LNEE 53, pp. 229.
springerlink.com

Index

MIX
Papier aus verantwortungsvollen Quellen
Paper from responsible sources
FSC® C105338

FSC
www.fsc.org

If you have any concerns about our products,
you can contact us on
ProductSafety@springernature.com

In case Publisher is established outside the EU,
the EU authorized representative is:
Springer Nature Customer Service Center GmbH
Europaplatz 3, 69115 Heidelberg, Germany

Printed by Libri Plureos GmbH
in Hamburg, Germany